AF443277

Ergonomics in Rehabilitation

Ergonomics

in

Rehabilitation

Edited by

Anil Mital

Ergonomics Research Laboratory
University of Cincinnati, USA

and

Waldemar Karwowski

Centre for Industrial Ergonomics,
University of Louisville, USA

Taylor and Francis
Philadelphia · New York · London
1988

USA	Taylor & Francis Inc., 242 Cherry St, Philadelphia, PA 19106–1906
UK	Taylor & Francis Ltd. 4 John St, London WC1N 2ET

Copyright © Taylor & Francis Ltd 1988

British Library Cataloguing in Publication Data

Ergonomics in rehabilitation.
   1.   Physically handicapped persons. Rehabilitation. Ergonomic aspects
I. Mital, Anil   II. Karwowski, Waldemar
362.4′ 475

ISBN 0-85066-456-X

Library of Congress Cataloguing in Publication Data

Ergonomics in rehabilitation/edited by Anil Mital and Waldemar
Karwowski.
   Includes bibliographies and index.
   ISBN 0–85066–456–X
   1. Rehabilitation. 2. Human engineering. I. Mital, Anil.
II. Karwowski, Waldemar, 1953–
RM735.E74   1988
617—dc19

Cover design by Russell Beach
Phototypesetting by
Chapterhouse, The Cloisters, Formby, England.
Printed in Great Britain by
Redwood Burn Limited, Trowbridge, Wiltshire.

Contents

Preface

Historically, ergonomics has aimed at establishing working capacities of healthy individuals and it is in this context that the overall objective of ergonomics has been defined: fitting the task to the worker. Very little attention, however, has been paid to the problems of disabled persons or to making them a productive part of society. In our zest for improving the quality of life and working conditions for the masses, many of us have overlooked a rather substantial part of the population; namely, the disabled. The preponderance of studies dealing with the capabilities and limitations of healthy individuals, published in the literature, underlies this fact.

Our objective in undertaking this project was twofold: (1) to point out that ergonomics can contribute, very effectively, in overcoming disabilities of individuals and assist in making the disabled a productive part of the workforce and (2) to stimulate interests of fellow ergonomists in rehabilitation research. If we are able to persuade even a small number of our colleagues to divert some of their attention to the problems of the disabled, we would consider this effort successful. Here, we are not suggesting that ergonomists ignore their concern for injury prevention and elimination of job-worker mismatches. Rather, we are hoping that the goal of putting the disabled employees back to work will become an integral part of their overall mission.

In the course of compiling this volume, it became evident that just as ergonomics is interdisciplinary in nature, so too is rehabilitation; and perhaps more so. Not only that, rehabilitating disabled individuals is a global priority for individuals from many diverse disciplines. The papers contained in this volume attest to the diversity of the authors' backgrounds. Rehabilitation of the disabled is of as much concern to a surgeon in India as it is to a biomedical engineer in Israel or a physiotherapist in Canada.

This edited volume contains 19 papers and one appendix. The papers have been divided into six sections: Introduction (1 paper); Physical disability: Prevention and re-habilitation (3 papers); Mental disability: Prevention and rehabilitation (2 papers); Job placement of disabled workers (3 papers); Occupational biomechanics (6 papers); and Assistive devices (4 papers). While some of the papers review the state of the art, others deal with theoretical and practical developments. Together, the collection presents a variety of perspectives and provides different insights to rehabilitation problems and solutions in research and practice. We hope this book will provide the impetus necessary for undertaking projects that will lead to design data, work practice guidelines, techniques, devices, etc., for integrating the disabled into the workforce.

The preparation of this volume has been time consuming and occasionally demanding, not only for us but for the authors and reviewers as well. We are thankful to all the authors for their contributions and for revising manuscripts to comply with the reviewers' comments, and to the reviewers for their prompt and thorough reviews. We are also thankful to David Grist, a former editor at Taylor and Francis, and Malcolm Clarkson, the Director of Book Publishing at Taylor and Francis, for their foresight in sanctioning the preparation of this volume.

Finally, we recognize that the time required for completing this volume has come at the expense of our families (Aashi, Jessica, Anubhav, Mathew, Chetna, and Bernardette) and to them we dedicate this book.

Anil Mital and Waldemar Karwowski

Cincinnati and Louisville
February 1988

PART I
INTRODUCTION

*Table 1.	Disabling accidents in the United States in 1978, by place of accident (Belknap, 1985).*

Place	Disabling injuries in thousands	Injury rates/1 million hours of exposure
All accidents	5,400	9·8
At work	2,200	11·6
Away from work	3,200	8·8
Motor vehicles	1,000	28·9
Public (non-motor vehicles)	1,100	8·4
At home	1,100	5·5

Table 2 shows the trends in on-the-job and off-the-job accidents leading to disabling injuries in the United States between 1945 and 1978. Between 1945 and 1978, the ratio of off-the-job to on-the-job injuries increased by 5 per cent to 1·45.

An important source of employment injury statistics is the International Labour Office (Geneva, Switzerland) which compiles accident statistics of many countries. Table 3 provides the number of persons injured as a result of industrial accidents alone, classified by the type of industry, in selected countries. Both the number of disabling injuries resulting in loss of workdays and the total number of workdays lost are given (the presented data, however, do not reflect the number of persons ''exposed to risk'', as expressed in terms of either full-time workers or hours of work, in a given sector of industry and country and, therefore, must be interpreted carefully). Unfortunately, with the exception of a few countries, the exposure risk data are not currently available. Therefore, comparisons between periods and countries cannot be made directly. Rather, the data presented in Table 3 illustrate the extent of human suffering due to industrial accidents. These accidents result in temporary or permanent, partial or total disability to work and, consequently, create an urgent need for comprehensive vocational rehabilitation efforts.

*Table 2.	Trends in on-the-job and off-the-job injuries in the United States, 1945–1978 (Belknap, 1985).*

Year	Disabling injuries (In thousands)		
	On-the-job (1)	Off-the-job (2)	Ratio (2/1)
1945	2,000	2,750	1·38
1955	1,950	2,400	1·23
1960	1,950	2,250	1·15
1965	2,100	2,700	1·29
1970	2,200	3,250	1·48
1975	2,200	3,200	1·45
1976	2,200	3,100	1·41
1977	2,300	3,200	1·39
1978	2,200	3.200	1·45
Change (%) from 1945 to 1978	+ 10	+ 16	+ 5

Table 3. Number of persons injured (thousands) with lost workdays (A), and number of workdays lost (thousands) B, in selected countries (after Year Book of Labour Statistics).

		Total		Mining		Manufacturing		Construction	
		1982	1983	1982	1983	1982	1983	1982	1983
Argentina	A	61·436	70·586	2·866	2·746	49·010	57·299	2·829	2·649
	B	991·801	1224·940	43·727	170·480	775·603	861·661	40·636	48·676
Egypt	A	62·312	62·933	1·763	0·976	48·277	48·719	4·673	5·367
	B	14120·060	946·610	64·180	28·040	1126·380	665·740	84·630	86·080
France	A	950·520	*	*	*	449·070	*	229·410	*
	B	27319·200	*	*	*	11459·500	*	8191·000	*
Hong Kong	A	70·879	70·895	0·101	0·104	23·794	22·892	19·420	17·739
	B	658·586	569·963	*	*	*	*	221·547	176·451
India	A	322·473	287·776	*	*	319·582	285·369	*	*
	B	3277·250	3095·85	*	*	3248·030	3066·930	*	*
Switzerland	A[1]	114·551	114·655	*	*	52·185	51·059	46·439	47·389
	B	2862·210	3374·530	*	*	1191·170	1366·230	1272·770	1530·900
UK	A[2]	396·000	*	38·000	*	134·000	*	45·000	*
	B	10544·600	*	1010·800	*	3442·700	*	1263·800	*
USA	A	2182·400	2182·700	57·500	42·200	783·700	763·700	197·900	210·000
	B	36875·100	37201·400	1450·700	1178·900	13293·200	12983·000	3786·200	3944·900

*Data Not Available.
[1]Including Fatal Cases.
[2]Excluding Northern Ireland.

Rehabilitation: Definition and scope

As defined by Krol (1985), rehabilitation aims to reduce the impact of disabling conditions in individuals and to enable them to achieve social integration. Comprehensive rehabilitation includes three components: medical, social, and vocational rehabilitation. The comprehensive rehabilitation methods include:

(1) remedial exercise,
(2) occupational therapy,
(3) physiotherapy,
(4) psychotherapy,
(5) reconstructive surgery,
(6) technical aids, and
(7) social intervention (Krol, 1985).

Vocational rehabilitation refers to that part of comprehensive rehabilitation which involves the provision of vocational services, such as vocational guidance, vocational training, and selective placement, designed to enable disabled individuals to secure and retain suitable employment (ILO, 1978). Vocational rehabilitation aims at evaluation of the disabled person's residual capacities for work (vocational assessment) and accommodating these capacities through training, application of technical aids, job modification, and selection and placement. It is in this context that ergonomics can play a significant role in facilitating rehabilitation efforts.

Figure 1 illustrates the concept of comprehensive vocational rehabilitation where all disciplines (i.e. medicine, engineering, ergonomics, and other human sciences, such as sociology and psychology) must interact to produce tangible results in both disability

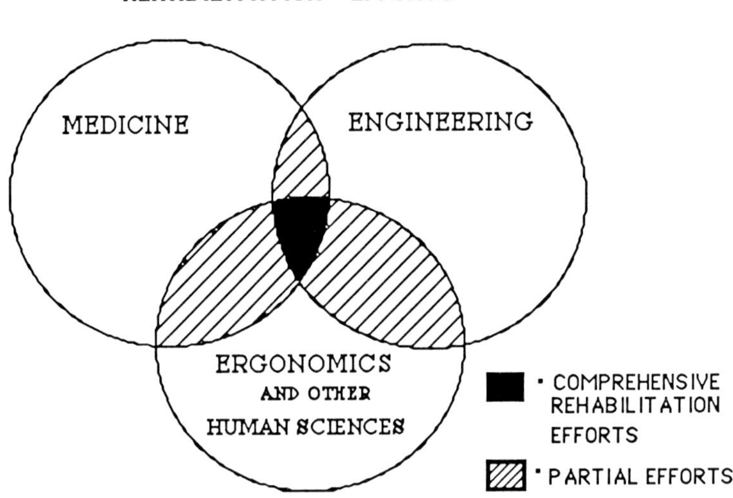

Figure 1.   *Comprehensive vocational rehabilitation efforts.*

prevention and rehabilitation. Although not always apparent, the application of ergonomic principles to remove incompatibility in the human-machine-environment system (Karwowski *et al.*, 1988), by matching human capacities and job requirements, is critical to long range success of all efforts aimed at disability prevention and vocational rehabilitation.

Disability definition

Disability is any restriction, or lack of ability, resulting from an impairment, to perform an activity in the manner or within the range of the non-disabled persons. Examples are communication problems due to hearing loss, mobility problems due to stiffness of joints, or incapacity to control own work. Disability, in such context, is seen as a disadvantage for an individual, resulting from an impairment, that prevents or limits a person's role compared with that of a non-disabled person (Kailes, 1985).

Impairment, according to Krol (1985), can be defined as any loss or abnormality of psychological or physiological functions or anatomical or physiological structures; for example, hearing loss or stiffness of a joint. Inpairments are the result of diseases, accidents or congenital disorders.

The role of ergonomics in rehabilitation

Ergonomics, the science of work, has the preservation of health and improvement of working efficiency as its main objectives (Singleton, 1982). It interacts closely with many other applied and life sciences, such as engineering and medicine, in order to implement effectively its findings and recommendations for job design or redesign. Historically, ergonomists have emphasized injury prevention and preservation of workers' health (Oshima, 1987). It is equally important, if not more so, that equal emphasis be placed on restoring disabled workers to useful employment, so that they too can contribute to the growth of society and enjoy a productive and satisfying life.

From the ergonomic standpoint, rehabilitation is a process of recognising functional limitations of people with disabilities, and designing the external environment around these limitations for the benefit of a given individual and society as a whole. The main aim of such a process is to reduce potential losses in productivity and wages due to incapacity of disabled workers to contribute fully, if at all, to their previous occupations.

According to Rohan (1985), most non-fatal injuries in North America occur at home and in domestic surroundings. The main causes of these accidents are:

(1) various manual and mechanical tasks related to running a residence,
(2) domestic duties,
(3) care of persons, goods, household linen, family clothes, furniture and carpets,
(4) cleaning the residence,
(5) handling combustibles and toxic products,
(6) kitchen work,
(7) shopping and delivery of food, and
(8) commuting to and from work.

In view of the above, consumer and home ergonomics can play a significant role in educating the general public, manufacturers and users, about human capacities and limitations such that these could be included in home design.

The role of ergonomics in disability prevention

Ergonomic designs can play an important role in preventing disabling injuries. Although each industry has its own characteristic types of accidents, it is widely accepted that the general causes of such accidents are the same (Evans, 1985). The following were identified as the leading causes of industrial accidents in the United Kingdom in 1978:

(1)   manual handling (25·5%)
(2)   falls and slips (18·1%)
(3)   machinery in motion (15·2%)
(4)   striking against objects (8·1%)
(5)   work transport accidents (7·5%)
(6)   struck by falling objects (5·6%)
(7)   hand tools (5·8%)
(8)   others (14·1%)

Ergonomic intervention, being a prerequisite to ensuring safe systems of work, is critical in analyzing cause of accidents and preventing accidents and related disabling injuries. Only by examining all human factors relevant to the job, the operator, the equipment, and their interaction, under given environmental conditions, can the disability prevention program work.

The role of ergonomics in vocational rehabilitation

Although rehabilitation in industry was pioneered by H. E. Moore in early 1925, its advances in many countries have been relatively slow and spasmodic (Guthrie, 1982). In the United States, the 1973 Vocational Rehabilitation Act of the United States Congress (1985) mandates all employers receiving federal contracts to employ and advance the employment of qualified disabled individuals. In this context, qualified disabled individuals are those who are capable of performing a particular job with ''reasonable'' accommodation to their physical and mental limitations. Only when, in spite of reasonable accommodation, physical and mental conditions result in an inability to perform a given job in a manner that is efficient and safe for the worker and those around him/her, are they recognized as activity limiting disabilities (Armstrong and Kochar, 1982). The same applies when the job aggravates a worker's impairment. Making ''reasonable'' accommodation for workers with specific disabilities becomes the critical factor in reducing the number of otherwise productive people who are removed from the working environment. Ergonomics can help disabled individuals overcome their limitations by analyzing and matching both the physical and mental work requirements, expressed in terms of attributes such as task frequency and duration, and their capabilities in terms of attributes such as gross body strength, reach location, hand action and force, perception, sensory abilities (vision, hearing, speech, touch, orientation, and balance), psychomotor and cognitive abilities (motor coordination and feedback), etc. The final accommodation may take the form of task modification or design and use of special purpose assistive devices. In both cases, ergonomic principles of matching human abilities and job requirements can be applied.

Table 4. Ergonomics research and application areas relevant to rehabilitation efforts

Classification of areas[1]	Disabling conditions[2]					
	Musculoskeletal conditions (39%)[3]	Circulatory conditions (25%)	Respiratory conditions (14%)	Metabolic conditions (14%)	Mental conditions (3%)	Senses (6%)
MAN AS A SYSTEM COMPONENT						
1. Perpetual input processes					*	*
2. Central processes					*	*
3. Basic motor processes (Psychological Comp.)	*				*	
4. Perceptual motor performance	*				*	
5. Factors affecting 4.					*	
6. Basic physiological processes		*	*	*		
7. Work capacity	*	*	*	*	*	
8. Basic anthropometric data	*					
9. Body mechanics data	*	*	*	*		
10. Basic data on sensory physiology						*
11. Factors affecting physiological and biomechanical functions	*	*	*	*	*	*
THE DESIGN OF MAN-MACHINE INTERFACE						
1. Visual displays						*
2. Auditory displays						*
3. Kinesthetic and tactile displays	*				*	*
4. Controls	*				*	*
5. Specialized input devices	*				*	*
6. Display-control dynamics	*				*	*
7. Workspace layout	*	*	*	*	*	*
8. Equipment design	*				*	*
9. Physical factors affecting human performance and processes	*	*	*	*	*	*
10. Specialized and protective clothing and equipment	*	*	*	*		

[1]After *Ergonomics Abstracts*, Taylor and Francis, Ltd.
[2]After Armstrong and Kochar (1982).
[3]Percentage of Restricted Persons Affected.

Final remarks

The most disabling conditions resulting in limitations of activity in the United States population under 65 years of age, and the related areas of ergonomic research and applications, are shown in Table 4. It can be seen that there is an enormous potential for ergonomics to contribute to the rehabilitation process more fully than it does at present.

Nagi (1969) pointed out that the potential for rehabilitation of a disabled individual depends upon three factors:

(1) characteristics of impairment (courses of underlying pathologies if any, functional limitations imposed, residual capacities, and prognosis for remediability of conditions and for other of restoration),

(2) the definition of the situation by the disabled, their reactions, and motivations, and

(3) the environmental conditions, especially the definition of the situation by significant others, such as those in the family, health, and work settings.

Nagi (1969) also indicates that the third group of factors have a pronounced effect on prospects for rehabilitation and, therefore, increased attention should be given to means of including favourable changes in the environment of the disabled. Undoubtedly, ergonomics can play a significant role in solving many of the above issues for the benefit of humanity.

References

Armstrong, T. J., and Kochar, D. S., 1982, Work performance and handicapped persons. In *Handbook of Industrial Engineering*, edited by G. Salvendy (New York: John Wiley & Sons), pp. 6.7.1–7.18.

Belknap, R. G., 1985, Accidents off-the-job. In *Encyclopedia of Occupational Health and Safety*, 3rd Edition, (Geneva: International Labnour Office) pp. 31–32.

Evans, D. J., 1985, Accidents, in *Encyclopedia of Occupational Health and Safety*, 3rd Edition, (Geneva: International Labour Office), pp. 21–33.

Guthrie, D. I., 1982, Rehabilitation in industry. In *Ergonomics and Occupational Health*, Proceedings of the 19th Annual Conference of the Ergonomics Society of Australia and New Zealand, edited by R. Rawling, pp. 44–53.

INTERNATIONAL LABOUR OFFICE, 1978, *Basic Principles of Vocational Rehabilitation of the Disabled*. 2nd Edition, Geneva.

YEAR BOOK OF LABOUR STATISTICS, 1984, International Labour Office, Geneva.

Kailes, J., 1985, Watch your language, please!, *Journal of Rehabilitation*, 51, 68–69.

Karwowski, W., Marek, T., and Noworol, C., 1988, Theoretical basis of the science of ergonomics. *Proceedings of the 10th Congress of the International Ergonomics Association*, Sydney, Australia. (In Press).

Kroll, J., 1985, Disability prevention and rehabilitation. In *Encyclopedia of Occupational Health and Safety*, 3rd Edition, International Labour Office, Geneva, pp. 648–652.

Nagi, S. J., 1969, *Disability and rehabilitation: Legal, clinical, and self-concepts and measurements.* (Columbus, Ohio: The Ohio State University Press).

NATIONAL SAFETY COUNCIL, 1978, *Accident Facts*. Chicago, Illinois.

Rohan, P. 1985, Accidents, Occupational domestic. In *Encyclopedia of Occupational Health and Safety*, 3rd Edition, (Geneva: International Labour Office), pp. 28–30.

Singleton, W. T., 1982, *The body at work: Biological ergonomics*, (Cambridge: Cambridge University Press).

UNITED STATES CONGRESS, 1985, *Preventing illness and injury in the workplace*. Office of Technology Assessment, Washington, D.C., Publication OTA-H-256.

PART II
PHYSICAL DISABILITY: PREVENTION AND REHABILITATION

2.

Driving for handicapped people

Christine M. Haslegrave

Department of Production Engineering and Production Management
University of Nottingham
University Park, Nottingham NG7 2RD
UK

Abstract

Physical handicaps often make it difficult to use an ordinary production car. Access can be a major problem for both driver and passenger, while the design of controls has to be matched to the functional capabilities of the individual driver. Provision of an adequate driving posture and storage for a wheelchair also have to be considered. Many of these problems can be overcome by suitable adaptations to the vehicle. Recent work in this field is reviewed to summarize the ergonomic guidelines which are available.

The driving task also requires cognitive skills and adequate speeds of reaction as well as physical skills and capabilities. Better methods of assessing these are gradually being developed, but much further work is necessary in this area. Disabled drivers need instructors who are skilled in training for their special needs.

New developments have occurred in powered controls and in their mode of operation, allowing greater numbers of very severely disabled people to drive. Steering controls can now be produced which require virtually zero force or very little movement to operate, and driving with a single multi-function control can be expected in the very near future.

Introduction

The ability to travel and take part in everyday activities is one of the most important concerns for disabled people, whether they wish to do the shopping, get to work, visit friends or go on holiday. Their travel needs are much the same as those for the population as a whole, depending of course on their age and other factors such as employment, but access to public transport can frequently be very difficult. People with walking difficulties have problems in reaching a bus stop, while many elderly people find it difficult to climb up steps to board a bus or train, or even to reach their seat in a moving vehicle. Those using a wheelchair have even greater difficulties with access to many forms of transport. Many of these problems have been discussed in detail by Brattgård (1978) and Shepard (1978) among others, while Brooks *et al.* (1974, 1978) have looked at the ergonomic implications for the design of buses.

The mobility provided by a car can obviously give much greater independence and opportunities for work and social activities for people with physical handicaps, although they may still find it difficult to use an ordinary production car. The design of mass-production cars very rarely takes account of the needs of drivers outside the normal design limits (whether they are very tall, very small, overweight or have other physical handicaps), since car manufacturers are inevitably constrained by economic considerations. Space in the vehicle is also often at a premium because of pressures for fuel economy and improved aerodynamics. Difficulties may therefore be encountered by both drivers and passengers who are handicapped, but most of the problems can be overcome by suitable adaptations to the car. The proper application of ergonomics can play a large part in the design of such adaptations for the layout and controls, and also in the training of drivers.

Problems of access and driving controls

In a sense, the constraints mentioned present very traditional ergonomic problems. These frequently mean that the conventional layout of the car has to be altered for a driver who is physically handicapped, so that he can reach the controls or to ensure that he has an adequate view of the road. Both drivers and passengers may also need special seating to give them proper support or a more comfortable driving posture.

Access to the car is often a major problem for both drivers and passengers. People with stiff or immobilized joints can have difficulty in getting their legs into the car, while others find it difficult to bend sufficiently for their heads to clear the top of the door. Others lack the strength to support themselves while transferring to the car seat, and wheelchair users have special problems since they also need adequate storage for a wheelchair, as do those using a walking stick, crutches or other aid. The provisions needed by a driver will depend upon whether he wishes to be independent, or whether he will have assistance in getting in and out of the car. Passengers often have assistance from others, but may be more severely disabled than drivers, and it may be necessary to provide assistance for their helpers in terms of lifts or hoists. They may also need special safety belts to give them support while sitting during the journey.

Most importantly the driver has to control the vehicle safely, which means that the controls must be of a type which he can operate despite his handicap, taking into account direction and range of movement as well as force of operation. On production cars the control forces for steering, service brake and parking brake controls are often too high for drivers with severe physical handicaps. Other controls also have to be considered since it is just as important to be able to operate the minor controls such as turn signals and light switches and even the door handles and locks.

This lists just some of the considerations related to the use of cars by disabled people. A study by the Motor Industry Research Association (MIRA) covering a large number of disabled people in Britain showed that the main problems they encountered in using cars were: spatial accommodation for access, an adequate driving posture, suitable mode of operation of controls, the strength needed to operate the controls, and storage of equipment (MIRA, 1979).

The study however showed that for the great majority these problems could be overcome by relatively simple solutions using ordinary production cars. A small number of drivers needed major modifications to the car controls, and the study showed that for these the control forces were the most important problem. The most difficult aspects of access were found to be the dimensions of the car doorway and the leg room which was available. These could necessitate more extensive modifications to the vehicle structure where people were not satisfactorily accommodated in a standard car. It may be stating the obvious to add that a driver needs to choose his model of car carefully, to minimize the extent of modifications which have to be carried out, since layouts and packages differ considerably. However, there are also quite subtle differences in design even between different versions of the same car model, and these can have major effects on their suitability for adaptation, so that it is necessary for anyone contemplating the conversion of a car to consider the design in some detail.

Conversion of vehicles

One possible solution to the difficulties could be the design of a special car for handicapped people. This concept has often been debated and various prototypes have been constructed, while one or two have in fact been produced and marketed. There are however good arguments against this approach. Firstly, it is relatively expensive to produce a car for a small market, and many of the special features which would have to be included (for instance facilitating transfer from a wheelchair) would not be needed by the majority of drivers who had less severe disabilities. There are too social and psychological reasons which make a special vehicle undesirable, because it makes the user's handicap conspicuous. Finally, it was shown by the MIRA study (1979) that a special car would not even be a practical solution because the handicaps which have to be catered for are so diverse, differing widely in degree and severity. A single car design could not satisfy all the requirements simultaneously. However, some of the features which are needed could easily be incorporated by production car manufacturers at the design stage, and a study by the Institute for Consumer Ergonomics (ICE, 1985) showed that such improvements would benefit many able-bodied users as well.

In most countries, therefore, physically handicapped drivers use ordinary production cars which have been modified for their particular needs. Adaptations range from the simple relocation of switches to the very complex conversion of the steering mechanism for foot operation. The most common type of conversion is the replacement of the brake and accelerator pedals by controls that can be operated by the hands instead of the feet. This is adequate for many drivers with leg disabilities, such as paralysis or amputation of the lower limbs. In fact, those with a single leg disability would normally just use a car with automatic transmission which avoids the use of a clutch pedal. Other simple conversions include such devices as quick release mechanisms to avoid the need for strength in releasing a parking brake, or a handgrip attached to the steering wheel to make it easier to grasp. The range of types of adaptation which are available in Europe have been described by Haslegrave (1986), while more detailed catalogues have been published by motoring organizations and assessment centres in Europe and the United States (see for example the Veterans Administration, 1978 or Less *et al.*, 1978b).

Most conversions are carried out by small workshops or specialist firms, although some common adaptations (such as hand controls for brake and accelerator) may be manufactured in standard form in relatively large numbers or even as kits for fitting by local garages. However, even these should be fitted by mechanics with some training since the importance of fine adjustments is often not appreciated by garage mechanics. A good ergonomic layout is very important in matching the adaptations to the needs of the individual drivers. For instance, a brake lever placed below the steering wheel is frequently used in conversion to hand-operation, and the driver has to be able to operate both simultaneously without taking his hand off the steering wheel. In this situation, finger reach and the distance of travel of the lever are critical, and it is important that the fitter is aware of this, as well as of the more technical aspects of modifying safety critical systems. Some of the technical considerations are described in Haslegrave (1985).

The key factor in such conversions is that the modifications must be chosen to suit the capabilities of the individual driver, particularly for severely disabled drivers who may be working very much closer to the limits of their strength and capability than is required of the able-bodied driver. Ergonomists have rarely been involved in the design of products or equipment to be used by single individuals (with the exception perhaps of space research), and it is an area in which they can make a large contribution.

The modifications which are needed are not necessarily very complicated or expensive, although often requiring imagination from the engineer or fitter, but more severely handicapped drivers may require radical changes to the controls and occasionally to the structure of the car. The MIRA study (1979) showed that about 10 per cent of people with walking difficulties need adaptations to the controls which are more complex than the simple conversion from foot to hand operation. Very severely handicapped people (such as high tetraplegics or those with weakening diseases of multiple sclerosis or muscular dystrophy) are now being enabled to drive and the possibilities with new technologies are enormous, limited mainly by the cost.

Entry and exit from the car

The first priority is actually getting in and out of the car, and access covers several features of the use of a car. Although principally concerned with the means of transfer to the driver's or passenger's seat, other functions very closely associated are the comfort and support given by the seat, the space available within the vehicle and the positioning of the safety belt or other restraint system. They are all affected by the structural design and layout of the car. Some other more detailed aspects which might be considered are the opening and closing of car doors (design of handles amd locks) and the putting on or release of the safety belt. All these are vital for the motorist to be able to use the car. Another consideration is the ability to store a wheelchair or other aid and the availability of equipment to help in lifting these into the car.

A study of the problems of access was carried out in Britain by the Institute for

Consumer Ergonomics (ICE, 1985) and some aspects were also discussed in the report from MIRA (1979). The features and dimensions of the car which have the greatest influence on ease of access are:

(a) height of seat (and the effective height varies depending on whether the car is parked on flat ground or beside a kerb)
(b) height of doorway, relative both to the ground and to the seat
(c) distance between seat and front edge of doorway (A-pillar)
(d) height of sill above the ground or kerb
(e) lateral distance between seat and outer edge of sill
(f) depth of footwell (height of sill above car floor)
(g) angle at which door opens
(h) obstacles such as door pockets or wheel arches intruding into the footwell
(i) availability of suitable handholds.

These factors may well interact with each other. For instance, less space may be required for leg room as the seat height is raised.

The study by the Institute for Consumer Ergonomics identified (through a postal survey and interviews) many of the problems which disabled people experience with access. Fifty-one per cent reported difficulty in getting into the car and sixty-seven per cent in getting out of it. They followed the surveys with practical trials with some 60 car users: wheelchair users, elderly people with limited strength and agility, and others whose disabilities included lower limb limitations, upper and lower limb limitations and neck/trunk/back limitations. In their report they describe the techniques which their subjects used for getting in and out of the car. The most common method of entry was first to sit sideways on the seat and then to swing or lift the legs into the car. On exit, their subjects reversed this procedure, if necessary using aids such as sticks to ensure stability before rising from the seat. Other techniques were used by people with different handicaps. The various techniques differed from the procedure followed by able-bodied people, who usually moved sideways, put one leg in, and then sat down and brought the other leg in. The observers noted that the able-bodied technique requires "twisting, ducking and transference of weight while balancing on one leg, grasping various parts of the car for support".

Various parts of the car were used for support during entry and exit: top edge and side of door, window ledge, arm-rest, A- and B-pillars (edges of doorway), door sill, cant-rail, grab handle, seat or head-rest, facia panel, steering wheel, gear lever or parking brake. Many of these structures were never intended by the manufacturers to be load-bearing, and so were likely to break or become worn.

The experiments were continued, using a car buck, in a series of trials to establish the most appropriate dimensions for the layout of the door and seat area which are critical for people who have access difficulties. They recommended limits for these dimensions which would be acceptable for 90% and 100% of their sample, and their figures are given in Table 1. The door angle could not be specified for all users because there is a conflict between the position of the door needed to give clearance during entry and the ease with which it can be reached to close it. Similarly, the preferred height of seat depends very much on the disability of the person. They concluded that a

between the seats inside the car. The structure of the car makes both these options difficult. One method often used by drivers (if they have sufficient strength) is to transfer first to the front passenger seat and then move across to the driving seat, pulling the wheelchair into the car behind the passenger seat. Alternatively, the front passenger seat may be removed to provide space for the wheelchair.

If assistance is available the wheelchair may be stowed in the boot of the car or on the roof rack. Powered hoists can be fitted to lift the wheelchair, and if the person has difficulty in lifting himself to transfer out of the chair body hoists can also be attached to the roof or door post of the car.

Some wheelchair users would like to be able to drive without transferring, using the wheelchair as a driving seat. A few conversions are carried out to permit this, although it is essential that the wheelchair is securely and rigidly locked in place. Ordinary wheelchairs are not sufficiently strong and need modification for this. Such conversions are obviously more suited to vans than to production cars, but a few special cars have been designed to be driven from a wheelchair.

Vehicle control

The primary controls in a vehicle are the steering wheel, brake and accelerator. All three functions are vital for safety and the driver must be capable of operating all three. They do not however have to be controls of the conventional form, and can be modified or replaced to compensate for the particular handicaps of individual disabled drivers (subject to the specific legal requirements in any country). Thus a person who has lost the use of his arms can have the steering adapted for foot control (and a number of different technical solutions have been devised for this). Similarly, a paraplegic can have the brake and accelerator pedals converted to levers operated by hand. Other secondary controls (such as parking brake or turn signals) are also of course very important, and in a conversion it is essential to ensure that they are all within reach and can be operated by the driver. Reach can be affected by the use of a prosthesis. Rozier (1977) for instance showed that amputees can have their workspace reduced by 45 per cent with a below-elbow prosthesis or by 83 per cent with an above-elbow prosthesis.

It would be generally agreed that ability to operate a control is determined by the various physical and psychomotor factors of strength, range of movement, co-ordination, stability, speed and reaction time (MIRA, 1979, Less *et al.*, 1978a, Mitchell *et al.*, 1975). These capabilities therefore have to be assessed for a handicapped driver in order to allocate the various control functions. The driver's capabilities should be used to full advantage, but the functions distributed so that none are overloaded. For instance someone with an above-knee amputation and artificial limb might be judged able to use this to operate one of the pedals while converting the other to hand operation. If a driver has the use of only one arm to operate several functions, switches such as turn signals can be operated by movements of the head or shoulder.

Re-design of the controls or layout can improve a driver's response time, as for instance when the brake and accelerator are allocated to pedals operated by the two different legs so that movement time is eliminated. Changes in layout may also help to improve co-ordination. A suitable posture is very important for operation of the controls as well as to provide stability and reduce fatigue. Mitchell *et al.* (1975) note that the reactive muscle spasm which occurs in some neurological conditions can be reduced with a more upright posture with legs at least partly folded. Spasm associated with upper motor lesion is often more likely in reclining postures where the limbs are extended.

The general ergonomic principles which apply to the layout and operation of controls and displays for driving are well summarized by Mitchell *et al.* (1975). They give some detailed advice on location, direction of operation and shaping of controls. After categorizing the different movements available in the body and limbs, they list the possible types of car controls which may use these actions or combinations of them to produce rotary, arc or linear translations.

In assessing an individual's ability to drive a car, the first consideration must be the demands of the standard unmodified car. A range of British car models was measured by MIRA (1979). This gives some indication of the braking and steering forces which are necessary, although power assistance has since become more easily available as an option on cars in Europe. The report however pointed out that the car controls need to be considered in all aspects of their normal use. For instance, steering forces are highest during low speed manoeuvres, as when parking a car, and both force and movement are quite low for most of the time spent driving on the road. The driver may therefore have to be capable of exerting the higher forces, but not need to sustain them over long periods. On the other hand, the force required to operate the accelerator pedal has to be sustained over long periods (although assistance can be provided during motorway driving by using devices such as cruise controls).

The MIRA report showed that braking forces on the foot pedals varied considerably between cars, with forces at the pedal of between 105 N and 210 N to give $0.8\,g$ deceleration. Steering forces without power assistance could be as high as 20 N at each hand when cornering in a car, while forces during parking are even higher. Drivers steering single handed will have to be able to apply double the force. Full lock-to-lock movements of the steering wheel take about $2\frac{1}{2}$–3 turns of the steering wheel, although this is only likely to be needed in parking manoeuvres. While driving in traffic the steering movements needed to maintain the vehicle's heading are in general within half a turn. If the steering system is modified for a driver with a more restricted range of movement, a reduction in the lock-to-lock travel will make the steering heavier unless power assistance is added.

In the study of people with walking difficulties MIRA found that 9.9 per cent of the drivers would need conversions which were more complicated than adaptation of foot controls to hand operation. Their disabilities were mostly generalized joint disabilities or muscular weakness, and localized joint or limb disabilities. The most frequent causes were rheumatoid arthritis, muscular dystrophy, poliomyelitis affecting both upper and lower limbs, and multiple sclerosis. Some of the people studied had disabilities which resulted in restricted arm movement or poor grip in the hands, but many of these were still able to use steering wheel controls. It is possible to change the normal steering method and use techniques such as ''feeding'' the wheel rim through the hands in small movements instead of making large arm movements, providing the driver can make the movements sufficiently fast. Other drivers are able to use a hand or arm guide attached to the wheel, so that the arm is used to wind the wheel from above. Many different techniques and adaptations can be used, so that it is impossible to generalize on suitable types of conversions, particularly for the more severe handicaps. Each driver needs to be assessed with care to identify whether the muscle groups they are able to use give them adequate range and direction of movement for the type of control being proposed.

MIRA found control forces were one of the principal problems in the use of production cars. The drivers with least strength were found to be those suffering from the diseases such as muscular dystrophy, multiple sclerosis and rheumatoid arthritis, as well as tetraplegics and hemiplegics. These would need controls with a very high degree of power assistance, and it is now technically possible to produce conversions with steering forces which are virtually zero. However some drivers may find these fatiguing to use if the controls are so light that they need to make constant corrections. Ultra-light steering conversions are commercially available, so that even tetraplegics with very high lesions are now driving.

Future developments in controls

Many new developments have occurred during the last ten years in powered controls and in mode of operation, several of which are described in Haslegrave (1986). Some of the most complex are remote steering systems. Microprocessor controls are now quite commonly installed on vehicles for many system functions, so that the direct mechanical or hydraulic linkages can be eliminated. Thus the conventional controls can more easily be replaced by others in different locations in the vehicle, making it much easier to choose a position and mode of operation tailor-made for the individual driver.

Over a period, the Transport and Road Research Laboratory in Britain collaborated with the University of Reading in a project (Feaver *et al.*, 1976) which has developed a small joystick control capable of operating the steering wheel through an electrically controlled potentiometer, and this is now manufactured for foot steering. It could however be placed in any convenient position in the vehicle, say for finger operation. A remote-control hydraulic steering system has also been developed in Britain by Steering Developments Ltd (1986), one of the first units being built for a thalidomide victim who needed to have a steering control placed at shoulder level. In this system, the operating force on the steering lever can be adjusted to suit the driver, and may be very close to zero if necessary. They have also developed a remote electronic brake control. A few manufacturers can provide power assistance conversions which reduce the force on the ordinary steering or braking system to zero.

Another recent development is the introduction of voice-operated controls, initially (for safety reasons) for ancillary functions such as the parking brake or seat adjustment. Most excitingly several research projects are aiming at the development of systems which can control steering, acceleration and braking all with a single multi-function control. There are described in Haslegrave (1986), while some of the earlier projects have already been reported by Micaleff and Rabishong (1983) and Mayyasi *et al.* (1973a, b). In the very near future there should be the possibility of driving with the use of a single limb.

While these developments in control design are very much to be welcomed in allowing greater numbers of very severely disabled people to drive, they are posing new problems in assessing the capability of the drivers (Haslegrave, 1985). These include deciding the optimum method and level of feedback of the handling information which is needed from the road. For instance, a design for a speed sensitive variable steering characteristic has already been proposed as an alternative to the conventional characteristic (Nishikawa *et al.*, 1979) and force feedback on the steering might be considered. One problem causing discussion is whether ultra-light controls (with high power assistance) give adequate feedback of handling information, either for able-bodied drivers or for people with very weak arms.

Mitchell *et al.* (1975) cite references in the literature commenting on the advantages and disadvantages of different dynamic characteristics for tracking tasks. There are three basic alternatives: static (isometric) controls which do not move and are force-sensitive, freely moving (isotonic) controls which are position-sensitive, and controls with variable dynamics which may be sensitive to steering angle or vehicle speed. They were forced to conclude that the evidence on the best performance is conflicting.

The cognitive performance requirements of the driving task also need study. There is much scope for further ergonomic research in these fields, and the new developments may well initiate some new fundamental research into the driving control system.

Measurement of driving capabilities

The next question is how the driver's capabilities can be measured and matched to the demands of the standard car or the modified controls, but relatively little research has been done in this area. This type of assessment is not concerned with a driving test as such, which is a legal requirement which in most countries is identical or very similar to the test for able-bodied drivers. Assessment should be primarily concerned with determining the most suitable adaptations and training which a driver will need to compensate for his handicaps and to enable him to drive safely and efficiently.

Bøgh and Poulsen (1967) studied the "demand/ability relation" for 50 handicapped motorists at the Polio Institute in Copenhagen by measuring their maximum isometric force while performing movements similar to those used during actual driving. They also measured the same movements in their subjects' own cars and found that "the handicapped motorists in a considerable number of cases are subjected to a very high degree of stress". For instance, in de-clutching able-bodied motorists used only 6 per cent of their maximum strength, while some of the handicapped motorists required as much as 65 per cent of their maximum strength, and two-thirds had to use between 25 and 50 per cent. The measurements in this study were carried out with strain gauge dynamometers attached to the controls in both the car seating buck and the individual cars.

MIRA (1979) also assessed the capabilities of groups of severely disabled people with various types of handicap, measuring the maximum isometric forces which they could apply to a steering wheel and brake control applied by foot or by hand. Those already able to drive were divided into three groups: those with access or space problems, those with modified controls and those with strength problems, to investigate the degree of interaction between the different requirements for the car design. It was found that, as might be expected, force capability for both steering and braking was highest among the group who reported access problems, but the variations within each group were considerable. The drivers who had the least strength were those with weakening diseases of multiple sclerosis, muscular dystrophy and rheumatoid arthritis. In general, they had greater strength in pulling a hand control towards them than in pushing it away, but during heavy braking in a car their body weight would assist in the pushing action, so that there would not necessarily be any reason for overriding the normal stereotype of pushing a control forwards to brake. Operation of the parking brake was also studied, and several drivers (with handicaps such as tetraplegia or muscular dystrophy) were unable to use the push button release.

The driving buck at the Polio Institute is still used to measure isometric strength of drivers for both steering and braking (MIRA, 1979). Many of the driving simulators used nowadays at assessment centres permit measurements to be made in the same way. Care has to be taken in these assessments, since with many disabilities strength and other capabilities vary from day to day, so that a single test session may not be sufficient. With controls such as the steering wheel, measurements are commonly made at intervals throughout the full range of movement to identify the most limiting conditions for operating the control. In this way it is possible to determine that the driver is capable of the full range of movement which is required, but no criteria have yet been suggested for the speed of movement which is required for the driving task (which is of course dependent on the type of control being used). This is generally determined by subjective judgement of an experienced assessor, and confirmed during road trials.

During driving, a driver's strength may be reduced through fatigue, particularly if a load or grip has to be sustained over a long period, or if the control has to be applied frequently (as with the clutch or brake during town driving). A study was made of this by Molbech (1963), who concluded that ''short lasting isometric contractions can be endured indefinitely if only 75 per cent of the maximum strength is used and, repeated 24 times per minute, if 65 per cent of maximum strength is applied''. His observations of rush hour traffic in cities showed that the clutch may be operated up to 12 times per minute, and the brake less frequently.

Reaction time is very important for safe driving and can be easily measured for both visual and auditory stimuli, but there is as yet no consensus on a safe criterion nor research evidence on which to base this (Haslegrave, 1986). Reports quoted there for the UK indicated that 0·7 second was taken as an adequate pure reaction time, although response to road situations involving decision-making would be longer, and a limit of twice reaction time was suggested for a normal driving response. Less *et al.* (1978a) in America use 0·75 second as a criterion for normal time to move from accelerator and push down on the brake. Bøgh and Poulsen (1967) measured reaction times for operating the controls, but found no difference between their able-bodied subjects and the handicapped drivers tested (who all had walking difficulties). Richter and Hyman (1974) found that reaction times were improved by over 25 per cent with hand-operated controls, because the driver did not have to move from accelerator to brake. They also tested a trigger switch, with which reaction time was more than halved.

Other functions are sometimes included in assessments. Less *et al.* (1978a) suggest criteria for assessing co-ordination, where subjective evaluation is considered adequate although various standardized tests can be used. In their view the important factors are eye–hand, eye–leg, head–hand, hand–leg and leg–leg coordination. Thompson *et al.* (1980) have proposed using a Critical Tracking Task in a driving simulator to evaluate the driver's perceptual, central processing and motor response delay time. This monitors the score for a subject carrying out a steering test likened to "steering a car on a slippery surface while the speed of the car gradually increases".

Less *et al.* (1978a) describe the methods used to assess potential drivers at the Human Resources Center in New York, where they used both functional and in-car measures. These are typical in scope (if not in documentation) for many assessment centres. Both evaluations are intended to be administered by a physical therapist and an experienced driving instructor. They are concerned with rating the performance of different muscle groups in terms of strength, range of motion, coordination, speed and reaction time, and the tests are carried out after the controls have been suitably adapted. They note that a low score on one task in the battery of tests could be compensated by a high score on another task. The example quoted is the use of good peripheral vision, which is balanced against low speed of neck rotation. During the in-car evaluation, they also assess motor perception, depth and spatial relations as well as general driving behaviour. They summarize the assessment by saying 'The effectiveness of the evaluation depends largely on the ability of the evaluator to observe performance and make intelligent assessments of functional limitations. At all times it must be remembered that the purpose of the evaluation is not to "pass" or "fail" the driver trainee, but rather to serve as a tool which would allow driving instructors to make realistic decisions concerning the potential of the physically disabled student.'

One issue which is often raised is the question of safety both for the driver and for other road users, sometimes used as an argument for requiring some form of objective measures to assess and test drivers. The review of practice in Europe (Haslegrave, 1986) found no evidence to suggest that drivers with modified controls have types of accidents which are different from any other driver, but there was some evidence to suggest that they may have a lower accident rate than able-bodied drivers (although no official separate statistics are collected from this category of driver).

Classification of handicaps

It is in fact very difficult to gauge the scale of the need for adaptations to cars for disabled drivers. The European Conference of Ministers of Transport (ECMT, 1986) noted that there is little standardization of data collected by different countries, and such data is rarely collected specifically with mobility handicap in mind. Haslegrave

(1986) tentatively estimated that in Europe 0·5 per cent of drivers may be physically handicapped, and that about 0·2 per cent may require adaptations to their cars. In Britain for instance, this would correspond to 70 000 drivers under the age of 65 years.

The information on the numbers of people who are disabled is usually collected from medical statistics based on classifications of injuries and diseases. This does not give a reliable guide to the handicaps experienced when performing a particular task. The physical handicaps resulting from a disease or accident can vary widely in type and severity. There is a great difference between the medical classification and assessment of severity for the purposes of treatment, and classification for the ordinary needs and activities in daily life. Information on the first of these is the most easily obtained. As Agerholm (1975) has said "although we have an effective and comprehensive classification and nomenclature of the diseases, disorders and defects which handicap, we have nothing comparable for the handicaps themselves." Haber (1971) has shown (in an analysis of statistics from the US national survey of the non-institutionalized civilian population in 1966) that the relationship between disease and severity of disablement was not strong. He could find no consistent pattern in the relationship.

It would be useful to have a classification of functional handicaps which could be related to specific design problems. This would help designers in producing aids and adaptations, as well as in planning provisions on a larger scale. Harris (1971) developed a classification scheme for handicaps experienced in various activities of everyday life, as part of a very detailed survey of handicapped and impaired people in the British population. Her analysis showed that approximately 65 per cent of those very severely, severely or appreciably handicapped were 65 years of age or older. These would therefore be considered as car passengers but would be less likely to want to drive. Similar literature in the United States was reviewed by Shepard (1978) to assess transportation travel demand for handicapped persons.

However, although such overall statistics give guidance on the total numbers of people likely to have special needs, they cannot give the fine detail needed for the design of particular equipment. At present, the information available in terms of medical classification of diseases or impairments must be used as far as possible and interpreted in the light of other information on the effects of different disabilities. The principal problems which affect the driving task are caused by locomotor, sensory and mental impairments, and these may be temporary, permanent or progressive. Armstrong and Kochhar (1981) for instance compiled a list of the major aspects of performance (such as strength, dexterity, motor co-ordination or cognition) which are affected by the most common types of disorder. Mitchell *et al.* (1975) also listed the abnormalities of movement (in amplitude, force and control) which occur with different types of impairment.

Although a particular disability can arise from many causes, these may be grouped in broadly related categories which result in similar types of handicap. The MIRA study (1979) developed a classification of functional disabilities which could be related to the requirements of the different elements of the driving task. The classification was used in analysis to identify correlations between the types of disability and the modifications needed to the drivers' vehicles, since for this particular study functional information could be obtained from medical records.

Table 2 shows the classification which was chosen, where diseases and other causes of handicaps were divided into seven categories according to the principal effect likely to influence the ability to drive or to enter and leave the car. It should be noted that the list in Table 2 was derived for the particular population in the study and is cited here as an example, not as a comprehensive list. There will of course be some overlap between the groups, since any disability may have multiple aspects.

Table 2. Scheme of disability classification (MIRA, 1979)

Group 1. Loss of function
   e.g.   Amputation of a limb
          Paralysis and paresis
          Cerebral palsy

Group 2. Localized joint or limb disability
   e.g.   Osteoarthritis
          Arthrodesis
          Spinal disc lesion
          Trauma

Group 3. Generalized joint disability and muscular weakness
   e.g.   Rheumatoid arthritis
          Multiple sclerosis
          Muscular dystrophy

Group 4. Spinal disabilities or size/reach limitation
   e.g.   Ankylosing spondylitis
          Small stature

Group 5. General weakness
   e.g.   Cardio-respiratory disease
          Malignant disease

Group 6. Lack of coordination
   e.g.   Ataxia
          Parkinson's disease

Group 7. Disorder of consciousness
   e.g.   Epilepsy

The first three groups relate to disabilities of the limbs, which may be the loss of a limb, restricted movements of joints or muscular impairment affecting one or more limbs. The fourth group includes disabilities of the spine which affect the ability to reach or turn, as well as other disabilities related to the space needed in the vehicle or the positioning of the controls. Diseases of the spine (e.g. spinal disc lesions) have however been placed in Group 2 since they are frequently associated with leg disabilities and their predominant effect is often a severe loss of power or movement of the legs with a less severe restriction on spinal movement. The fifth group contains diseases causing general weakness which may severely affect walking ability and way of life, but have less effect on the limb functions related to driving a car other than the strength needed to operate controls.

Agerholm (1975) went further and produced a classification relevant to the problems of handicap, in which the basic structure consisted of nine Key Handicaps: locomotor, visual, communication, visceral, intellectual, emotional, invisible, aversive and senescent handicaps. These were further subdivided into 41 components of handicap, with further increasing detail as necessary. This scheme should prove very useful in assessing the driving task if it is possible to define the categories in sufficient detail, but as yet we do not have the handicap profiles and national statistics based on this type of classification.

Table 3.   Measures of mobility (MIRA, 1979)

	Level of mobility
Ambulant	– Able to walk for at least 60 metres, with or without the use of a stick
Semi-ambulant	– Unsteady, unable to walk more than 60 metres, or relies on one or two sticks or a crutch to stand or move, but can walk around the car, position himself for access and lift objects in and out
Assisted ambulant	– Relies heavily on crutches or two sticks to support weight, or uses walking frame, but can with difficulty walk around the car, position himself for access and lift objects in and out
Non-ambulant	– Able to move only very short distances with crutches or walking aid, or confined to a wheelchair, and cannot walk around the car and may need assistance in entering
	Ease of entry and exit
Obstructive	– Rigidly extended shoulder, hip or knee – Arthrodesed spine
Non-functional	– Flaccid paralysis or absence of limb (where artificial limb is not used) – Rigid shoulder, elbow or knee not in extension, but unable to support body weight independently – Unable to maintain posture of trunk
Semi-functional	– Moderately or severely reduced power or restricted mobility of limb, such that body weight could not be supported independently by that limb – Pain in bending spine
Functional	– Normal or only slight reduction in power and mobility – Limb able to assist in support of body (may include artificial limb)

Table 4.   Levels of limb functions (MIRA, 1979)

Movement	
	– Normal or slight restriction – Fixed joint (not hip or shoulder) including drop foot with spring extension, below-knee or below-elbow prosthesis or caliper (but with no other restrictions in the leg or arm) – Restricted – Total loss of use
Power	
Arm or leg	– Normal or slight reduction – Moderate reduction – Severe reduction
Hand	– Normal or slight reduction – Reduced grip
Control	
	– Normal – Loss of control/coordination or instability of joint

Analysis of functional ability in the driving task

Where more information is available, it is possible to continue the analysis in more detail and to consider specific abilities and to classify these in terms of the functions which can be used for a particular task. A set of functional measures was devised in the MIRA study (1979) for the assessment of some of the elements of the driving task. These are shown in Tables 3 and 4. They were based on capabilities rather than handicaps — for instance on the degree of strength and range of movement in each arm and leg or whether the person was sufficiently mobile to store his walking aid in the back of the car and then walk round to the driving seat. This study had the advantage that the population of interest was well defined and both disabilities and levels of limb functions were well documented in the records. It was therefore possible to consider which limbs had sufficient function for particular aspects of the task (considering arms, hands and legs in turn). The measures could then be matched to the requirements of the driving task in terms of ability to operate the controls, mobility in reaching the car, ease of manoeuvring for getting in and out, and difficulties with seating space and position.

In a restricted workspace such as a car, the ease of entry and exit is mainly affected by the functions of the legs and spine and also by the flexibility of the spine. Thus the functions of each leg and the spine can be rated on the scale given in Table 3, and the need for special modifications can be related to the level of leg and spine functions. The ability to perform motor tasks — to operate controls — was assessed by categorizing the levels of limb functions, as shown in Table 4, in terms of movement, power and co-ordination (or control). A scale is given for each of the three aspects.

A number of analyses were made using the measures described here to study a sample of over 1000 disabled people with walking difficulties, to identify what adaptations they would need to be able to drive an ordinary production car. Their functional disabilities were determined from a study of their medical records and were then related to the requirements of the different elements of the driving task. This covered questions such as the suitability of the height and width of car doorways, the need for special seating or support, the types of controls needed and the need for assistance in lifting a wheelchair into a car. The measures were also correlated with types of injuries and diseases grouped according to the handicaps they might be expected to cause. This provided some guidance on the types of adaptations which would be needed by different groups of handicapped people. The study proved very clearly that medical categories of disability could not be used to specify an individual's requirements for vehicle adaptations, except for very straightforward handicaps such as the amputation of a limb.

Other researchers have used different schemes and scales of functional ability. Feeney (described in Mitchell *et al.* (1975)) categorized arm movement outputs in terms of number of limbs available, their force output as greater or less than 15 N, and movement as greater or less than 5 cm. Mayyasi *et al.* (1973) assessed each limb segment simply in terms of whether it had full function or reduced function (which they regarded as having no function and equivalent to amputation). Less *et al.* (1978a) use a more complex scheme in assessing drivers and make a functional evaluation covering many aspects of performance, each judged on the following scale:

4–normal   performance without difficulty
3–fair   some difficulty
2–poor   inability to perform at an acceptable level without assistive devices
1–weak   inability to accomplish the task even with assistive devices

Finally, although not specifically related to driving, Jefferys *et al.* (1969) devised a series of standardized tests to measure motor impairments and limitations for daily living activities. These scored performance on tests of reaching, grasping and manipulating, as well as strength in the upper limbs and actions such as standing, walking, stepping and bending.

Assessment and training

The various research studies cited above give useful information on the strength capabilities and mobility of disabled people, but the driving task requires cognitive skills and adequate speeds of reaction as well as physical skills. Better methods of assessing these are gradually being developed, but much further work is needed in this area.

Fitness to drive is determined by three main aspects: medical condition, mental condition and functional capacity, while functional capacity itself has five aspects: eyesight, mental faculty, control of posture, and amplitude and power of movement (Mitchell *et al.*, 1975). The criteria used to determine fitness to drive and the way in which these are tested varies to some extent from country to country. There are as yet few objective criteria to assess driving skills. The most commonly applied is a minimum visual capability, which can be measured quite simply although there is still debate over the relationship between different aspects of visual capability and performance in the driving task.

Certain medical conditions may disbar a person from driving, but in most countries the severity of a physical handicap does not prohibit a person from driving, providing suitable adaptations are available for the car and he can demonstrate his ability to drive in a road test. Medical organizations have also published guidance to doctors and other medical practitioners concerning the effects on driving of different medical conditions (as for instance Raffle, 1985).

However, the effects of physical handicaps need to be considered together with the adaptations to the vehicle. In the final analysis this needs to be assessed in a road trial, but many assessment centres are now using simulators to help both in the choice of suitable control adaptations by measurement of force capability and in the assessment of driving performance. These simulators vary greatly in complexity from a simple static seating buck which represents the layout of the car and controls, through controls which are instrumented to measure the force applied, to highly sophisticated systems able to present road scenes and monitor the driver's hazard awareness and reaction time (Cornwell, 1986).

It is important to note that there is as yet little objective data for standards of performance which could be measured in these simulators. There has also been little research to investigate the degree of correlation between measures in a simulator and actual performance on the road. This is an area in which research should be concentrated to gain a better understanding of the driving process, and the simulators themselves should certainly provide a valuable research tool. Two preliminary studies have been reported by Kochhar *et al.* (Kochhar *et al.*, 1981, Boydstun and Kessel, 1980) and Thompson *et al.*, (1980). Both indicate that simulators do have a potential for evaluation of current driving performance, for identification of difficulties with psycho-motor skills, and possibly also for training. Their driving simulators were of the fixed base (or stationary) type with a computer-controlled roadway display. Steering wheel, brake and accelerator were used to control speed, heading and lateral position of the vehicle relative to the roadway display. Both studies were concerned with the visual, decision-making and motor control skills involved in driving. Kochhar and his colleagues studied lateral position control and heading angle control during slow and fast driving in the presence of wind gusts, and they identified differences between a group of subjects with neuromuscular or sensory dysfunctions and a control group.

The complete driving task integrates many skills. Drivers may well be able to compensate for some of their deficiencies by modifying their driving techniques, and experienced practitioners who use simulators constantly stress the need to confirm such assessments with final road trials in an actual vehicle. There is always a danger that disabled drivers will be rejected or discouraged at too early a stage by using a questionnaire approach or set battery of tests, when alternative driving techniques, control modifications or extended physical training would permit them to drive.

Disabled drivers need instructors who are skilled and understand their special needs. An assessment of their difficulties is important at an early stage because many disabled drivers cannot contemplate purchasing or adapting a car until they are sure that they will be capable of driving. Some driving schools as well as assessment centres now have adapted cars available for teaching, although the most severely disabled drivers will still need individually adapted cars. It may also be necessary for a disabled person to improve the strength or range of movement of poorly used muscle groups by physical training over a period of time before he is able to drive. The instructor needs to be able to advise on these matters, and training may take a very much longer time than for an able-bodied driver.

More fundamentally, many severely disabled drivers have cognitive, perceptual or sensory handicaps which have to be considered during driver training, particularly in the case of progressive disabilities (Simms, 1986a, b, Bardach, 1969, Hofkosh *et al.*, 1969). Bardach describes the problems resulting from an altered or faulty body image following brain damage, which may cause difficulty in following spatial instructions. Some hemiplegics for instance tend not to scan the side of their body which is disabled, and others with brain damage have difficulty in shifting attention or in selecting the most relevant visual cues to which they should respond. Hemiplegia in particular presents problems during the early period after the person becomes disabled. Later when the handicap is stabilized and the patient has begun to adjust to his new body image, he may be taught to compensate for the deficiencies in spatial awareness.

Such special skills for instructors are now being developed in a more systematic fashion and their own training is beginning to be studied in both America and Europe (Colverd *et al.*, 1978, Less *et al.*, 1978b, Wider undated).

References

Agerholm, M., 1975, Handicaps and the handicapped — a nomenclature and classification of intrinsic handicaps, *Journal of the Royal Society of Health*, **1**, pp. 3–8.

Armstrong, T.J., and Kochhar, D.S., 1981, Work performance and handicapped persons. In *Handbook of Industrial Engineering*, edited by G. Salvendy, (New York: John Wiley & Sons), pp. 6,7,1–6,7, 18.

Asmussen, E., Poulsen, E., and B ogh, H.E. 1964, Measurements of the muscular strength necessary for driving a motor car, Communications from the Testing and Observation Institute Nr. 19, Danish National Association for Infantile Paralysis, Hellerup, Denmark.

Bardach, J.L., 1969, Psychological factors affecting driving behaviour in the disabled, *The Medical Clinics of North America*, **53**, **3**, 692–696.

Bean, L.E., 1969, Car transfer technique of a patient with quadriplegia, *Physical Therapy*, **49**, 602–4.

B ogh, H.E., and Poulsen, E., 1967, Investigation of the demand/ability relation in handicapped motorists, *Communications from the Danish National Association For Infantile Paralysis*, 26, (Hellerup, Denmark: Danish National Association for Infantile Paralysis).

Boydstun, L.E. and Kessel, D.S., 1980, Techniques for assessment and training of perceptually disabled drivers, Center for Ergonomics Report, (Ann Arbor, USA: Center for Ergonomics, The University of Michigan).

Brattgård, S–O., 1978, Swedish experience in modifying vehicles and infrastructure. In *Mobility for the Elderly and Handicapped*, Proceedings of the International Conference on Transport for the Elderly and Handicapped, Cambridge, 4–7 April 1978, edited by N. Ashford and W.G. Bell, (Loughborough, UK: Loughborough University of Technology), pp. 51–60.

Brooks, B.M., Ruffell–Smith, H.P., and Ward, J.S., 1974, An investigation of factors affecting the use of buses by both elderly and ambulant disabled persons, British Leyland (Truck and Bus Division) Contract Report, (Leyland, UK: Leyland Vehicles).

Brooks, B.M., Edwards, H.M., Fraser, C.R. and Levis, J.A., 1978, Passenger problems on moving buses, Leyland Vehicles Contract Report, (Leyland, UK: Leyland Vehicles).

Colverd, E.C., Less, M., DeMauro, G.E. and Young, J., 1978, Teacher's preparation course in driver education for the physically handicapped: a sample course, Human Resources Center Publication, (New York: Human Resources Center, Adapted Driver Education Unit).

Cornwell, M., 1986, A study of the car control and seating requirements of 908 disabled people assessed by the Banstead Place Mobility Centre between May 1982 and April 1985, *Proceedings of the 4th International Conference on Mobility and Transport for Elderly and Disabled Persons*, July 1986.

Deyoe, F.S., and Anderson, W., 1970, Wheelchair-car transfers for quadriplegics, *American Corrective Therapy Journal*, **24**, 130–132.

European Conference of Ministers of Transport (ECMT), 1986, Transport for disabled people — international comparisons of practice and policy with recommendations for change, (Paris, France: OECD Publications Service).

Feaver, J.L., Penoyre, S. and Stoneman, B.G., 1976, A drive-by-wire vehicle control system for severely disabled drivers, I.E.E. Conference Publication 141, *Proceedings of I.E.E. International Conference on Automobile Electronics*, July 1976.

Haber, L.D., 1971, Disabling effects of chronic disease and impairment, *Journal of Chronic Diseases*, **24**, 469–487.

Harris, A.I., 1971, *Handicapped and impaired in Great Britain*, (London: HMSO).

Haslegrave, C.M., 1985, Car controls for physically handicapped drivers, *Proceedings of the Tenth International Conference on Experimental Safety Vehicles*, Oxford, 1–4 July 1985.

Haslegrave, C.M., 1986, Car control conversions for disabled drivers, TRRL Research Report RR29, (Crowthorne, UK: Transport and Road Research Laboratory).

Hofkosh, J.M., Sipajlo, J. and Brody, L., 1969, Driver education for the physically disabled — evaluation, selection, and training methods, *The Medical Clinics of North America*, 53, 3, 685–689.

Institute for Consumer Ergonomics (ICE), 1985, Problems experienced by disabled and elderly people entering and leaving cars, TRRL Research Report RR2, (Crowthorne, UK: Transport and Road Research Laboratory).

Jefferys, M., Millard, J.B., Hyman, M. and Warren, M.D., 1969, A set of tests for measuring motor impairment in prevalence studies, *Journal of Chronic Diseases*, **22**, 303–319.

Kochhar, D.S., Miller, J.M. and Boydstun, L.E., 1981, An approach to training the disabled driver, *I.E.E.E. Transactions*, 375–379.

Less, M., Colverd, E.C., DeMauro, G.E. and Young, J., 1978a, Evaluating driving potential of persons with physical disabilities, Human Resources Center Publication, (New York, U.S.A.: Human Resources Center, Adapted Driver Education Unit).

Less, M., Colverd, G.E., DeMauro, G.E. and Young, J., 1978b, Teaching driver education to the physically handicapped, Human Resources Center Publication, (New York, U.S.A.: Human Resources Center, Adapted Driver Education Unit).

Mayyasi, A.M., Pulley, P.E., Hyman, W. A. and Swarts, A.E., 1973, Categorization of disabilities and functional limitations imposed in the driving task, *SAE Paper 730466*, SAE Automobile Engineering Meeting, Detroit, 14–18 May 1973.

Mayyasi, A.M., Pulley, P.E. and Swarts, A.E., 1973a, A complete one-handed pistol-grip controller, *SAE Paper 730469*, SAE Automobile Engineering Meeting, Detroit, 14–18 May 1973.

Mayyasi, A.M., Pulley, P.E. and Swarts, A.E., 1973b, Complete single-footed automobile controller, *SAE Paper 730470*, SAE Automobile Engineering Meeting, Detroit, 14–18 May 1973.

Micaleff, J.P., and Rabishong, P., 1983, Motorized assistance for the handicapped: control systems, vehicles, walking machine, *Proceedings of the First European Workshop on Evaluation of Assistive Devices for Paralysed Persons*, Milan, 27–29 April 1983.

Mitchell, J., Feeney, R.J. and Corney, G., 1975, Car controls for the disabled — an analysis of the problems and indication for their solutions, ICE Report (Loughborough, UK: Institute for Consumer Ergonomics).

Molbech, S.V., 1963, Average force at repeated maximal isometric contractions at different frequencies, Communications from the Testing and Observation Institute 16, (Hellerup, Denmark: Danish National Association for Infantile Paralysis).

Motor Industry Research Association (MIRA), 1979, Personal transport for disabled people, *MIRA Project Report K44554*, (Nuneaton, UK: Motor Industry Research Association).

Nishikawa, M., Toshimitsu, Y. and Aoki, T., 1979, A speed sensitive variable assistance power steering system, *SAE Paper 790738*, SAE Passenger Car Meeting, Dearborn, 11–15 June 1979.

Raffle, A. (Ed.), 1985, *Medical aspects of fitness to drive: a guide for medical practitioners*, (London, UK: The Medical Commission on Accident Prevention).

Richter, R.L., and Hyman, W.A., 1974, Research note: driver's brake reaction times with adaptive controls, *Human Factors*, **16**, 1, 87–88.

Rozier, C.K., 1977, Three-dimensional work space of the amputee, *Human Factors*, **19**, **6**, 525–533.

Shepard, M.E., 1978, Handicapped persons in the U.S. and public transportation travel demand: a literature review and annotated bibliography, *General Motors Research Publication GMR-2513*, (Warren, USA: GM Research Laboratories).

Shotton, M.A., 1985, Belt up — if you can!, *Applied Ergonomics*, **16**, **2**, 127–133.

Simms, B., 1986a, Learner drivers with spina bifida and hydrocephalus: the relationship between perceptual-cognitive deficit and driving performance, Z. Kinderchir. (in press).

Simms, B., 1986b, The brain damaged learner driver: screening considerations, (in preparation).

Steering Developments Ltd, 1986, Information Leaflet, (Hemel Hempstead, UK: Steering Developments Ltd).

Thompson, R.R., Repa, B.S. and Leucht, P.M., 1980, Evaluating the driving potential of the handicapped using a simulator, *SAE Paper 800421*, SAE Congress, Detroit, 25–29 February 1980.

Veterans Administration, 1978, Add-on automotive adaptive equipment for passenger automobiles, Prosthetics and Sensory Aids Service Program Guide M–2 Part IX G–9, (New York, USA: Veterans Administration).

Wider, H., (undated), Korperbehinderte al motorfahrzeuglenker. (Zurich, Switzerland: Strassenverkehrsamt des Kantons Zurich).

3.

The partnership of ergonomics and medical intervention in rehabilitation of workers with Cumulative Trauma Disorders of the hand

Waldemar Karwowski

Center for Industrial Ergonomics
University of Louisville
Louisville, Kentucky, 40292, USA

and

Morton L. Kasdan

Surgery of the Hand
Louisville, Kentucky, 40202, USA

Abstract

This paper addresses both the treatment and prevention of cumulative trauma disorders (CTDs) of the hand, and discusses selected ergonomic aspects of rehabilitating workers affected by these ailments. The first part of the paper focuses on the most common problems of CTDs encountered in private practice of surgery of the hand. Conservative and medical management of nerve entrapments, inflammatory pathology of tendons, vascular disorders, and diseases of the joints are discussed. In the second part of the paper, several case studies are presented illustrating ergonomic intervention to reduce the prevalence of CTDs in the workplace.

Introduction

Economic competition in manufacturing today is so intense that human efficiency is an absolute necessity for survival. Workers are expected to produce, in larger volume, a better quality product than ever before. Modern industry has placed demands on its employees to perform at such a high level of efficiency that the human hand is wearing out (Armstrong, 1986).

Due to social and economic changes in the last twenty years, many contemporary factories employ a different character of workers than in the past. Most employees are no longer willing to tolerate job-related musculoskeletal strains, pain and physical suffering. At the same time, however, the number of workers totally or partially disabled because of job-induced cumulative trauma disorders (CTDs) has reached an all time high (Brown, *et al.*, 1984; Kelsey, *et al.*, 1980). Many major plants are having significant problems maintaining desired production volume and labour standards. In view of this, it is not surprising to find that the management of American industry is looking to both industrial engineers and physicians to keep the assembly lines going.

Unfortunately, there is no quick solution to these problems. Only a better understanding of the engineering aspects of job design by the physician can help the patients. Equally, only by a better understanding of the anatomy and pathophysiology of CTDs can the industrial engineer modify the job design to help both the worker and the employer.

This paper addresses both the treatment and prevention of the CTDs of the hand, and discusses ergonomic aspects of rehabilitating workers affected by these ailments. The first part of the paper focuses on the most common problems of CTDs encountered in private practice of surgery of the hand. A discussion of the conservative and medical management of each disorder is also presented. Four basic topics are discussed: (1) nerve entrapments, (2) inflammatory pathology of tendons (synovium), (3) vascular disorders, and (4) diseases of the skeleton (joints). In the second part of the paper, several case studies are presented illustrating ergonomic intervention to reduce the prevalence of CTDs in the workplace.

Job-related CTDs of the hand observed in private practice

Nerve Entrapments

Peripheral nerve entrapments which may result from work will be considered from proximal to distal. A cervical radiculopathy usually has some form of predisposing problem, namely cervical arthritis. The nerve roots leave the spinal column by the neural foramina. These spaces may already be narrowed because of arthritic spurs. Patients often perform jobs with the neck in an extended position, such as an overhead job on an automotive assembly line (working in a pit underneath the vehicles), or house painting. Treatment should first consist of a job modification. Neurosurgical intervention, a soft cervical collar, and cervical traction should be deferred until there is objective evidence of nerve entrapment (Bradley, 1984).

Thoracic outlet syndrome

Thoracic outlet syndrome has become an umbrella term for nerve and vascular entrapments at the level where the clavicle, first, and second ribs form an outlet tunnel. The neurovascular structures, leaving the chest, pass under the clavicle and over the first and second ribs on their way to the upper extremity. Movements requiring prolonged and vigorous activity with the arm hyperabducted above the head markedly narrow this space. There may be predisposing factors, such as a congenital cervical rib or fascial bands, that limit this space. The diagnosis is not always clear cut and treatment can be very controversial (Derkash, *et al.*, 1981; Machleder, *et al.*, 1985; Roos, 1985). A thorough medical work-up of the patient with radiological and neurological consultations is necessary. The single most important treatment should be a complete change of work activity.

Cubital tunnel syndrome

Cubital tunnel syndrome is entrapment of the ulnar nerve just at and distal to the notch of the elbow. Affected workers frequently perform a job that requires the elbows and forearms to rest against a table-top or work surface. For example, in the automobile assembly plant, pressure is exerted on the proximal forearm and elbow by employees reaching over hoods to install parts and screws. This can produce compression of the ulnar nerve and related symptoms of pain, numbness and tingling, radiating into the little and ring fingers, and decrease of grip power. This can progress to actual atrophy of the intrinsic muscles of the hand if the pathology is not corrected. Complete rest of the involved extremity, with an elbow splint, should be instituted as soon as possible. An explanation of the anatomical problem, and instructions to avoid pressure, i.e. leaning on the affected area, should also be given to the worker.

Operations for decompression of an ulnar nerve entrapment at the elbow consist of simple release of the fibro-osseous tunnel to transfer the nerve to a new anatomical position (Miller and Hummel, 1979). If the neuropathy has progressed to positive electrical diagnostic studies and muscle changes, the prognosis for complete recovery is guarded. When examining patients using the electrophysiological techniques, Payan (1969) found that sensory recovery was as satisfactory following conservative treatment as with surgical transposition of the nerve.

Pronator syndrome

Pronator syndrome is a proximal neuropathy of the median nerve. The term *pronator syndrome* relates to the entrapment or compression of the median nerve in the area between the distal third of the arm and the wrist (Hartz *et al.*, 1981; Howard and Hill, 1986; Johnson *et al.*, 1979). Four areas of compression have been documented. Progressing from proximal to distal, the nerve passes under the arch of the flexor digitorum superficialis muscle; secondly, the nerve may be compressed within the pronator teres muscle; the third area is just at the elbow where the nerve passes beneath the lacertus fibrosis; and, finally, the most proximal lesion can occur in the distal third of the humerus just beneath the supracondylar process in the ligament of struthers.

Any activity that will produce edema around the elbow or forearm can lead to pronator entrapment. Patients will often notice forearm pain and weakness of flexion of the index finger and/or thumb. There may or may not be isolated motor or sensory symptoms. This activity is frequently seen from strenuous flexion of the elbow and wrist (activities that resemble the weight-lifter performing wrist curls). Pronator syndrome is more commonly associated with an acute strain, as opposed to the usual cumulative trauma disorder. However, it may be seen in association with patients who have CTDs such as carpal tunnel syndrome or tendonitis, in addition to median nerve entrapment in the forearm.

The conservative treatment of median nerve entrapment in this area entails intensive follow-up evaluation. This should include the history of any progression or regression of symptoms, physical examination including testing of range of motion and muscle strength, and finally careful myoelectrical diagnostic studies. A resting wrist and elbow splint, restriction of activities, and a nonsteroidal anti-inflammatory agent may provide relief of symptoms. The patients must be cautioned to avoid any heavy lifting or gripping activities. It is worthwhile to wait 6–12 weeks with regular evaluations before considering surgical intervention. The operation is an extensive procedure requiring exploration of the nerve from above the elbow to the distal forearm. Patients will be on restricted activities following surgery for a minimum of three months.

Carpal tunnel syndrome

Carpal tunnel syndrome is the best known of the peripheral nerve entrapments and CTDs. This nerve compression results in a symptom complex described by patients as numbness and tingling of the thumb, and the index, long and ring fingers. Symptoms may vary from pain in the palm of the hand to a sensation of coolness or a sleepiness of the hand. Typically, symptoms will be aggravated by certain activities, such as repetitive or prolonged gripping or pinching. Nocturnal symptoms are also frequent. The patients describe awakening at night with the hand numb, requiring vigorous shaking of the extremity. A multitude of systemic diseases may be associated with carpal tunnel syndrome. There is also a long list of systemic diseases or neuropathies that can mimic carpal tunnel syndrome. A thorough evaluation by an experienced neurologist is necessary for making the correct diagnosis (Ditmars and Houin, 1986; Kasdan and Janes, 1988). Numbness and tingling in the fingers can be a manifestation of a neurological disorder from the brain to the distal peripheral nerves.

The treatment of carpal tunnel syndrome can vary from the most conservative, such as job design changes, to surgery and decompression of the nerve, internal neurolysis, and radical synovectomy. An approach combining all of the conservative methods of management offers the best chance for a resolution without surgery. This would consist of a job change or modification of activities, weight loss (if necessary), treatment of systemic disease, wrist splinting (usually at night), and prescription of Pyridoxine (Ditmars and Houin, 1986; Kasdan and Janes, 1988).

Cumulative trauma disorders of the digital nerve are somewhat unique in that the most frequent cause is a non-occupational problem. Repetitive trauma to the ulnar digital nerve of the thumb is most frequently related to the hobby of bowling (Marmor, 1966; Minkow and Bassett, 1972; Siegel, 1965). As the bowling ball is grasped, the digital nerve is trapped between the proximal phalanx of the thumb and the edge of the hole in the ball. The avid bowler may grasp the bowling ball in excess of 100 times a week. The tremendous force necessary to hold a 12–16 pound bowling ball can produce excessive pressure on the very tender digital nerves. This repeated trauma will eventually produce a fibrosis (or scar) surrounding the nerve. The patient will frequently report a tumour mass in addition to numbness and tingling in the distribution of the involved nerve. This ''tumour'' is the nerve enlarged by the surrounding scar tissue.

The treatment of choice is, of course, to stop bowling. However, some patients will reduce the frequency, but find it necessary to continue the sport as a psychological outlet. In this case, the treatment is a modification of the way the ball is held, and, perhaps, beveling the hole in the ball. This same pathology can be seen in the patient who has a job holding scissors or shears over the ulnar side of the proximal phalanx of the thumb. However, any digit can be involved if the digital nerve is exposed to repeated compression. Surgical management will only be successful if the patient does not return to the same harmful activity.

Tendons

Medical terminology used in describing CTDs of the upper extremity varies. Terms such as focal dystonia, tenosynovitis, tendonitis, repetition strain injury, occupational cramp, and overuse syndrome, have all been used (Fry, 1986). For the sake of brevity, we will use the term most frequently referred to by physicians, namely, tenosynovitis, in the description of this common cumulative trauma disorder.

One of the most prevalent disorders of this kind is the stenosing tenosynovitis, trigger finger (see Figure 1). The tendons develop a restriction in motion where the digits junction with the palm of the hand. This can be related to some form of repetitive gripping activity. The initial symptom may simply be localized pain, progressing to clicking and then sticking of the tendon on flexion and extension, and, finally, loss of motion of the digit.

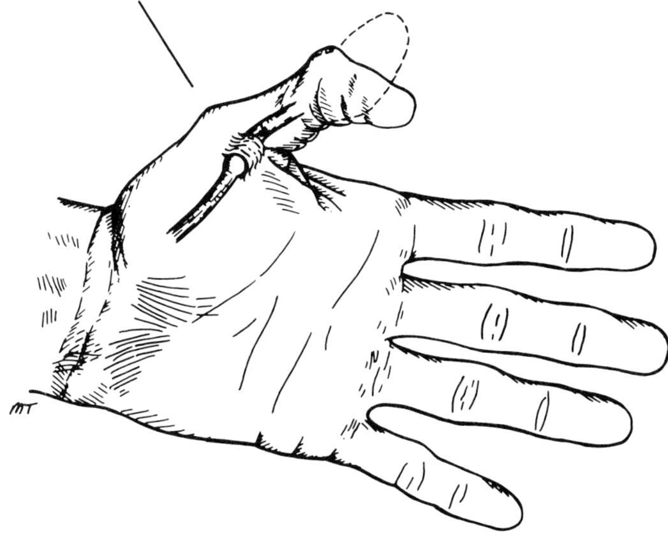

Figure 1. Stenosing tenosynovitis in trigger finger can occur at the A-1 ligament of all five digits.

The second most common form of tendonitis is usually manifested by a rubbing or grating sensation with extreme pain on flexion and extension of the wrist and digits. The area affected is usually the forearm and wrist. The patient is often engaged in an occupation (or hobby) that requires heavy lifting, gripping, or pinching activities with simultaneous motions of flexion and extension of the wrist. This can be treated by immobilization in a splint and restricted activities, depending on the level of the patient's cooperation, and nonsteroidal anti-inflammatory agents. Some success has been reported with the controlled use of vitamins B6, C, and E.

The third form of tenosynovitis frequently encountered is DeQuervain's disease (De Quervain, 1985). This is a painful and sometimes disabling stenosing tenosynovitis on the radial side or thumb side of the wrist (see Figure 2). Job restrictions require avoiding repetitive twisting motions of the wrist, such as wringing out a cloth or sanding a dowel.

Conservative management of tenosynovitis consists of anti-inflammatory agents and a splint to immobilize the symptomatic area. Injection with corticosteroids and a change in the way the patient uses his hands is also helpful. If the patient has not responded in three to six weeks, then surgery should be considered. Surgery is usually curative and the patient may resume the occupation that brought on the problem after about eight weeks. Hypersensitivity of the scar and distal paresthesias are frequently encountered and may be permanent.

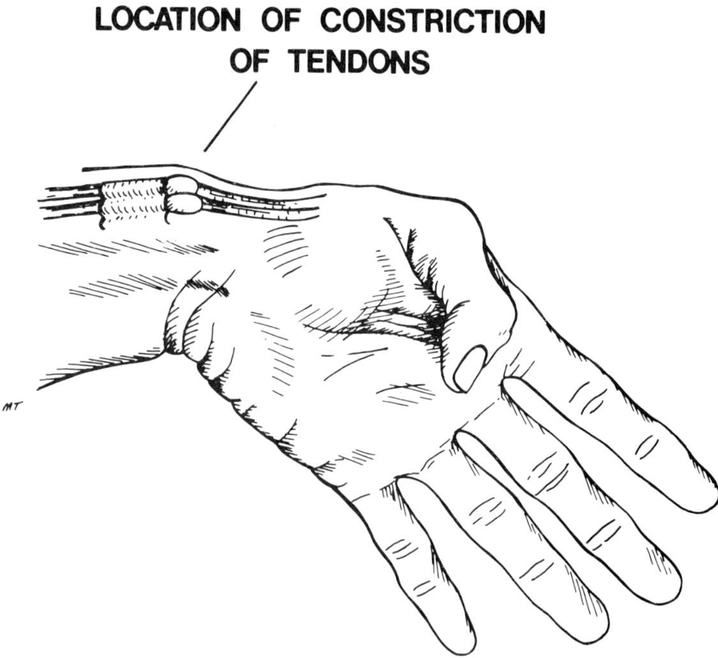

Figure 2. Stenosing tensosynovitis of the abductor pollicis longus and extensor pollicis brevis on the radial side of the wrist is commonly called DeQuervain's.

Lateral epicondylitis

Tennis elbow, or lateral epicondylitis, is typically reported as pain over the lateral aspect of the elbow and proximal forearm. The condition is aggravated by wrist motions and typically has tenderness over the lateral aspect of the elbow. This condition is also described as developing from repetitive activities of the forearm extensor muscle group. Conservative treatment will require restricting activities, a firm wrap around the proximal forearm, and nonsteroidal anti-inflammatory agents. Injection of steroids can provide permanent relief if the patient does not resume the aggravating activities. Various surgical procedures have been reported; most authors describe success. However, it is imperative that the patient is well motivated and cooperative before considering surgical intervention.

Kienbock's disease

Kienbock's disease is a deterioration of the lunate (Taleisnik, 1985), i.e. one of the eight carpal bones between the hand and the long bones of the forearm. This lunatomalacia can be related to either circulatory problems to the bone or repeated compression. The small, very essential, bone is susceptible to tremendous forces, especially when a heavy tool is gripped which produces tense and heavy pounding to the hand transmitted, in turn, to the wrist and forearm. The treatment, initially, is that of job change, rest, and splinting. In more advanced stages, with complete collapse of the bone and a painful wrist, surgical intervention may be necessary. Surgical procedures for this are many and controversial, and the treatment may vary from simple excision of the bone to total wrist fusion.

Ulnar artery occlusion

An aneurysm of the ulnar artery in the proximal palm can result from repeated use of the hand as a mallet. This is frequently seen in automobile mechanics and individuals working on an assembly line who use the heel of their hand to push or pound (Koman and Urbaniak, 1985; Kornberg *et al.*, 1983). The ulnar artery passing through the small Guyon's canal is vulnerable to injury. A wall of the vessel becomes weakened and a false aneurysm can develop. The adjacent ulnar nerve can be compressed, developing into an ulnar neuropathy. Surgical intervention is necessary to remove the aneurysm, which in turn relieves the pressure on the ulnar nerve. This is a preventable lesion. The worker should be trained to avoid repeated pounding or pushing with the heel of the hand. In situations where such actions are necessary, a padded glove should be provided and some type of instrument should be devised to prevent this problem from developing.

Approaches to diagnosis and treatment of CTDs

First step

Evaluating the patient with a possible cumulative trauma disorder involves more than just a visit to the physician's office. The patient may start out by seeing the plant physician, who will, in turn, try an approach of conservative management. This may include rest, nonsteroidal anti-inflammatory agents, and, if no response, a referral to a qualified hand surgeon. Unfortunately, the treatment cannot stop here, and it is necessary to bring in other specialists to work on the problem.

The team approach

The team approach to evaluating and treating CTDs can be developed by building on the relationship between the medical and engineering specialists. This approach, through the combined expertise of several specialists, facilitates the identification of an occupational problem versus a systemic disease. It is not uncommon for a systemic disease to be manifested through a group of symptoms that can resemble a cumulative trauma disorder. By collaborating in the thorough evaluation of the patient, a more exact diagnosis can be established. The interdisciplinary group should be composed of a plant physician, hand surgeon, radiologist, neurologist, clinical psychologist, physical therapist, industrial engineer and an ergonomist. Once a diagnosis is established, a line of communication between the patient, employer, and all specialists involved is essential. This must take the form of written as well as direct verbal communication.

CTDs in the workplace: Review of case studies

Occupational risk factors of CTDs

Cumulative trauma disorders (CTDs) of the upper extremities have been identified as major occupational health and safety concerns in modern manufacturing industries (National Safety Council, 1984; Habes and Putz-Anderson, 1985). According to Armstrong *et al.*, (1986), some of the occupational risk factors of these disorders include: (1) repetitive or sustained exertions, (2) uncomfortable postures of the shoulder (elbows raised above mid-torso, reaching down and behind the torso), forearm (inward or outward rotation with a bent wrist), wrist (palm flexion or hyper-extension, or ulnar or radial deviation), or hand (forced to pinch instead of grip), and (3) mechanical stress concentration near the base of the palm, or on the bottom or sides of the fingers. Incessant mechanical vibrations, cold, or habitual use of gloves may also contribute to the risk of the job-induced CTDs.

As indicated by Habes and Putz-Anderson (1985), an ergonomic evaluation of repet-itive tasks, and subsequent workplace redesign, can prove cost-effective when measured against the time lost, medical treatment, and rehabilitation required for workers suffering from job-related CTDs as a result of performing repetitive, stressful and unnatural movements and/or muscular contractions in the course of their normal workdays.

The case studies reviewed in this paper focus primarily on how hands and arms are affected by peculiarities of the workplace design and task requirements in four modern industrial environments, i.e.: (1) an auto assembly plant, (2) a food-processing plant, (3) a furniture factory, and (4) a clothing factory.

Information gathered during interviews with workers and management, as well as related medical records, helped to identify jobs with a high rate of CTD-related complaints. On these jobs, all tasks involving repetitive upper body or hand movements, especially flexion or extension of the hand in combination with forceful exertions and use of hand tools, were studied. Task analyses were based in data obtained from review of the available job descriptions provided by the plants' industrial engineering departments, on-site visits to these plants, and photographic and video-recordings of the workplaces. In each case, the identified existing CTD-inducing tasks were analysed and recommendations made suggesting specific improvements in the workplace.

Analysis of selected assembly operations in an auto assembly plant

Karwowski (1987) studied several manual operations performed at a light truck assembly plant. These tasks included: (1) tie-down of the seat panel and front grill assembly, (2) steering rod assembly, and (3) an outside electric wire hook-up operation. Analysis of the seat tie-down task (see Figure 3a) disclosed that this job involved excessive bending of the upper body forward, and flexion and extreme extension of the worker's wrist. In addition, the inflexible nature of an assembly line set-up and layout hampered worker movements.

Although the effective height of the workplace could not be altered, the following changes in work design were recommended: (1) redesign of the tool (a drive gun which causes ulnar deviation of the hand), (2) suspending the gun so that the workers' hands and arms are not unduly strained by its weight, (3) tilting the bolt bins towards the worker in order to reduce wrist flexion when reloading the gun, and (4) periodically transferring the workers to opposite sides of the assembly line to prevent accumulation of stressful effects on one side of the body. Similar recommendations were made with respect to the front grill assembly task (Figure 3b).

The steering rod installation task (see Figure 3c) was found to be extremely demanding, embodying numerous CTD risks. Causative factors included the tight quarters within the truck cab requiring that assembly be performed in awkward positions for the worker's upper body, arms and hands. The task required exertion of large arm and shoulder forces due to the steering rod's weight and geometry (length), high task repetitions and forceful pinching (Figure 3d). Because of the layout limitations and excessive stresses to the worker, it was recommended that the steering rod operation be divided into two separate tasks, and that the rod be pre-positioned by another worker whose primary duty was outside the main assembly line.

The third task, installation of the wiring harness (see Figure 3e) required repetitive pinching, pushing, and pulling of wires through a narrow opening. It was determined that the problem was in the design of small plastic grommets inserted in the openings after the wires are positioned. The grommets were made of hard plastic and, therefore, were difficult to install. Experienced workers revealed that formerly grommets of softer material were used which required much less force than arm-jerking motions to install. It was also observed that during the wire loom operation, workers must walk along with the moving assembly line. Given the significant precision requirement of the operation, it was recommended that moving walkways be installed at this particular workstation.

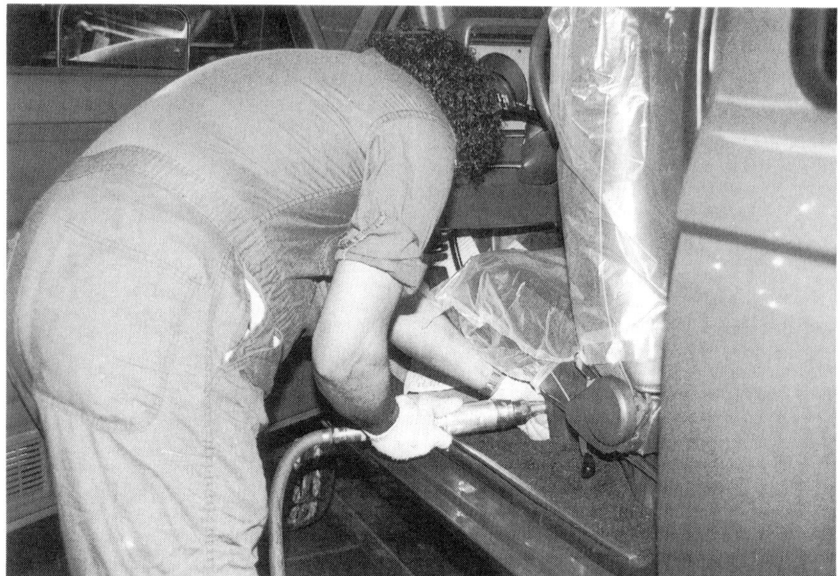

a

b

Figure 3.　Operations at an automotive plant:
(a)　Seat tie-down;
(b)　Front-grill assembly;

c

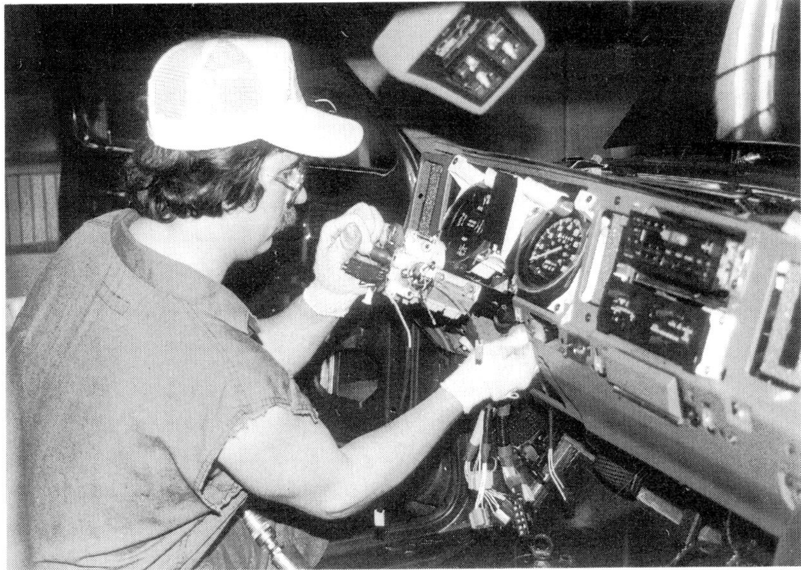

d

(c)   Steering column assembly;
(d)   Adjusting the steering column;

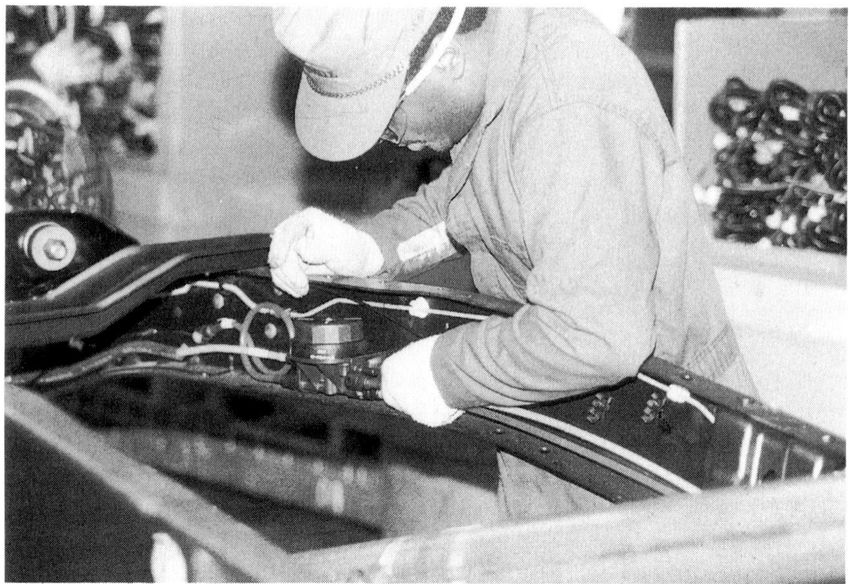

Figure 3. Cont'd.   *(e)   Electric wire hook-up.*

Analysis of lettuce cleaning at a food service plant

In this study (Karwowski, 1987), the female operators worked in teams on two different conveyor lines cleaning lettuce (see Figure 4a). Workers performed repetitive motions and exertions of arms and hands, cutting with commercially available knives with thin, non-oval handles (Figure 4b). A temperature of about 40°F was maintained in the building. Operators wore rubber gloves, for hygiene rather than safety reasons.

The height of the first conveyor line was too high with respect to the operators' stature, forcing them to keep their arms and elbows raised (see Figure 4c). The edge of the barrier enclosing the conveyor on the side facing the operator was sharp, not providing suitable forearm support. It was observed that during cleaning operations, "incidental" contacts of worker's arms with the conveyor edge occurred rather frequently. Also, the operators would often rest their forearms on the edge of the conveyor (see Figure 4d), while they waited for vegetables to be loaded into the storage container. On the other hand, the work surface of the second line was much too low, forcing the workers to bend forward and extend their arms excessively in order to pick up lettuce from the conveyor.

As a result of the task analysis, it was recommended that the first line's conveyor surface be lowered by 1·5 inches, while the height of the conveyor's edge facing the operators be lowered by 1 inch. In addition, the barrier's sharp edge should be eliminated by bending its edge outward, towards the operators, in a 1 inch radius curve, and covered with a soft material. The work surface of the second line should be raised approximately 3 inches to eliminate excessive, repetitive bending from the waist for the operators. Both conveyors should be tilted towards the operators by approximately 30 to 40 degrees.

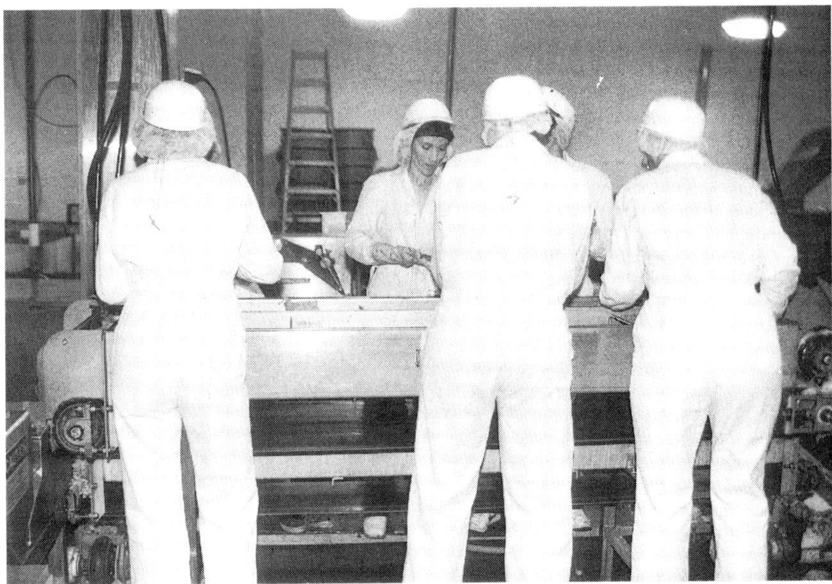

a

b

Figure 4.   *Operations during lettuce-cleaning:*
(a)   The conveyor line;
(b)   Typical knife design;

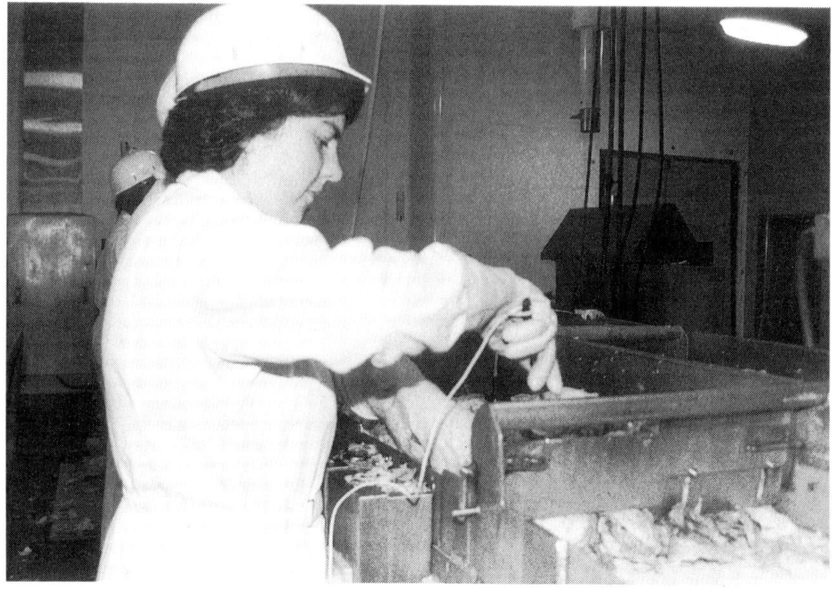

c

d

Figure 4. Cont'd.   *(c)   Arm abduction;*
                                    (d)   The sharp edge of the conveyor line.

In addition, knives with thicker rectangular handles with a width-to-height ratio of about 1 : 2 and a covering of some thin, pliant material to reduce stress to the user's palms, should be provided. The operators should be given a choice of a few different knives, in order to select a knife whose handle will fit comfortably in their hands. It was also recommended that a variety of sizes and styles of cold-protective gloves be made available to workers on a trial basis, so that they could individually select the glove best suited to them.

Analysis of the Chair Pad-making Operations

In the chair-pad sewing task (see Figure 5a), female workers performed a series of repetitive motions with the hands and forearms, most requiring application of force, as in pinching and pressing down. Such motions were routinely done with ulnar or radial deviation of the wrists, inward or outward rotation of the forearms, and excessive abduction of the arms and shoulders. These unnatural motions were combined with simultaneous stressful bending and twisting of the upper body. Two tasks subjected to detailed analysis (Karwowski, 1987) are described here: (1) stuffing of foam filler into the pad-case, and (2) sewing the edge of a large chair pad.

During the foam-stuffing operation, the operator's forearms were abraded by moving constantly against the fabric of the pad-case while the hands are pinching and repositioning the foam pillow in order to fit it into the envelope. Therefore, it was recommended that the foam be compressed to a lesser degree than is presently done. This could be accomplished by adjusting the metal feed chute to close, and consequently compress the foam moving through, to a width of 9 inches (a width of only 8 inches was allowed prior to the analysis). This way, the amount of force required to reposition the foam inside the pad would be greatly reduced. In addition, the width of the feed chute in open-mode should be increased from 15 to 16 inches, so that abrasive contact (see Figure 5b) between operator's hands and the metal sides of the feeder is reduced.

The operator was also advised to change the method of removing the filled pad from the feeder. The recommended, more natural position for the hands, which reduced stresses to the hands and forearms, was one in which the operators would grab the back edges of the fabric (away from the body) on each side of the pad with their hands, and allow the pressing mechanism to push the pad out of the feeder. It was also suggested to the plant management that the operators would also benefit greatly from semi-chair/standing supports.

With respect to the chair pad's edge-sewing operation (see Figure 5a) the following recommendations were made. The operators should be provided with adjustable chairs providing adequate back support. The chairs should rotate, to eliminate some stressful twisting of the upper body and allow for easy changes in seated work posture. The tables on which the sewing machines are mounted have sharp edges, which should be rounded off and, if possible, covered with a thin layer of pliant material. The entire assembly should be repositioned by turning the table approximately 15 degrees

a

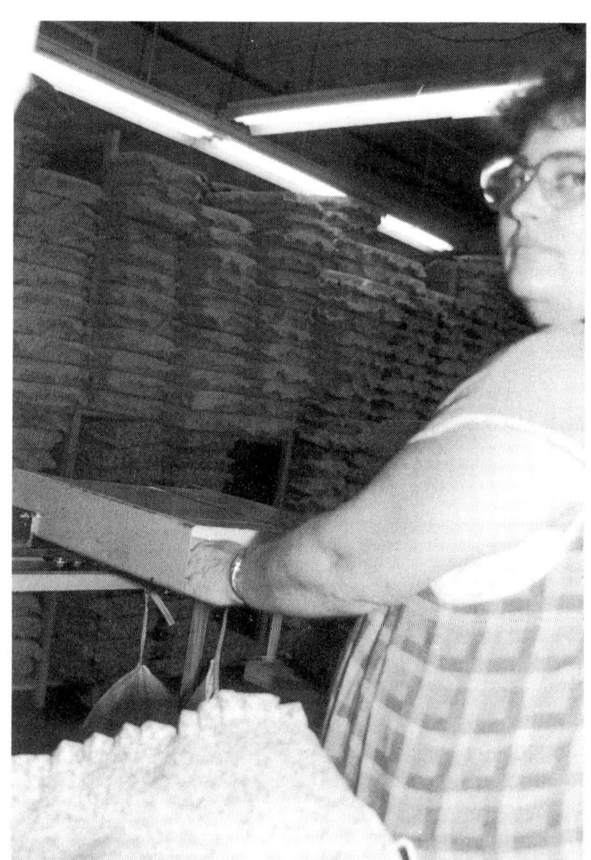

b

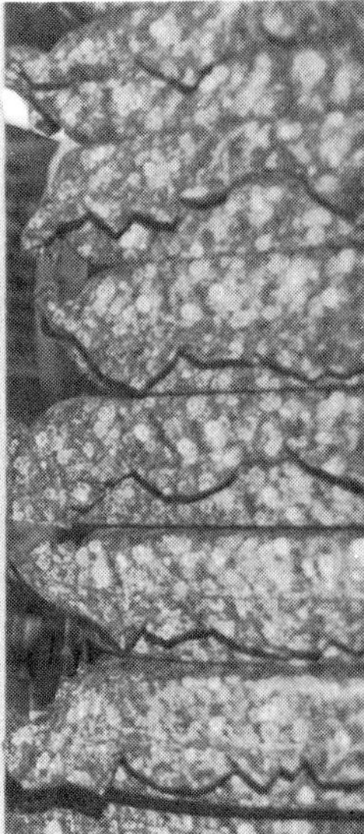

c

Figure 5. *Operations during fabrication of chair-pads:*
(a) *Edge-sewing task;*
(b) *Foam-stuffing operation;*
(c) *The chair-pad design.*

counter-clockwise (for right-handed operators) to reduce the presently excessive ulnar deviation of the right hand. A support should be provided for the operator's right forearm. All sewing machines should be tilted towards the operators, to relieve stress on the upper body caused by bending forward and in order for the operators to see their work more clearly. The sewing machines should be individually adjusted for sitting height of the operators.

It was also recommended to change the chair pad's foam design (Figure 5c) in order to reduce the need for pinching and stressful hand exertions while sewing the edge of the pad. Specifically, pre-cutting the back edge of the pad-filling foam to triangle forms and/or extending the top of the fabric edge, so that the sewing could be done along the bottom edge of the pad, was suggested.

Analysis of the fabric code-marking task at a clothing plant

The female operator, seated at table, was using a "code marker" tool to impress numeric codes on previously-cut shirt parts. The operator would use her right arm to press down on the tool's handle to make each mark. These forceful, repetitive exertions were particularly affecting the palm of the hand as the pinched grip style was used for this procedure. The hyperextended wrist was subjected to considerable pressure in this operation. The resistance of the handle when pressing down, and the weight of the marker itself, created unnecessary stresses to the forearm and wrist of the operator.

The marker used was not designed for prolonged, repetitive use, but rather for occasional coding only. It was suggested that management contact the makers of coding and marking equipment, to determine whether ergonomically-improved devices are commercially available. Additionally, the operator was to be moved to another table where there would be no restrictions in assuming a proper, comfortable sitting posture. Before, a control rope located below the table top interfered with the position of the operator's legs, and prohibited changes in seat height.

Conclusions

Although implementation of ergonomic principles of job design may not always be an easy task, it is very important to convince management that even small changes in the nature of repetitive jobs would be cost-effective in the context of improving worker comfort while performing thousands of repetitions per shift, and have pronounced effect on the prevention of CTDs of the upper extremities. In our experience, reductions in the amount of force per exertion necessary to perform a task, the elimination of one or more exertions-per-cycle, decreasing movement distance, (even slightly) or reducing the length of the workshift have all proven to increase work efficiency and job productivity while reducing worker time-lost due to CTDs.

It is essential that the ultimate goal of all those involved in treating and managing workers with a cumulative trauma disorder be his/her return to gainful employment. This requires a combined effort of cooperation between management and the professionals treating the individual patient. It is important to emphasize to management as well as the labour organizations that, in general, the longer the injured employee remains out of work, the more difficult it is to obtain a satisfactory result and ultimate recovery.

To place a worker completely off-duty because of, for example, a restriction from repetitive gripping or pinching activities, can only lead to a patient who is physically and psychologically disabled. The longer the worker stays out of work, the more difficulties he/she will have in returning to work. It was observed that, in many cases, the impaired worker behaves like a small child missing school because of an illness.

With each day away from the job, it is more difficult for the worker to return to the work environment. Therefore, management should implement specific incentives for prompt return to some kind of restricted duty. There should certainly be no great financial gain from being out of work, and there should be some "carrot" for returning to work. It is in the best interest of everyone concerned that the employee with a cumulative trauma disorder be returned to some worthwhile employment. Finally, only with an interdisciplinary team approach to prevention and treatment of CTDs can the effective return of a patient to a productive workforce be accomplished.

References

Armstrong, T.J., 1986, Ergonomics and Cumulative Trauma Disorders. *Hand Clinics*, **2**, 553–565.

Armstrong, T., Radwin, R.G.J., Hansen, D.J. and Kennedy, K.W., 1986, Repetitive Trauma Disorders: Job Evaluation and Design. *Human Factors*, **28**, 323–336.

Bradley, W.G., 1984, Diseases of the Spinal Roots. In *Peripheral Neuropathy*, Volume II, edited by P.D.Dyck and P.K.Thomas (Philadelphia, W. B. Saunders), pp. 1368–1382.

Brown, C.D., Nolan, B.M. and Faithfull, D.K., 1984, Occupational Repetition Strain Injuries. *The Medical Journal of Australia*, **3**, 329–332.

De Guervain, 1985, Ueber eine Form von chronischer Tendovaginitis. *Correspondenz-Blatt f. Schweizer Aerzte*, **25**, 389–394.

Derkash, R.S., Golberg, V.M. and Mendelson, H., 1981, The Results of First Rib Resection in Thoracic Outlet Syndrome. *Orthopedics*, **4**, 1025–1029.

Ditmars, D.M. and Houin, H.P., 1986, Carpal Tunnel Syndrome. *Hand Clinics*, **2**, 535–532.

Fry, H.J.H., 1986, Overuse Syndrome, alias Tenosynovitis: The Terminological Hoax. *Plastic and Recontructive Surgery*, **78**, 414–417.

Habes, J.D. and Putz-Anderson, V. 1985, The NIOSH Program for Evaluating Biomechanical Hazards in the Workplace. *Journal of Safety Research*, **16**, 49–60.

Hartz, C.R., Linscheid, R.L. and Gramse, R.R., 1981, The Pronator Teres Syndrome: Comprehensive Neuropathy of the Median Nerve. *The Journal of Bone and Joint Surgery*, **63–A**, 885–890.

Howard, F.M. and Hill, N.C., 1986, Entrapment of the Anterior Interosseous Nerve. *Contemporary Orthopaedics*, **12**, 43–46.

Johnson, R.K., Spinner, M. and Shrewsbury, M.M., 1979, Median Nerve Entrapment Syndrome in the Proximal Forearm. *The Journal of Hand Surgery*, **4**, 48–51.

Karwowski, W., 1987, Prevention of Cumulative Trauma Disorders of the Upper Extremity Through Job Redesign: Case Studies. In *Trends in Ergonomics/Human Factors IV*, edited by S.S. Asfour (Amsterdam: North Holland), pp. 1021–1028.

Kasdan, M.L. and Janes, C.J., 1988, Carpal Tunnel Syndrome and Vitamin B-6. *Plastic and Reconstructive Surgery* (in press).

Kelsey, J.L., Harris, P. and Kreiger, N., 1980, *Upper Extremity Disorders: A Survey of Their Frequency and Cost in the United States*, (St. Louis: C. V. Mosby Company) pp. 1–71.

Koman, L.A. and Urbaniak, J.R., 1985, Ulnar Artery Thrombosis. *Hand Clinics*, **1**, 311–325.

Kornberg, M., Aulicino, P.L. and DuPuy, T.E., 1983, Ulnar Arterial Aneurysms and Thromboses in Hand and Forearm. *Orthopaedic Review*, **12**, 25–33.

Machleder, H.I., Moll, F. and Verity, M.A., 1985, The Anterior Scalene Muscle in Thoracic Outlet Compression Syndrome. *Archive Surgery*, **121**, 1141–1144.

Marmor, L., 1966, Bowler's Thumb. *The Journal of Trauma*, **6**, 282–284.

Miller, R.G. and Hummel, E.E., 1979, The Cubital Tunnel Syndrome: Treatment with Simple Decompression. *Annals of Neurology*, **7**, 567–569.

Minkow, F.B. and Bassett, F.H., 1972, Bowler's Thumb. *Clinical Orthopedics*, **83**, 115–117.

National Safety Council, 1984, Cumulative Trauma Disorders (CTDs) and Ergonomics: Causes, Treatment and Preventive Measures. *Satellite Video Conference Seminar*, May 16, 1984.

Payan, J., 1969, Electrophysiological localization of ulnar nerve lesions. *Journal of Neurology, Neurosurgery, Psychiatry*, **32**, 208.

Roos, D.B., 1985, Transaxillary First Rib Resection for Thoracic Outlet Syndrome: Indications and Techniques. *Contemporary Surgery*, **26**, 55–62.

Siegel, I.M., 1965, Bowling and Thumb Neuroma. *Journal of the American Medical Association*, **192**, 263.

Taleisnik, J., 1985, *The Wrist* (New York: Churchill Livingstone).

4.

Rehabilitation in Hansen's Disease

Atul Shah
Honorary Assistant Professor of Plastic Surgery
and
Kumkum Saluja
Formerly Registrar in Plastic Surgery

*J.J. Hospital
Bombay, 400 008
India*

Abstract

Hansen's disease leads to deformities of the lower and upper extremities. Eventually, these deformities become a serious handicap and limit the functional capabilities of the afflicted. This paper defines the characteristics of individuals suffering from Hansen's disease and demonstrates that disabilities caused by Hansen's disease can be overcome by: (i) use of splints, (ii) reconstructive surgery, and (iii) modification of the articles and tools of daily use, and (iv) proper job-worker matching.

Introduction

Rehabilitation is required in Hansen's disease because the disease process makes the afflicted individual a handicapped person, and also because of the social stigma attached to the disease. In spite of the change in nomenclature of the disease in some countries from Leprosy to Hansen's disease, social ostracization of the afflicted still continues. Given opportunities in the proper environment, with the aid of modified tools, equipment and machinery as well as restoration of the altered anatomy of the individual with the help of reconstructive surgery, all leprosy affected individuals can be rehabilitated to lead a life of gainful employment. For example, a young son of a carpenter who has learnt some carpentry but has taken to begging after contracting the disease, can be offered an opportunity to carry out carpentry in a rehabilitation centre. If necessary, the handle of a chisel or hammer can be modified to suit the hand grip. Many voluntary institutions offer such opportunities by paying daily wages which are generally more than the individuals can earn from begging. We have also come across examples in which a patient is living within a family but is unable to work due to his handicap. The handicap results from impairment of function which, in turn, is the result of nerve damage. The nerve damage is quite amenable to various plastic surgical corrections. Tendon transfer procedures in which a functioning tendon is employed for the restoration of function of paralysed muscles, restore the altered anatomy and make a person fit from an ergonomic standpoint. However, realization of this concept of offering proper environmental setting, modified tools and reconstructive surgery as a combined mode of rehabilitation, for almost all patients of Hansen's disease, is yet to be achieved.

Objectives

(1)   To define the anatomical, physiological and behavioural characteristics of the afflicted.

(2)   To compensate the functional disabilities of affected persons by the use of splints, surgery or by modification of articles in daily use and those used at work, in order to optimize safety, health, comfort and efficiency.

(3)   The ultimate objective is to rehabilitate the affected individual physically and mentally by proper use of principles of ergonomics so that he can once again be an independent and useful member of society.

Defining the problem and the handicap

The stigma of leprosy is due to the irreversible nature of the visible, physical deformities caused by the disease process *per se*, and also secondary (but preventable) deformities which result from neglect of the affected parts. Deformities of the feet can be hidden completely, those of the hands partially, but those of the face not at all.

Facial deformities occur mainly in lepromatous leprosy, or in the bacteriologically positive cases, except lagophthalmos. The primary tell-tale signs on the face are the loss of eyebrows, depressed nose and facial wrinkling. Eyebrow loss tends to identify such individuals within a crowd. The ulceration which may occur in the interior of the nose gives rise to rhinitis and a typical smell in the breath. Encrustation, destruction and occasional bleeding may also invite other bacterial infections to occur and induce more suffering. The ultimate destruction of cartilage and depression of the nose pronounces the change in the patient's personality. Concomitant with it are changes in the facial skin, which sags down. The patient looks much older than his age. In short, his normal youthful life style is disturbed both physically and psychologically.

On the other hand, in the tuberculoid spectrum of the disease there are deformities like the claw hand, foot drop and claw toes. The majority of the deformities arise due to the involvement of nerves at various levels. The involvement of facial nerve branches to the orbicularis occuli and/or the frontalis muscle, which are especially prone (Antia 1966) result in lagophthalmos. The loss of blinking action, corneal drying due to constant exposure, accidental introduction of dust particles with resultant conjunctivitis, and keratitis may all culminate in damage to eye-sight, and even more so if corneal sensation is also affected.

The handicap associated with the loss of intrinsic action, due to ulnar nerve paralysis, is not great except in a country like India where people use their hands directly for eating rice. This activity involves the action of cupping of the fingers and the thumb and cannot be effected in the absence of intrinsics. Besides this, the grasp may be affected to a certain extent, particularly the diagonal grasp necessary for proper use of a hammer or screw driver, handling of levers, hand carts, ploughing implements etc. The span of the hand necessary for lifting spherical objects or discs, also made by intrinsic muscles, is reduced. Dexterity is lost and simple tasks such as sewing, closing buttons etc. become difficult or impossible. This is much more pronounced if there is absence of sensation. The total claw hand following a high ulnar nerve lesion with a low median nerve lesion affects all five digits. In such a hand, only the key pinch is possible. The thumb lies in the plane of the palm and thus the deformity is described as the 'ape thumb' or 'ape hand' deformity. When the mechanism of grasp by a paralysed hand is compared with a normal grasp, the following observations can be made. To grasp a spherical or cylindrical object with the palm of the hand, the claw hand opens out more than required because of the hyperextension at the metacarpophalangeal joints. When it starts embracing the object, the metacarpophalangeal joint does not flex and stabilize the fingers, but the flexion starts from the distal interphalangeal joint, followed by proximal interphalangeal joint. Hence, if the object is more than 5 cm in diameter the grasp cannot be effected and the object will slip out of the palm. In a normal hand, metacarpophalangeal joint stability and movement at the transverse arch, along with extension of proximal and distal joints, allow the grasp of the object. A similar effect is seen in the pinch, where abduction and rotation of the thumb are essential. In the absence of this movement in the claw hand, only the key pinch remains.

The damage caused by the loss of sensation is much worse and far greater in the lower limb than in the upper limb. During standing and locomotion, the plantar surface of the foot is subjected to repeated stress and strain of shearing forces and pressure from the weight of the body. In the absence of the protective sensations of pain, these uncontrolled forces and inadvertent, unnoticed injuries lead to plantar ulceration at pressure points. As these ulcers are painless, the person continues to bear weight on them and thus a simple ulcer is converted into a complicated one with added osteomyelitis, which leads to mobility impairment. When this handicap is considered quantitatively, there are 45 716 cases with deformities in Maharashtra state alone, and the deformity rate is 16·14 per cent. It is also estimated that at present 30 000 patients in Maharashtra require rehabilitation (statistics from the ''Report of the study team on leprosy, Maharashtra State, 1982''), i.e. they require instruction in health education, proper use of insensitive limbs, availability of reconstructive surgery, and offer of jobs, food, shelter and clothing on a regular basis to enable them to lead gainful and productive lives. In advanced, smear-positive lepromatous cases it is not morally and scientifically possible to allow the patients to continue their jobs, because it is believed that they are the carriers of the disease and may spread the disease. It is beneficial, though, to persuade the employer to allow a person a long leave so that he may resume his job

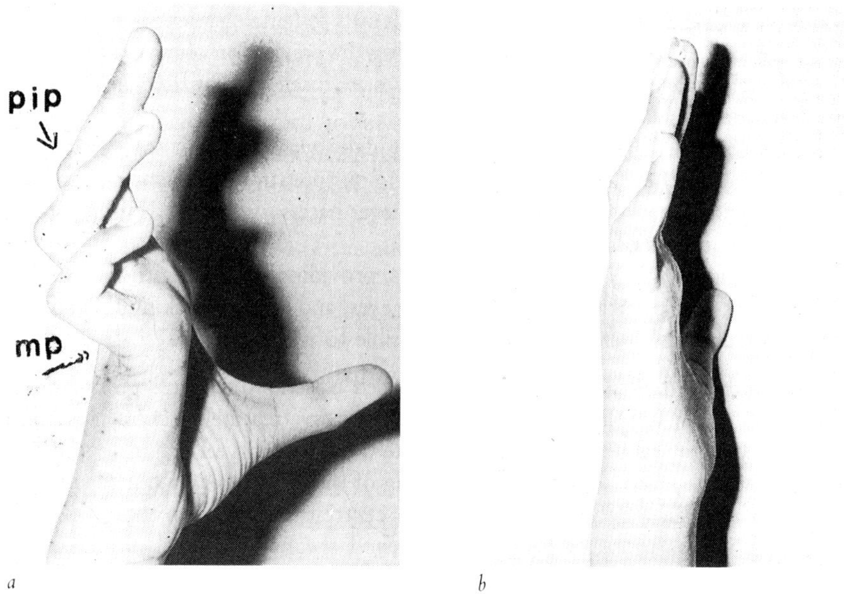

Figure 1.(a) A typical (pre-operative) claw hand deformity, showing hyperextension at the metacarpophalangeal joint (MP) and flexion at the proximal interphalangeal joint (PIP).
(b) The straightening of fingers obtained (post-operative).

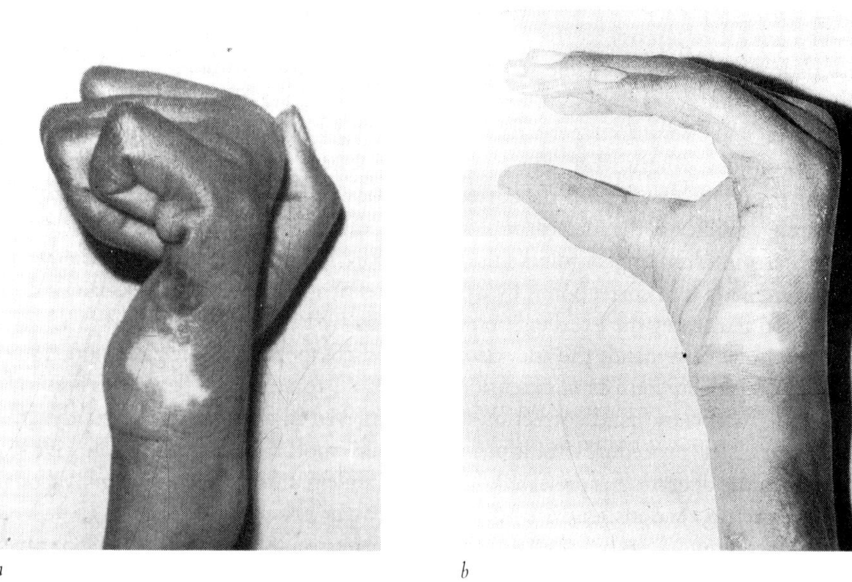

Figure 2.(a) A total claw hand cannot have an effective grasp unless metacarpophalangeal joints are stabilized by an operation to enable the fingers to straighten. Note too, the position of the thumb in adduction.
(b) Post-operative effective finger straightening (lumbrical position) obtained for the normal grasp. The thumb is also brought into abduction rotation by opponensplasty.

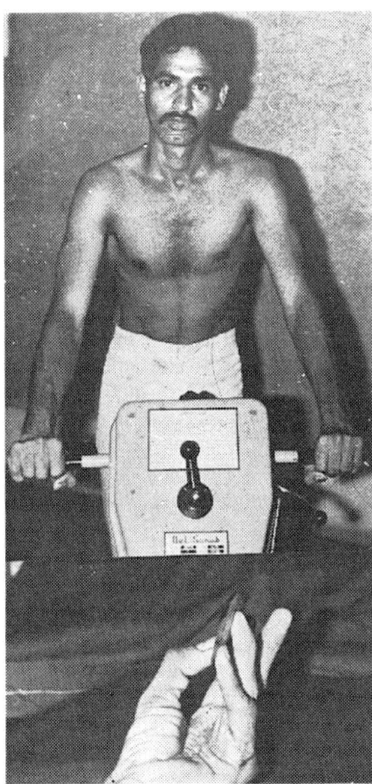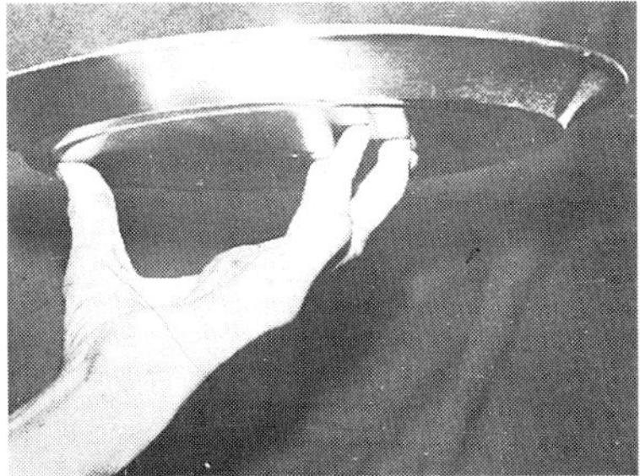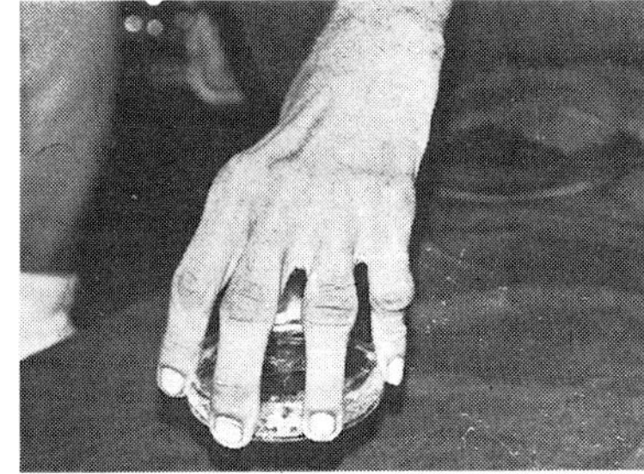

Figure 3. *Various activities possible after the correction of transverse metacarpal arch and claw hand by the authors' technique.*

to enable one to eat rice with the hands, as is the custom in India, require not only the correction of stability but also the movement of the ring and little finger towards the volar aspect. This oblique movement towards the thumb base is carried out by the movement of the transverse metacarpal arch. Generally, it requires a separate operation but the first author's unique procedure (Shah, 1987) corrects the reversal of transverse metacarpal arch along with correction of ulnar claw hand and enables the patient to eat rice with hands and to perform all activities as shown in Figure 3. Deep plantar ulcer with osteomyelitis, which do not heal with conservative management, need to be tackled surgically. Various local flaps have been described. The successful resurfacing of heel ulcers with the flexor digitorum brevis myocutaneous flap has been reported by the author (Shah and Pandit, 1985). For resurfacing metatarsal head ulcers, the author has been using a neuro-vascular pedicle island flap taken from the lateral aspect of the great toe. An example of this is shown in Figure 4a.

The sensory motor deficit due to nerve lesion in leprosy is reversible if nerve decompression is performed within six months of the onset of leprous neuritis. In a detailed study (Shah, 1986) it was found that all modalities of sensation tend to recover. The recovery rate is better when the decompression is followed by corticosteroid therapy for a period of 4–6 months (personal observations, unpublished). Figures 5a and 5b show the return of hypothenar abduction following ulnar nerve decompression and corticosteroid therapy.

We have observed that correction of the depressed nose has sometimes been more satisfying to the patient than functional reconstruction of the hands. It cannot be too strongly emphasized that any rehabilitation programme must provide for the tackling

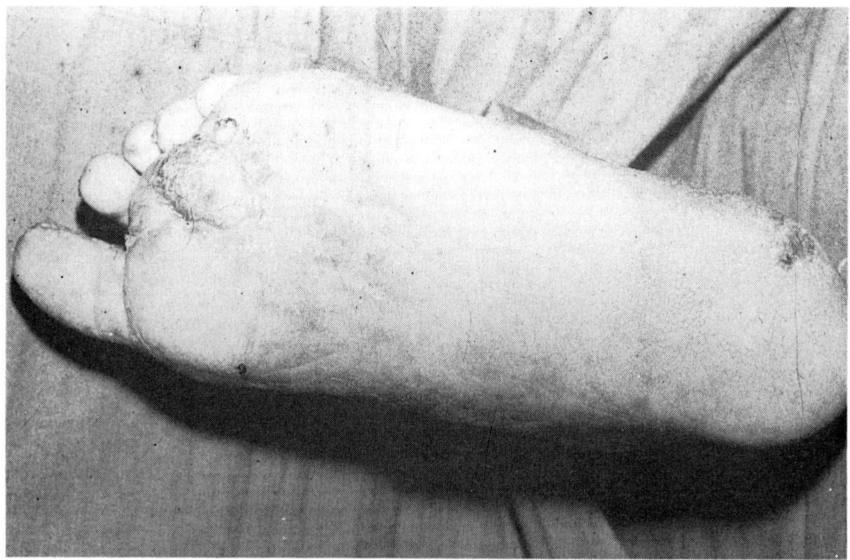

a

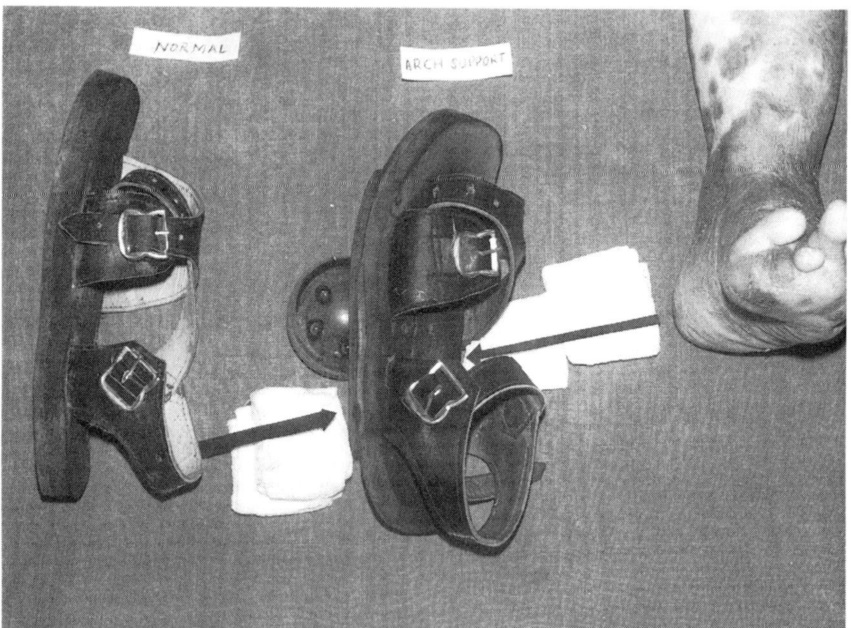

b

Figure 4. (a) Neurovascular island pedicle flap for ulcer on the metatarsal head taken from the medial side of the great toe to cover skin deficiency with the same type of skin.
(b) The microcellular rubber foot-wear (left) for prevention of ulceration. The same footwear is modified by providing medial and/or lateral arch support where the foot is grossly deformed.

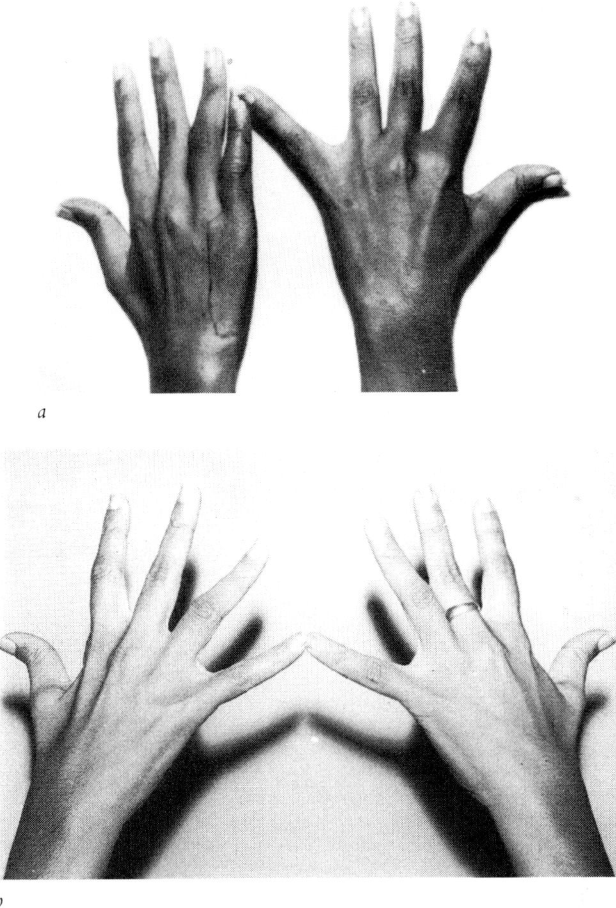

*Figure 5.(a) Weakness of the little finger for abduction and sensory loss before neural decompression.
(b) Normal power of abduction and complete sensory recovery.*

of the functionally insignificant (but cosmetically important) stigma ever present in lepromatous cases. Loss of the eyebrow, depressed nose and facial wrinkling constitute the major deformities. Fortunately, most of these deformities can be corrected by simple plastic surgical procedures. The total personality change by such multiple operations may take a year or more, but is very gratifying. The eyebrow reconstruction is generally done by the island flap based on the anterior branch of the superficial temporal artery (Antia, 1964). Nasal reconstruction is done by providing the inner lining by nasolabial flaps (Antia and Büch, 1972). Facial wrinkles are eliminated by nasolabial excision of skin (Antia, 1974). Figures 6a to d illustrate an example of a 'personality change' operation. This patient is now employed as a compounder at a dispensary.

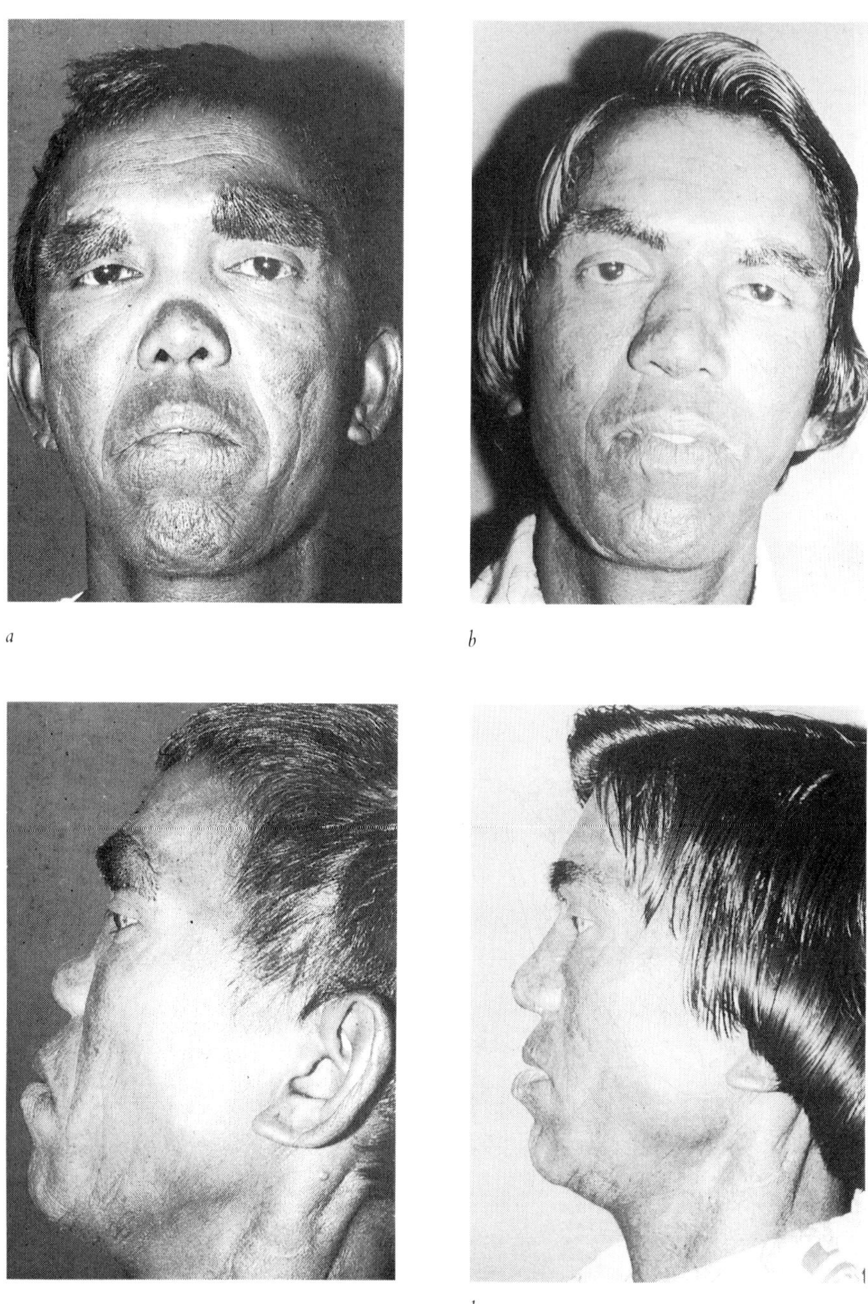

a

b

c

d

Figure 6.(a) The 'stigma' of Hansen's disease, the depressed nose and the facial wrinkles (eyebrows have been reconstructed from the scalp).
(b) The dramatic change in personality after plastic surgery, leading to his re-employment and rehabilitation.
(c) Profile of the same patient before, and (d) after plastic surgery.

Instrumentation and equipment

The goverment and voluntary agencies have played a major role in increasing public awareness, early detection, ensuring regular treatment, recognising and referring cases which will benefit from surgery, etc. Patients with established disabilities are rehabilitated to perform their activities of daily living by the modification of clothing, utensils, footwear etc. For example, persons who have lost all the fingers of both hands (Figure 7) are advised to wear shirts with magic tape instead of buttons, and trousers are provided with elastic at the waist. Another patient has only a 'lobster claw' type grip possible in his thumb web. He was provided with a spoon with a long handle which allows class III lever system principle to lift load with minimum effort, and with which he was able to eat his meal comfortably without external assistance. Spoons and the handles of other utensils are modified into a hook into which the patient can slip his palm. All handles should be insulated. In the work-place, the patient's span of the hand is sometimes insufficient to grip an instrument. For example, a carpenter could not hold a conventional metal plane. The metal plane was modified and made longer with a rounded projection at the back, which can be firmly grasped, and a side bar at the front provided stability while using it. Figure 8a illustrates the ordinary and modified metal planes and Figure 8b the patient's mechanism of grasp.

Figure 7. Three cases of advanced deformities and mutilated hands who can be helped only by design modifications of various articles for the activities of daily living.

a

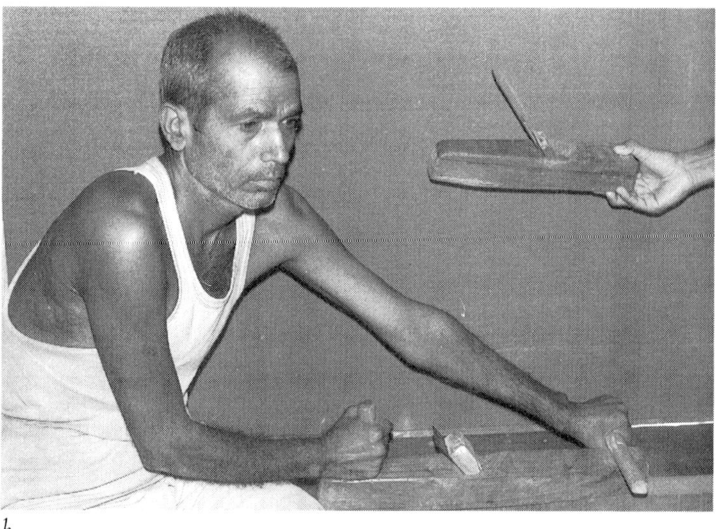

b

Figure 8.(a) The modified metal plane for wood carving.
(b) The alteration allows for comfortable working.

Patient-occupation matching

The administrative officer in charge of the resettlement project generally decides on the type of occupation that a patient can undertake. Preference is given to the occupation he held prior to his diseased state. Due importance is given to the patients desire as to whether he wishes to return to the same occupation, the availability of the job and the extent to which his disability interferes with his job. All efforts are made to suit the job to his physical state by alterations to machinery and tools, with or without prior surgery. The aim is to increase his efficiency, to restore his body image and thus stabilize his psychology to facilitate social acceptance.

In order to maximize the gains to the patients and the employer, the following jobs are proposed which cover the majority of occupational areas and available opportunities.

1. Carpentry
2. Horticulture
3. Farming
4. Mechanical repairs
5. Handloom weaving
6. Cobbler (shoe and chappal making)
7. Supervisor, watchman and clerical jobs.

The criteria adopted for suitability of the occupation were as follows:

(1) Knowledge and experience of a job.
(2) Physical suitability and willingness.
(3) Allotment of job for temporary period to observe the work output.
(4) Opinion of medical personnel for reconstructive surgery.
(5) When the matching is final and the deformities uncorrectable, modification of the equipment/machinery for long term benefit.

However, our experience suggests that anatomical restoration by reconstructive surgery, and physiological operations such as nerve decompression, lead to a better and easier rehabilitation. This is in view of the availability of the surgical procedures which not only correct the deformity, and the associated handicap, but also enable the patient to function like a normal individual. The cosmetic correction is equally good. However, this is only feasible when the patient is seen early in the course of the disease. The modification of articles or equipment and machinery is required for insensitive limbs and when the joints are fixed in certain positions of contracture (where surgery cannot be of much use). Out of two hundred cases seen by us, one hundred and forty underwent the surgery and were able to regain their occupation with almost normal performance. It is only in a few cases that we have to adjust the occupation to their handicap.

Splintage

Splints play an important role in facilitating surgery and rehabilitation of the patients. The first step in correction of a physical deformity is to apply the appropriate splint. Plaster-of-paris splints, moulded in such a way as to correct the deformity, to elongate the shortened skin and release the tendon contraction gradually, have been employed for many years. However, this splintage needs to be changed often and can be quite cumbersome for large scale application. Therefore we have prepared splints from other materials including rubber, felt, aluminium strips, leather, elastic and rubber bands. The main advantage of the splint is that it can be adjusted by the patient himself, and needs review only once weekly, thereby decreasing the workload of physiotherapists.

Basic design of splints

Splints can be classified as either static or dynamic. The static splints are used for support. The dynamic splint not only provides support but also allows the movement. The spring or elastic bands incorporated in the design (as described below) enable movement to occur while the splints provide stability, aiding movement and allowing the opposing group of muscles to act against the elastic resistance of the splint. Each splint has an aluminium strip as the base, the thickness of which depends on the rigidity required. Felt is emplaced on this strip as padding so that it does not rub against the skin. The elastic straps are held in position at a desirable site with a rivet or button clamped on to the leather cover of the splint. Velcro strips are attached to fasten it to the finger, palm or forearm. Figure 9 illustrates a number of the varieties of splints which are currently being used:

(1) Splint for mobile ulnar claw hand, consisting of a wrist strap and the finger straps which are joined by rubber bands so as to flex and stabilize the metacarpophalangeal joint, and enable the patient to extend the interphalangeal joint with the help of the long extensors. The regular use of this splint prevents proximal interphalangeal joint contracture.

(2) Splint for the mobile total claw hand, consisting of the same design as above, but where another buckle and rubber band are incorporated in a diagonal direction to enable the abduction of the thumb so as to prevent the adduction deformity of the thumb and thumb web contracture.

(3) Gutter splint for independent proximal interphalangeal joint extension without correcting metacarpophalangeal hyperextension is applied as a temporary splint in some cases.

(4) For dorsal skin contracture in the region of the thumb web, a spring action splint with an aluminium cup is applied in such a way that one end will rest on the index finger and the other end on the thumb. This particular splint is very useful in opening out the skin contracture in the region of the thumb web and to stretch the adductor fascia.

(5) An embroidery ring splint is used to strengthen the flexor profundus after the author's preferred method of reconstruction by lasso technique.

(6) For fixed, proximal interphalangeal joint contracture we have tried a dorsal splint with rubber traction on the fingers but have not found it as satisfactory as moulded plaster-of-paris splints.

Results

The results were evaluated in terms of patients leading an independent gainful life in society or in a resettlement colony. The first group included those patients who did not

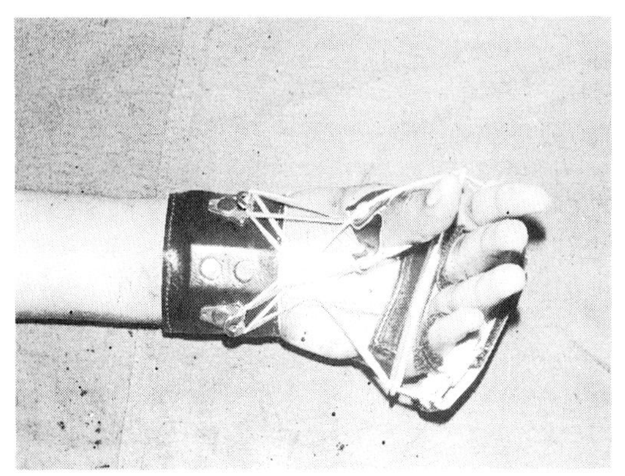

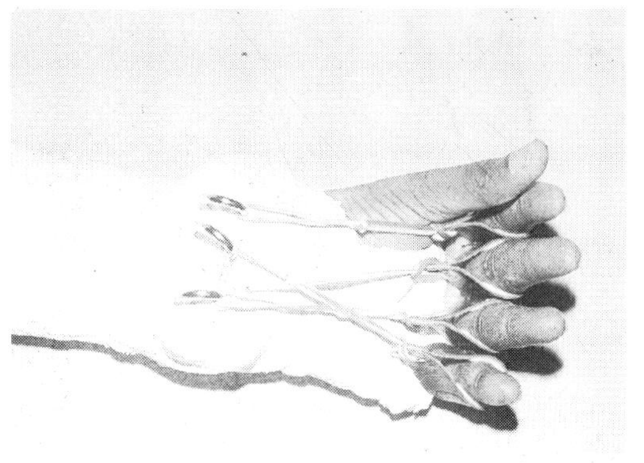

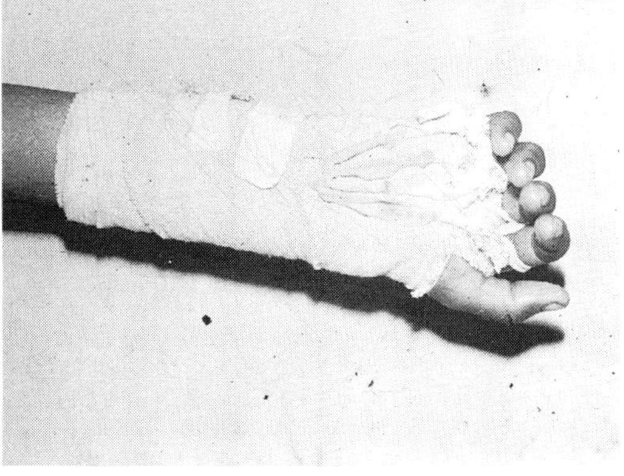

Figure 9. Simple splints used for re-education exercises (a) Elastic bandage splint; (b) P.O.P. splint; (c) Regular splint.

have any tell-tale signs and those whose stigmata had been masked by surgery, and were thus not recognized as cases of Hansen's disease. In the last two years 40 claw hands were operated upon by the authors' technique. Of these, 35 were successful, 2 were unsatisfactory and 3 were lost to follow up. All of the successful cases have been leading independent, gainful lives in society. The other group included those persons who were once beggars, and the ostracized individuals who preferred to stay among the afflicted. Our results of rehabilitative surgery at a rural leprosy colony have been published (Shah and Bhagat, 1984). To elaborate and bring our statistics up to date, we have performed 139 operations. Of these, 130 have been satisfactory. Of the remaining 9 unsatisfactory cases, 6 were of plantar ulcers which had recurrence and the other 3 were actually re-operations on the hand. On average, 2 operations were required per patient. The change in personality and attitude was adjudged to be extremely good.

Conclusion

In view of our experiences described above it can safely be concluded that:

(1) Rehabilitation in Hansen's disease should have a many fold approach covering psychological awareness of the nature of the disease, its effects on function and the prevention of dehabilitation.

(2) Incentives for modified machinery and tools, to return the afflicted to gainful employment, should be made available.

(3) Creation of social awareness where a person will be accepted back in society is a necessity.

(4) Rehabilitation settlements for the physically handicapped with facilities for re-constructive surgery should exist in adequate numbers.

(5) Ultimate eradication of stigma should be a primary aim for optimized environmental resettlement.

Acknowledgements

We wish to thank Miss Nilofer Bhagat for overall assistance and excellent photography and to Miss Virgil Fernandes for typing the manuscripts. We are grateful to the Trustees of the Shram Mandir Rural Leprosy Rehabilitation Colony for carrying out our study. We also wish to thank LEPRA for supporting our project on standardised splints described in this study.

References

Antia, N.H., 1964, Reconstructive surgery of the face. In *Leprosy in theory and practice*, edited by Cochrane (Bristol: John Wright & Sons), 2nd edition, pp. 497–509.

Antia, N.H., 1966, The 'Temporalis Musculo-facial sling' for the correction of Lagophthalmos. *Indian Journal of Surgery*, **28**, 389–396.

Antia, N.H., 1974, The scope of plastic surgery in leprosy; A ten year progress report. *Clinics in Plastic Surgery*, **1**, 69–82.

Antia, N.H. and Buch, V.I., 1972, Correction of nasal deformity with nasolabial flaps. *Indian Journal of Plastic Surgery*, **5**, 23–25.

Brand, P.W., 1958, Paralytic claw hand, with special reference to paralysis in leprosy and treatment by the sublimis transfer of Stiles and Bunnell. *The Journal of Bone and Joint Surgery*, **40–B**, 618–632.

Report of the study team on leprosy, Maharashtra State, India, 1982.

Shah, A., 1984a, Correction of ulnar claw hand by a loop of flexor digitorum superficialis motor for lumbrical replacement. *The Journal of Hand Surgery*, **9–B**, 131–133.

Shah, A., 1984b, Multiple superficialis motor for opponens and lumbrical replacement: One stage correction of leprous claw hand. *The Journal of Hand Surgery*, **9–B**, 285–288.

Shah, A., 1985, One in four flexor superficialis lasso for correction of claw hand. *The Journal of Hand Surgery*, **10–B**, 404–406.

Shah, A., 1986, Evaluation of nerve function deficit: Its improvement by nerve decompression or corticosteroid therapy. *Indian Journal of Leprosy*, **58**, 216–224.

Shah, A., 1987, A new technique for correction of reversal of transverse metacarpal arch and ulnar claw hand. *Transactions of IXth International Congress of Plastic and Reconstructive Surgery*, edited by R.J. Maneksha, (New Delhi: Tata McGraw Hill), 533.

Shah, A. and Nilofer, B., 1985, Rehabilitation surgery for leprosy handicapped at a rural leprosy colony. *International Journal of Rehabilitation Research*, **8**, 345–347.

Shah, A. and Pandit, S., 1985, Reconstruction of the heel with chronic ulceration with flexor digitorum brevis mycutaneous flap. *Leprosy Review*, **56**, 41–48.

Shah, A. and Saluja, K., 1985, Biomechanics of reconstruction of claw hand in leprosy. *Proceedings of the international conference on Biomechanics and Clinical Kinesiology of hand and foot*, I.I.T. Madras, 73.

PART III
MENTAL DISABILITY:
PREVENTION AND
REHABILITATION

5.

A study of correctional movements of mentally retarded persons

Tadeusz Marek and Czeslaw Noworol
Jagiellonian University
Department of Work Psychology
Institute of Psychology
Golebia 13
31–007 Krakow
Poland

Abstract

This study dealt with the description of correctional movements of mentally retarded persons. Twenty one normal and twenty one mildly mentally retarded persons were examined. The Meile Ball Test was used. The stabilization phase of directional manual acts was analysed. It was found that normal and retarded persons differ in the character and range of correctional movements in this phase ($P < 0 \cdot 05$). These findings are potentially useful for the ergonomical design of work-stations for the mentally retarded as well as for planning appropriate rehabilitation therapy.

Introduction to the problem

Mentally retarded persons perform very simple actions during their work. They perform simple manual movements which are mostly directional (the movement is directed towards given objects). Following our previous investigations and review of the literature (Whiting, 1984; Stelmach and Requin, 1980), the directed action of a movement consists of two phases. The first phase is the dynamic phase (long movement towards a given object), while the second one is the stabilization phase (precise and short correctional movements). Figure 1 shows schematically the two phases for action of directed movements. It is known that the structure of such movements is completely different for normal and mentally retarded persons (Kapecka and Marek, 1982). For retarded persons, disturbances in smoothness and anticipation of movement are very frequent.

The aim of this study was to describe the correctional movement space for mentally retarded and normal persons. The main question was whether the mentally retarded differed from normal persons, in number of correctional movements, shape of the correctional movement space, and volume of the correctional movement space.

75

Table 1.   *Mean and standard deviation of number of correctional movements appearing during the stabilization phase of action for normal and retarded persons.*

Examined persons	Number of correctional movements	
	Mean	Standard deviation
Normal	0·7	0·4
Retarded	6·5	2·3

characteristic for the correctional movements, particularly the shape and the volume of that space. Therefore, the projections of the space onto the three planes were analysed (Figure 2). These are the planes: yz, on which movements were recorded by camera X; xy, on which movements were recorded by camera Z; and xz, on which movements were recorded by camera Y.

For each person the shape of maximal ranges of correctional movements for the ten repeated tasks were drawn. The results for one examined retarded person are shown in Figures 3–5, and for one normal person in Figures 6–8. To show the differences, Figures 6–8 (representing the maximal ranges in the normal group) are presented. The shapes of ranges of actions shown in the figures for retarded and normal persons are typical for each of the investigated groups. Generally, mentally retarded persons are characterized by the maximal ranges of correctional movements in the quadrants (y_-z_+), (x_-y_-), and (x_-z_+). We can say that this is the overlapping of movement for a mentally retarded person. The shape of the range of correctional movements for retarded persons is asymmetrical, while for the normal persons is quite symmetrical.

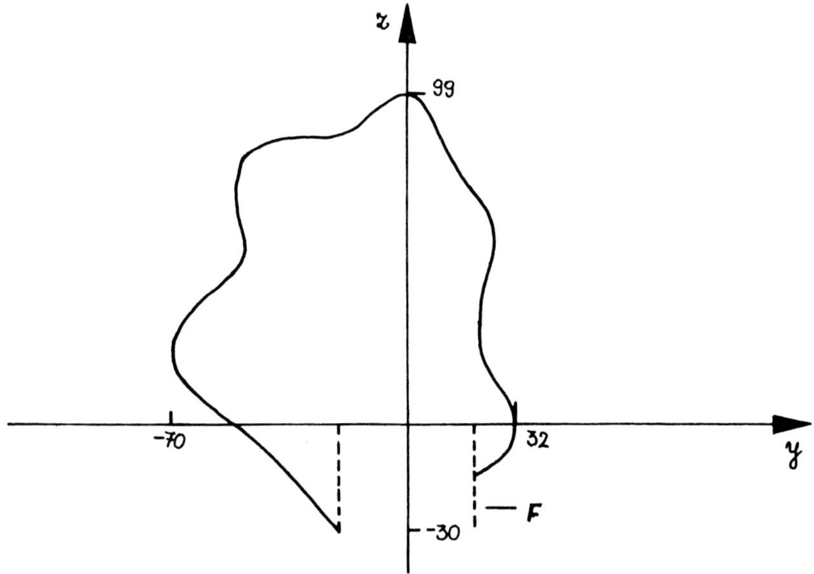

Figure 3.   *The projection of the range of correctional movements (in mm) into the yz plane. Example for one retarded person.*

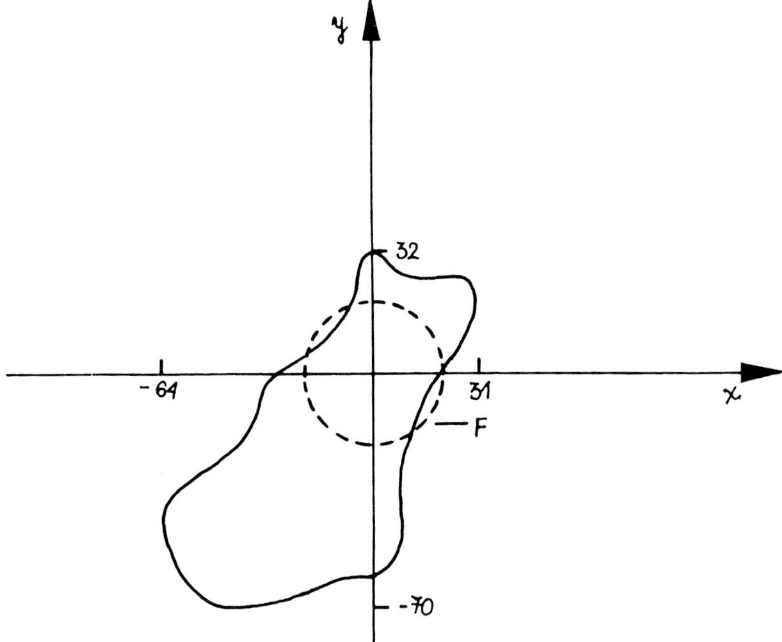

Figure 4.　*The projection of the range of correctional movements (in mm) into the xy plane. Example for one retarded person.*

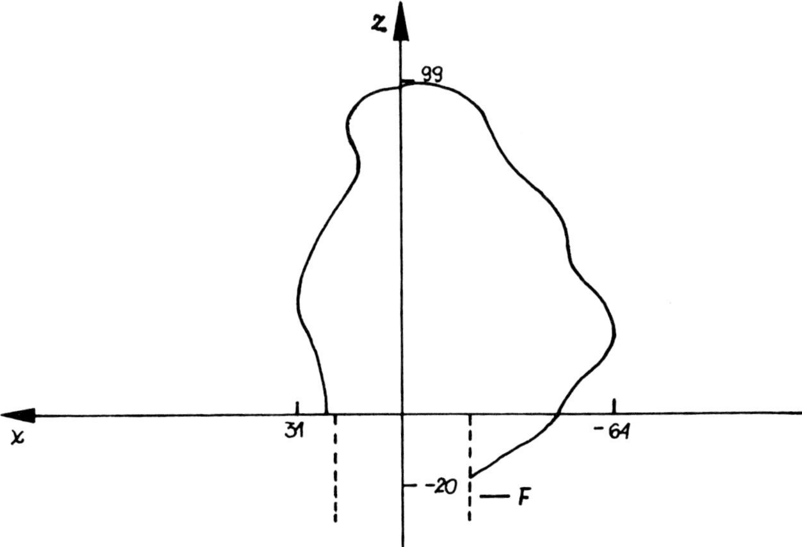

Figure 5.　*The projection of the range of correctional movements (in mm) into the xz plane. Example for one retarded person.*

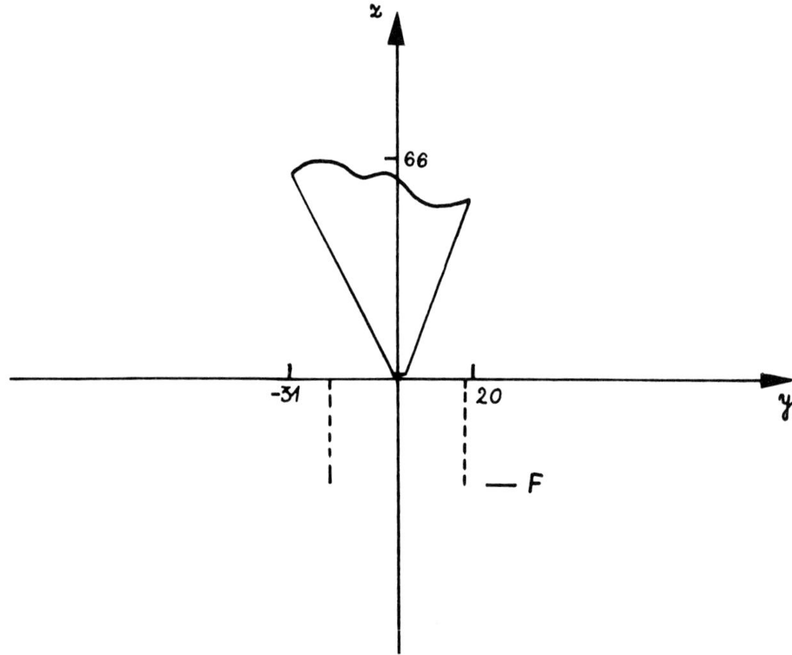

Figure 6. The projection of the range of correctional movements (in mm) into the yz plane. Example for one normal person.

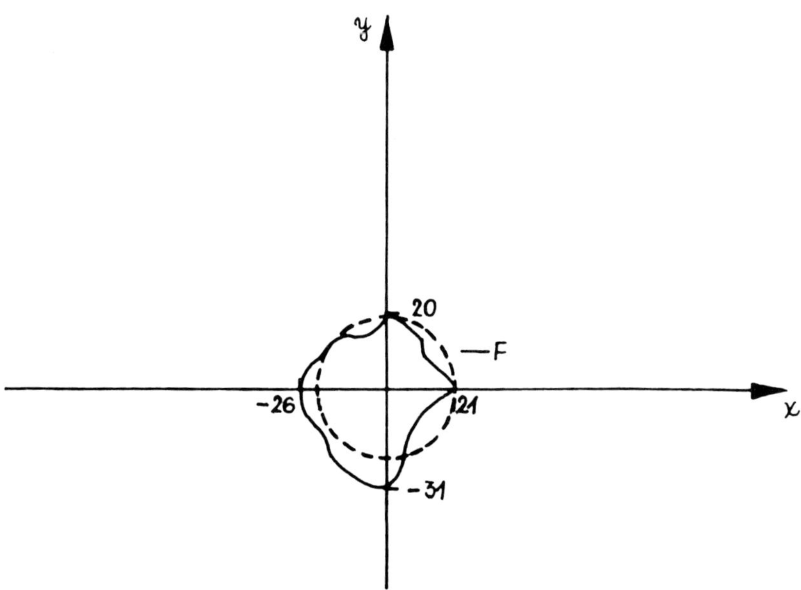

Figure 7. The projection of the range of correctional movements (in mm) into the xy plane. Example for one normal person.

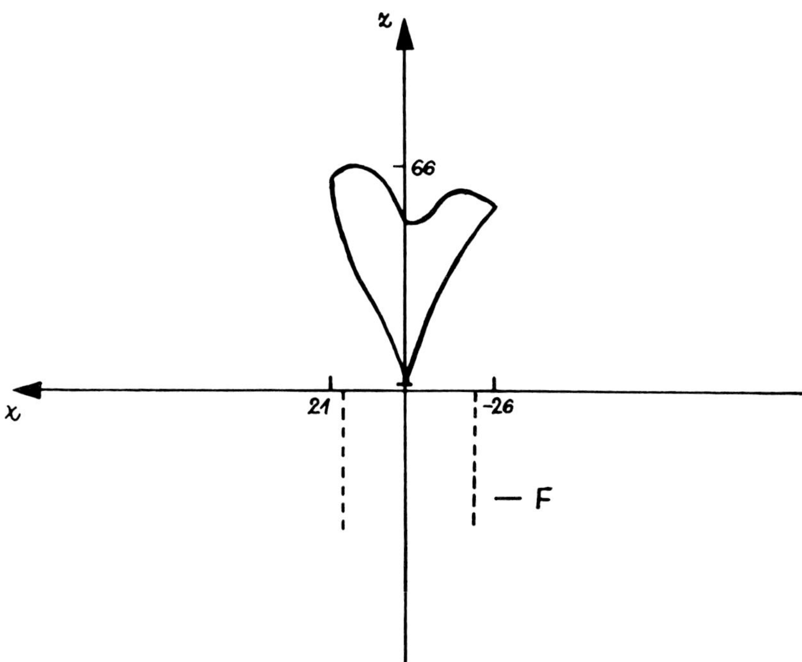

Figure 8.   The projection of the range of correctional movements (in mm) into the xz plane. Example for one normal person.

Table 2.   Mean and standard deviation for the area of correctional movements space projections into planes: xy, xz, and yz — for normal and retarded persons.

Planes	Quadrants	Areas for retarded persons (mm)		Areas for normal persons (mm)	
	Mean	Mean	Standard deviation	Mean	Standard deviation
xy	x^+y^+	11·3 +	4·4	3·7	0·8
	x^-y^+	4·9	1·2	2·0	0·5
	x^-y^-	47·5 +	17·9	2·6	1·1
	x^+y^-	5·1	0·9	1·9	1·0
xz	x^+z^+	27·8 +	6·5	5·4	1·9
	x^-z^+	35·7 +	12·3	4·8	2·1
	x^-z^-	4·3	2·7	0·0	0·0
	x^+z^-	0·6	0·5	0·0	0·0
yz	y^+z^+	16·4 +	3·7	3·3	1·2
	y^-z^+	52·3 +	16·4	6·0	2·6
	y^-z^-	4·4	2·1	0·0	0·0
	y^+z^-	1·1	0·8	0·0	0·0

Note: + = areas which differ significantly from others in the retarded group ($P < 0.05$).

The mean values and standard deviations for the three planes, yz, xy, xz, of area for each quadrant were calculated for both retarded and normal groups (Table 2). By using ANOVA and the Duncan Range Test it was proved that both groups differ significantly in the areas of all quadrants. In the case of the retarded group, the area of movement is larger than that for the normal group. Within each plane, the areas which differ significantly from others are marked for the retarded group (Table 2) with an asterisk. In the normal group, there are no differences between areas of movements within each plane. It can be seen (Table 2) that the area for correctional movements are asymmetrical in the case of retarded persons, and are symmetrical for normal persons, at least from a statistical point of view.

Discussion

For about the last 15 years Bernstein's ideas concerning the problems raised by the organization and control of motor activity and their reformulation have become more and more popular (Whiting, 1984). According to one of the laws of Bernstein's biodynamic structures which make adaptive movement possible, the regulation of movement, with consideration to the temporal as well as spatial aspects, in response to sensory feedback requires a central structure that can perform accurate anticipatory tracking of peripheral stimuli. Additionally, change in sensitivity of the receptors commanded by the centres may be at play. Perceptual anticipation intervenes when the subject learns the predictable characteristics of signals. On the basis of this knowledge, his anticipation of temporal and spatial events enables him to set up adaptations before the occurrence of the events. Anticipation is particularly important with respect to movements in which sensory motor activity is continuous and changes during the movements as in the present study. The fact that the area of correctional movement is larger for retarded persons than for normal persons depends probably on the disturbances in anticipation. As a result, when the sensory motor activity changes, in the second stabilization phase of directed action of the movement, errors appear. In other words errors appear when the dynamic phase transforms into the correctional phase.

When the spatial aspect of movement is analysed, errors appear in the overlapping of movements. In this case the prevention of errors depends on the perceptual anticipation as well as on the capacity to anticipate the effects of movement, i.e., on the capacity to anticipate the effects of both movement phases accurately (Whiting, 1984). Thus, from the theoretical point of view, rehabilitation should be directed towards improving the capacity of perceptual and effector anticipation.

Conclusions

The results of this study seem to be important for the ergonomic design of workstations for the mentally retarded as well as for planning for their rehabilitation therapy. The complex design of the work-station should be directed toward improving the correctional movements. The work space for mentally retarded persons should be prepared in such a way that the whole space of correctional movements lies in the optimal range of movements.

It is necessary to stress here that the model of rehabilitation therapy based on mastering the work movements can not be realized without ergonomic treatment. The work-station for rehabilitation therapy should be designed not only for the mastering of the work movement but also for mastering the systems of control and correction of movements. To reach this goal, it must be possible to make changes in the particular parameters of the work-station, so that the mentally retarded persons are able to make appropriate corrections in their movements. During progressive rehabilitation these changes should increase successively. Rehabilitation should be directed towards decreasing the range of movements in quadrants (y_-z_+), (x_-y_-), and (x_-z_+).

References

Kapecka, E. and Marek, T., 1982, Ergonomics for the mentally handicapped — motion study. In *Proceedings of the 8th Congress* of IEA, edited by K. Noro, Tokyo, 126–127.

Maloney, M.P. and Ward, M.P., 1979, *Mental Retardation and Modern Society*, (New York: Oxford University Press).

Stelmach, G.E. and Requin, J. (Eds.), 1980, *Tutorials in Motor Behavior* (Amsterdam: North-Holland).

Whiting, H.T.A. (Ed.), 1984, *Human Motor Actions* (Amsterdam: North-Holland).

6.

Assessment of the cognitive components of driving performance among disabled individuals

John R. Schweitzer, W. Drew Gouvier[1], Charles R. Horton, Margaret Maxfield, Mike Shipp and Paul N. Hale, Jr.

Center for Rehabilitation Science and Biomedical Engineering
Louisiana Tech University
P.O. Box 10426 T.S.
Ruston, Louisiana 71272–0046, USA

[1]Department of Psychology
Louisiana State University
Baton Rouge, Louisiana 70803–5501, USA

Abstract

The paper describes aspects of a research programme directed towards the development of a valid assessment battery for measuring perceptual and cognitive skills involved in driving.

Outcomes of three studies are presented: a preliminary case study, a controlled two-group comparative study and a three-group comparative study. The findings focus on the validation of a battery of 8 tasks, by examining the ability of these tasks to discriminate able-bodied individuals from those afflicted with head injury. Scores were also obtained for subjects driving a Small-Scale Vehicle (SSV) and driving a Full-Scale Vehicle (FSV), making it possible to study actual application of cognitive performance to driving. Results indicated that performance on the REACT and SEARCH tasks differentiated well among the groups. Two of the seven driving manoeuvres involving execution of left turns and assessed by the Tracking Simulator were also found to be significant discriminators. The findings are discussed in terms of the restricted generalizations made from these results, and the apparent promise in the use of computer assisted tasks for cognitive/perceptual assessment.

Introduction

There has been increasing interest in the use of cognitive and perceptual assessment techniques in evaluating disabled individuals who are seeking driving licenses. There is

85

good reason for this interest; traditional approaches to disabled driver evaluations have often relied on subjective judgements of occupational therapists and driver evaluators to decide whether a person has sufficient cognitive and perceptual capacity to drive. Although many therapists and evaluators incorporate some objective testing into their decision-making process, the relative lack, or disuse of, cognitive assessment procedures with established validity increases the likelihood that subjects with significant deficits may proceed to actual driving in the course of the assessment. This contingency may result in a dangerous situation for the subject, the evaluator, and others in their environment.

For example, patients with damage affecting the right cerebral hemisphere often misinterpret or fail to notice stimuli on their left. This phenomenon, which is known as left visual neglect or left hemi-imperception is observed in as many as 60% of patients with right brain injury (RBI) (Friedland and Weinstein, 1977). While very simple procedures such as visual confrontation can identify cases of gross imperception, the use of more sensitive procedures results in the identification of imperceptions among many cases who do not evidence gross deficits (Trobe *et al.*, 1981). Using complex tasks that require detailed searching to find stimuli that are identical to a matching stimulus, left-side imperception has been demonstrated among brain damaged patients who had no problem with any of the measures typically used to demonstrate these phenomena (Blanton and Gouvier, 1985). The task of driving an automotive vehicle involves a set of complex perceptual and cognitive operations (Schlesinger, 1972; Weaver, 1982); thus, the findings of Blanton and Gouvier (1985) and Trobe *et al.* (1981) would suggest that many subjects with subtle undiagnosed perceptual problems may be advanced to on-the-road assessment, and thus place themselves and others at needless risk of accident and subsequent re-injury.

Although research on cognitive and perceptual assessment of disabled drivers is limited, some significant contributions have been made in recent years. Work at the University of Michigan (Kewman *et al.*, 1985; Sivak *et al.*, 1981a, 1981b, 1983, 1984) has demonstrated that cognitive and perceptual measures can be valid predictors of driving performance among both disabled and able-bodied individuals, and that the deficits identified in such assessments may have direct implications for treatment and remediation.

In the initial study (Sivak *et al.*, 1981b) psychological tests were shown to predict driving skills among brain-damaged subjects and able-bodied subjects, although different tasks proved to be the best predictors for these two groups of subjects. Among the tests that predicted driving skills of brain damaged subjects, the Picture Completion subtest of the WAIS had the highest correlation ($r = 0 \cdot 72$) with driving performance. Next, these researchers conducted a pilot intervention targeting the perceptual abilities believed to be involved in the Picture Completion task (Sivak *et al.*, 1981a). Four brain-injured subjects received training on paper and pencil perceptual tasks, and two other brain-damaged subjects viewed driver education filmstrips as a control intervention. Both interventions led to improvement on the battery of predictive tests used in the preliminary assessment, although the improvement was greater for the experimental group. Both groups showed a comparable degree of improvement in their driving performance. Subsequent studies with different populations of brain-damaged subjects have supported the efficacy of providing training on paper and pencil perceptual tasks in improving driving skills (Sivak *et al.*, 1984). It was also demonstrated that computerized video tasks hold promise as a training medium (Sivak *et al.*, 1983). In the most recent report from the University of Michigan group (Kewman *et al.*, 1985) brain-damaged subjects were provided with a series of structured practice exercises in piloting an *AMIGO* three-wheel vehicle. Compared to subjects who were given comparable durations of unstructured practice on the *AMIGO* trainer, the subjects who received formal training made significant improvements in their driving performance, while the control subjects did not.

Other researchers have demonstrated that psychological variables can serve as useful predictors of driving performance (Jones *et al.*, 1983; Timmermans *et al.*, 1986; Stokx and Gaillard, 1986). The use of such off-the-road measures is intended to complement, rather than replace the findings of on-the-road assessment, but such predictive measures can be used to identify subjects who would be unsafe in a road assessment.

Multiple regression procedures have been used to identify the best predictors from a battery of psychological measures. In one study of 28 head-injured individuals (Timmermans *et al.*, 1986), 79 per cent of the variance in driving performance could be accounted for by the following five variables; night vision, depth perception, auditory reaction time, the WAIS-R Arithmetic subtest, and the WAIS-R Digit Symbol subtest. A larger study cross-validating the predictors is currently underway. (Timmermans, personal communication, 1986).

Jones *et al.*, (1983) report their experiences using a battery of cognitive and psychomotor tasks in assessing disabled drivers. They report that a tracking task included in their battery was the most efficient single predictor in differentiating among subjects who passed and subjects who failed the on-the-road assessment, although other tasks such as reaction time also contributed. The usefulness of reaction-time measures in predicting driving skills was upheld in a recent study reported by Stokx and Gaillard (1986). While these authors were not able to identify which specific links in the information processing chain (e.g. stimulus encoding, memory comparison, response selection and motor preparation) were most affected by head injury, they did note a high correlation ($r = 0 \cdot 69$) between simple reaction time and driving performance by TBI subjects through a slalom course.

The relationship between laboratory-based assessments and driving skills has also been examined by others. For example, Golper *et al.*, (1980) have demonstrated that post-stroke aphasics tend to be cautious in their decision about returning to driving. In their study of twenty left brain-damaged stroke patients, ten decided to resume driving and ten decided not to resume driving. Professional evaluations by physicians and therapists supported the decision to return to driving that was made independently by ten of the patients. Only 4 of the remaining ten patients (who had not chosen to resume driving) were considered by the professionals to be good candidates for driving. Related findings on left and right brain-damaged drivers are discussed by Bardach (1971) who notes the greater difficulty encountered in getting right brain-damaged patients safely behind the wheel, compared with left brain-damaged patients. Blaauw (1982) studied the validity of simulator tasks in predicting driving; he reports that such tasks are good predictors of 'lateral control' of a motor vehicle. Subjects who make left-sided errors on the simulator tend to make left-sided errors in driving; similarly, right-sided simulator errors were associated with right-sided driving errors.

These studies suggest that measures of reaction time, simulated driving, and some of the performance subtests of the WAIS offer useful predictors of driving skill. The present study examined these and several other predictors in relation to cognitive deficits associated with head injury. It is presumed that brain injury affects cognitive functions, and thereby leads to deleterious changes in driving skills. Thus, those tasks that can discriminate brain-injured from able-bodied individuals are likely to be useful predictors of driving performance among the brain-injured.

Preliminary case study

The specific tasks used in the preliminary studies described here were the REACT and SEARCH tasks from the Computer Programs for Cognitive Rehabilitation (Gianutsos and Klitzner, 1981) and performance of the seven manoeuvres assessed by the Louisiana Tech Tracking Simulator (Schubert and Irwin, 1985). These manoeuvres simulate starting and stopping, right turn from stop, left turn from stop, right turn while moving, left turn while moving, low speed serpentine tracking and high speed serpentine tracking.

Five individuals with histories of cognitive deficits volunteered to participate in the evaluation. The aetiologies represented included cerebrovascular accident (CVA), hypothyroiditis, developmental disability and cerebral palsy. All subjects were adults and were able to understand verbal task instructions accompanied by demonstrations. All five subjects were able to complete all parts of the tests; however, the SEARCH subtest of the Cognitive Rehabilitation software proved frustrating for several subjects.

The site for computer testing was the Driver Assessment Laboratory at Louisiana Tech University Center for Rehabilitation Science and Biomedical Engineering, with room darkened, monitor screen brilliance controlled and corridor traffic redirected.

Testing on the Tracking Simulator was supervised by a driving instructor and a re-habilitation engineer. One subject simulator record was lost due to a power surge in an electrical storm. No other technical difficulties occurred during testing.

Although statistical generalization was impossible from the pilot study, the feasibility of data collection was assured and expected interrelationships of variables appeared to exist. A formal controlled study was therefore proposed.

Controlled two-group comparative study

The assessment performance of individuals who had sustained head injuries with that of individuals with no history of head trauma was compared in this study. The injured group consisted of seven subjects in post-acute recovery phases from traumatic brain injury (TBI) and the able-bodied subjects were volunteers recruited from intro-ductory psychology classes at Louisiana Tech.

The objectives of this study were:

(1) To investigate correlations of SEARCH and REACT scores with scores on the seven manoeuvres measured by the Tracking Simulator.
(2) To determine which measures offer evidence of residual cognitive deficits.

Data Collection Procedures

The Mobile Assessment Laboratory (MAL) was used to obtain data from the head injured subjects. The SEARCH and REACT programmes were run on a Commodore microcomputer. A simplified version of the SEARCH task was developed for use on the Commodore computer; the simplified version contained fewer alternatives to the target shape than did the original test. The MAL implementation of the tracking simulator was used.

Subjects were screened for inclusion in the TBI group by the Occupational Therapy Clinic of the Rehabilitation Unit at a local hospital. To be eligible, volunteers had to have suffered traumatic brain injury but to have recovered past the acute stage and reached medical stability. They also had to be seeking to become licensed drivers. All seven had between one and seven years of previous driving experience.

All testing of head-injured subjects was conducted in the MAL. Evaluation staff consisted of an occupational therapist, with technical support provided by a rehabilitation engineer. The tests for the control group were conducted over the course of a week by an occupational therapist. All subjects were able to complete the testing without difficulty.

Analysis

Pearson product moment correlations and Spearman rank order correlations were computed between individual manoeuvre scores and the REACT and SEARCH scores. Although most variables showed no evidence of non-normal distributions, one of the tracking simulator manoeuvres yielded skewed scores. For this reason, the Spearman correlations are reported in preference to Pearson correlations. Differences between scores for the two groups of subjects were analysed by t-tests.

Results

The most promising parts of the correlation analysis are presented in Table 1. Significant correlations were observed between performance on two of the manoeuvres on the Tracking Simulator and performance on the REACT task ($p < 0.01$). The correlations were strongest with those parts of the REACT test that were presented to the client's left field of vision (REACT–LEFT). These two manoeuvres both involved making left turns, the left turn from a stop (MANVR2), and the left turn while slowing (MANVR4).

respond to the task. This yielded a motor-dependent measure (MOTWAIS) composed of the first three sub-tests, and a motor-free measure (NOMOTOR) composed of the latter two.

Subjects also received the Motor Free Visual Perception Test, MVPT, (Bouska and Kwatry, 1983) and the Baylor Adult Visual Perceptual Test (1980). These two tests examine different aspects of visual perceptual functioning using motor-free and motor-dependent response formats.

The Trail Making Test, Parts A and B, from the Halstead Reitan Battery (Reitan and Davison, 1974) and the Symbol Digit Modalities Test (Smith, 1968) were used as measures of information processing speed and efficiency. The Symbol Digit test is quite similar to the WAIS Digit Symbol sub-test, but the advantage of the Symbol Digit test is that it can be completed in a non-motor format, and thus can offer a purer test of cognitive function for quadraplegics. The final predictive measure was the Driver Performance Test, DPT (Weaver, 1982). This measure assesses knowledge of how to respond to potentially dangerous driving situations by using a series of videotaped scenes. The fast pace of its automated presentation makes it a very challenging and demanding test to complete, and in many respects, the DPT appears to measure capacity for information processing, integration, and motoric output, as well as the knowledge about how to respond while driving.

The criterion measure involved expert ratings of performance in driving an appropriately modified full sized vehicle about a closed course. Eight specific manoeuvres were evaluated. These were starting and stopping, left turn from stop, right turn from stop, left turn while slowing, right turn while slowing, slalom navigation at 10 m.p.h., slalom navigation at 15 m.p.h., and performance on an evasive manoeuvre in which the subject was given only several seconds to veer left, veer right or stop. Two raters independently rated these manoeuvres for eight disabled subjects; perfect agreement was reached on 86 per cent of the observations. The primary rater also rated performance on parameters of following instructions, tracking, braking, accelerating and appropriate speed control.

Procedure

Disabled subjects were individually evaluated either at a regional hospital, or at our Center in Ruston, LA. Closed course driving ranges were prepared for completion of the driving component of the assessment.

Initial screening involved completion of medical and driving history forms and interviews, as well as objective evaluation of mental status (Folstein *et al.*, 1975). Subjects received evaluation of vision and hearing functions, range of motion, muscle strength and assessment to identify the presence of hemispatial visual neglect. The cognitive processing component involved completing the assessment measures described in the preceding section, including REACT and SEARCH subtests from the Computer Programs for Cognitive Rehabilitation (Gianutsos and Klitzner, 1981). Administration of the Driver Performance Test in a group format supplied additional information on cognitive processing and knowledge together.

Subjects then completed evaluation on the Tracking Simulator (Schubert and Irwin,

1985) prior to advancing to driving the Small-Scale Vehicle (Hale *et al.*, 1987) on a closed course. The criterion assessment component involved driving the Full-Scale Vehicle on a closed course.

Analysis of variance was carried out for comparing all three groups. For comparing just the TBI and ABLE groups, *t*-tests were conducted, and stepwise discrimination procedures identified the best tasks for separating these two groups. Stepwise regression analysis was carried out for discovering which variables among the batteries of tests permit valid predictions of the driving scores. Factor analyses were carried out to cluster those tests that seem to be measuring similar dimensions.

Three-group study results

There were clear and significant differences between the groups in terms of driving performance. With a score of 100 being perfect performance, the mean performance of the able-bodied subjects was $94\cdot7$, for the SCI subjects it was $83\cdot6$, and for the TBI subjects it was $66\cdot4$. Thus, although the SCI subjects tended to have greater physical problems with motor control and postural stability, the cognitive deficits observed among the TBI subjects appeared to be an even greater impediment to safe driving.

All of the psychometric measures except the MVPT correlated significantly with driving performances at levels of $p < 0\cdot001$. Correlations between measures and the criterion (FSV = Full Scale Vehicle score) are shown below:

MOTWAIS	$0\cdot807$	NOMOTOR	$0\cdot706$
SYM-DIG (W)	$0\cdot782$	SYM-DIG (O)	$0\cdot839$
TRAILS	$0\cdot668$	MVPT	$-0\cdot382$ $(p < 0\cdot06)$
BAYLOR	$0\cdot632$	DPT	$0\cdot847$

Multiple regression analysis using only two predictors, the DPT and the NOMOTOR WAIS variables, yielded an R^2 coefficient of association of $0\cdot81$. Even when corrected for shrinkage (Pedhazur, 1982), the adjusted R^2 remained an impressive $0\cdot79$. Including knowledge of group membership in the regression equation made no significant addition to the level of prediction. A three-step multiple regression including the MOTWAIS, DPT and Tracking Simulator accelerating/braking error scores yielded an even higher R^2 coefficient of association with a value of $0\cdot85$.

The results that look most promising for contributing to eventual formalization of cognitive assessment were those from the discriminant analyses. Although several such analyses yielded worthwhile results, the most interesting, perhaps, used just one discriminator: the oral Symbol–Digit Modalities Test. The discriminant function obtained from the analysis assigned members of the TBI group and the ABLE group to their correct groups with no mistakes. When applied to the third group, SCI, the function found 2 of the 8 more like the TBI group and the others more like the ABLE group. In the light of findings that approximately one half of spinal-cord injured patients may also suffer brain injury as well (Davidoff *et al.*, 1985; Wilmont *et al.*, 1985), the discriminant analysis opens an exciting avenue for research, and suggests that cognitive assessment may be a useful means for differentiating brain-injured SCI patients from those who have not sustained brain injury.

Although the REACT test also discriminates perfectly between the TBI and ABLE groups, it does not contribute significantly in regression to explaining variance in full-scale driving scores (FSV). Prediction of FSV by Symbol-Digit Modality score alone yielded a coefficient of association (R^2) of 69 per cent. Several other combinations of regressor variables yield R^2 values of 84 to 85 per cent. It can be seen that the discrimination according to cognitive skills was not precisely limited to those skills that are applied in driving. Since the aetiology for driver evaluation candidates would ordinarily be known, it will become important to isolate those discriminating scores that actually pertain to driving ability.

Conclusion

The results of the studies reported, especially the three-group study, lead to optimism about eventual establishment of formal assessment procedures for disabled subjects who are seeking driver training or licensing. A brief screening battery could be developed which might permit the ready identification of subjects who can be advanced directly to on-the-road assessment, and allow comprehensive cognitive assessment to be reserved for those subjects who have difficulty on the screening assessment.

Acknowledgements

Supported by cooperative Agreement G0083C0097 from the National Institute on Disability and Rehabilitation Research (NIDRR, formerly NIHR) of the United States Department of Education, and also by grant funds from the Louisiana Division of Rehabilitation Services.

References

Bardach, J.L., 1971, Psychological factors in the handicapped driver, *Archives of Physical Medicine and Rehabilitation*, **52**, 328–332.

Baylor Adult Visual Perceptual Assessmant, 1980, (Dallas, TX: Occupational Therapy Department, Baylor University Medical Center).

Blaauw, G., 1982, Simulator and Intrumental Car: A Validation Study. *Driving Experience and Task Demands.* (Soesterberg, Netherlands: The Human Factors Society, Institute for Perception). pp. 473–485.

Blanton, P.D. and Gouvier, W.D., 1985, Sex differences in hemispatial performance following right-hemisphere cerebrovascular accidents, *Journal of Clinical and Experimental Psychology*, **7**, 619.

Bouska, M.J. and Kwatry, E., 1983, *Manual for the Application of the Motor Free Visual Perception Test to the Adult Population*, (Philadelphia, PA: author).

Carpenter, M.B., 1978, *Core text of neuroanatomy*, (Baltimore: Williams and Wilkins).

Davidoff, G., Morris, J., Roth, E. and Bleiberg, J., 1985, Closed head injury in spinal cord injured patients: Retrospective study of loss of consciousness and post-traumatic amnesia, *Archives of Physical Medicine and Rehabilitation*, **66**, 41–43.

Folstein, M., Folstein, S. and McHugh, P., 1975, 'Mini-Mental State': A practical method for grading the cognitive state of patients for the clinician, *Journal of Psychiatric Research*, **12**, 189–198.

Friedland, R. and Weinstein, E., 1977, Hemi-inattention and hemisphere specialization: Introduction and historical review. In E. Weinstein and R. Friedland (Eds.), *Hemi-inattention and hemisphere specialization*, (New York: Raven), pp. 1–32.

Gianutsos, R. and Klitzner, C. 1981, *Computer programs for cognitive rehabilitation.* (Bayport, NY: Life Science Associates).

Goldstein, G. and Shelley, C., 1981, Does the right hemisphere age more rapidly than the left? *Journal of Clinical Neuropsychology*, **3**, 65–78.

Golper, L.A., Rau, M. T. and Marshall, R.C., 1980, Aphasic drivers and their decisions on driving: An evaluation. *Archives of Physical Medicine and Rehabilitation*, **61**, 34–40.

Hale, P.N., Gouvier, W.D., Schweitzer, J.R. and Shipp, M., 1987. A small scale vehicle for assessing and training driving skills among the disabled. *Archives of Physical Medicine and Rehabilitation*, **68**, 741–742.

Jones, R., Giddens, H. and Croft, D., 1983, Assessment and training of brain-damaged drivers. *American Journal of Occupational Therapy*, **37**, 754–760.

Kewman, D.G., Seigerman, C., Kintner, H., Chu, S., Henson, D. and Reeder, C., 1985, Simulation training of psychomotor skills: Teaching the brain injured to drive, *Rehabilitation Psychology*, **30**, 11–27.

Leber, W., Jenkins, R and Parsons, O., 1981, Recovery of visuospatial learning and memory in chronic alcoholics, *Journal of Clinical Psychology*, **37**, 191–197.

Long, C.J. and Gouvier, W.D., 1982, Neuropsychological assessment of outcome following closed head injury. In R. Malatesha and L.C. Hartlage (Eds.), *Neuropsychology and Cognition, Vol. 1*, (Boston: Martinus Nijhoff).

Pedhazur, E.J., 1982, *Multiple Regression in Behavioral Research*, 2nd edition, (New York: Reinhart and Winston).

Reitan, R.M. and Davidson, L.A., 1974, *Clinical neuropsychology: Current status and applications*, (New York: Hemisphere Publishing).

Schlesinger, L.E., 1972, Human factors in driver training and education. In T.W. Forbes (Ed.), *Human factors in highway traffic safety research*, (New York: Wiley).

Schubert, R. and Irwin, E., 1985, A microprocessor based simulator for objectively evaluating prospective drivers of adapted personal licensed vehicles. In *Biomedical engineering IV: Recent developments*, edited by B. Sauer (New York: Pergamon).

Silber, S.M. and Olson, P.L., 1984, Improved driving performance following perceptual training in persons with brain damage. *Archives of Physical Medicine and Rehabilitation*, **65**, 163–167.

Sivak, M., Hill, C.S., Olson, P.L. and Henson, D.L., 1981a, *Preliminary testing of techniques to improve driving performance of persons with brain damage via perceptual/cognitive training.* (Ann Arbor, MI: University of Michigan, Highway Safety Research Institute), Document UM-HSRI-81-12.

Sivak, M., Olson, P.L., Kewman, D.G., Won, H. and Henson, D.L., 1981b, Driving and perceptual/cognitive skills: Behavioral consequences of brain damage. *Archives of Physical Medicine and Rehabilitation*, **62**, 476–483.

Sivak, M., Hill, C.S. and Olson, P.L., 1983, *Computerized video tasks as training techniques for driving-related perceptual deficits of persons with brain damage: A pilot evaluation.* (Ann Arbor, MI: University of Michigan, Transportation Research Institute), Document UMTRI-83-48.

Sivak, M., Hill, C.S., Henson, D.L., Butler, B.P. and Olson, P.L., 1984, Improved driving performance following perceptual training in persons with brain damage. *Archives of Physical Medicine and Rehabilitation*, **65**, 163–167.

Smith, A., 1968, The Symbol Digit Modalities Test: A neuropsychologic test for economic screening of learning and other cerebral disorders. *Learning Disorders*, **3**, 83–91.

Stokx, L.C. and Gaillard, A.W., 1986, Task and driving performance of patients with a severe concussion of the brain. *Journal of Clinical and Experimental Neuropsychology*, **8**, 421–436.

Timmermans, S., Bouman, J. and Reed, P., 1986, The role of assessment and driving outcome for head injured individuals. Paper presented at the annual meeting of the National Head Injury Foundation, (Chicago, Il: November).

Trobe, J., Acosta, P., Krisher, J. and Trick, G. 1981, Confrontation visual field technique in detection of anterior visual pathway lesions. *Annals of Neurology*, **10**, 28–34.

Weaver, J.K., 1982, *Driver Performance Test*. (Dunwoody, GA: SPA Productions).

Wechsler, D., 1955, *Wechsler Adult Intelligence Scale*. (New York: The Psychological Corporation).

Wilmont, C.B., Cope, N., Hall, K.M. and Acker, M. (1985), Occult head injury: Its incidence in spinal cord injury. *Archives of Physical Medicine and Rehabilitation*, **66**, 227–231.

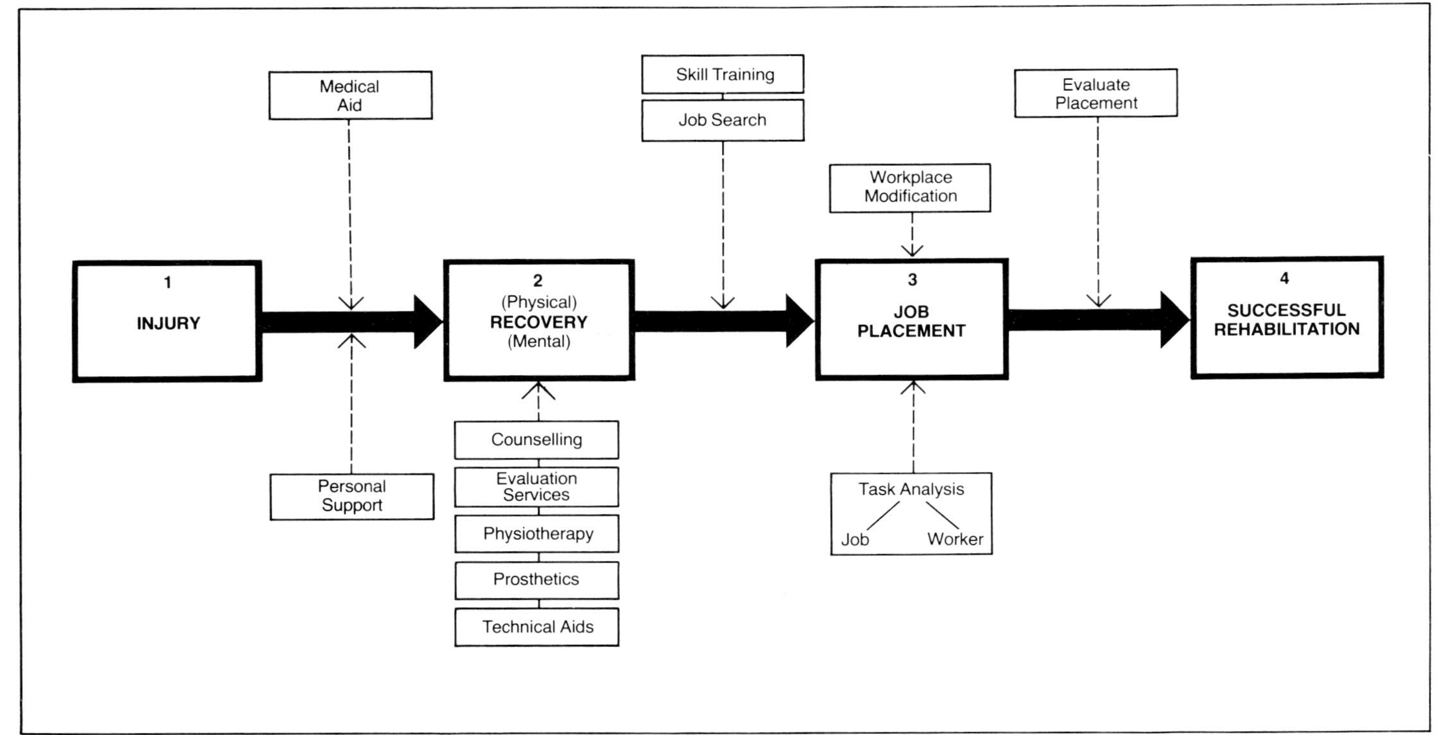

Figure 1: The rehabilitation process

work and around the workplace. Mobility related disabilities met with by Rehabilitation Services range between reduced endurance for walking, visual impairment, amputation, incontinence and disfigurement. Each of these has implications for job access. If the job is not accessible, then regardless of the person's ability to perform the work itself, job placement will not be successful.

Two approaches to worksite access are possible. Either a building can be assessed and modified so as to be accessible to anyone who might want access or routes taken by individuals can be assessed and buildings modified on a selective basis to solve immediate needs. General building assessment rather than individual route assessment is the most common and, at first sight, the logical choice, but results in a number of disadvantages from the point of view of occupational access.

Whole building assessments either have to be exhaustive in terms of the range of disabilities taken into account, or make certain limiting assumptions about the potential user population. A common simplifying assumption is to equate mobility impairment with wheelchair use yet there are many other forms of disability which impair access.

Another factor is that access problems are usually faced in the context of an existing facility. For a new public building, adequate wheel chair access should be a *sine qua non*, but for an existing private access building for, say 250 employees, there is a strong argument for responding to accessibility problems in the face of specific requests. Since most organizations are faced with limited budgets, site modifications are likely to be prioritized on the basis of demonstrated and not hypothetical need.

Even in the unlikely event that, for a particular building, all access and internal mobility problems have been eliminated for all potential users, the building may represent only one of many barriers to job access for a given person and some individual assessment scheme will always remain necessary. Successful job placement depends upon the surmounting of **all** barriers along relevant personal access routes.

From a largely client-oriented point of view such as that of the Workers' Compensation Board, modification of complete buildings based on hypothetical assumptions about future patterns of demand represents a distribution of resources away from the more pressing problems of existing case loads. Development of procedures for evaluating occupational access needs to focus on individual cases, not generalized building assessments and route evaluation should be considered as part of the overall process of matching clients to new jobs.

Application of ergonomics

Ergonomics is defined as the practical application of scientific knowledge about human characteristics to match task demands to worker characteristics within the context of the person–process–environment shown in Figure 2.

The person–process–environment model has several advantages for rehabilitation. It changes the focus from a worker as a disabled person to a focus on functional demands and their interactions within a particular setting. It makes setting criteria for successful rehabilitation more objective and emphasizes the need to concentrate on local task requirements while providing a conceptual framework which is easy to comprehend by

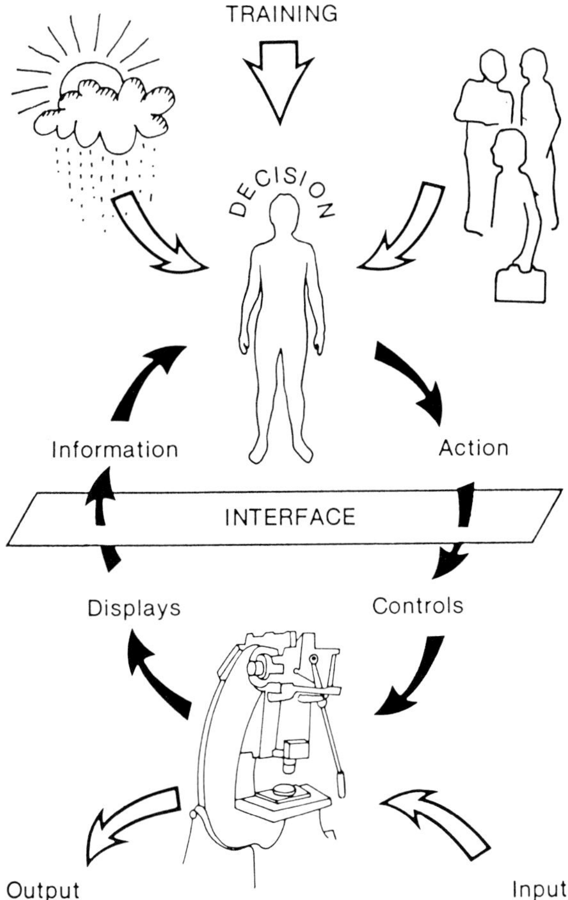

Figure 2: The person–process–environment model

analysts whose main experience is not ergonomics. Finally, it provides a basis for integrating the various sources of information available.

The model also makes it plain that the completion of any task requires activities which fall into some common categories, namely:

(a)　Information must be sensed . . .
(b)　. . . transmitted to the brain . . .
(c)　. . . interpreted on the basis of past experience . . .
(d)　. . . an action determined . . .
(e)　. . . the action carried into effect . . .
(f)　. . . and the consequences of the action evaluated by repeating steps 1–5.

This is true whether the task is lifting a box or driving a vehicle. The ability of any individual to perform specific functions within these categories depends upon three factors:

(a) His or her own abilities – sensory, physical, or intellectual.
(b) The demands of the task itself.
(c) The additional constraints imposed by the physical and social environment.

The problem for Rehabilitation Services is to first identify any mismatch for the worker in question and then to devise a strategy for eliminating the mismatch.

Anticipated benefits for vocational rehabilitation of applying ergonomics include: improved job retention following placement; reduced second injury and medical absence rates; and improvements in job performance and the general quality of work life for individual workers. Availability of task related information in systematic form was also expected to benefit follow-up evaluation.

Definitions

In order to set up the procedures it was necessary to agree on some common terminology. A *job* was defined as all the work expected of an employee fulfilling a specified job role. Each job comprised a number of *tasks* and each task a number of *activities*. Occasional or informal tasks need to be taken into account since it is actual demands placed on a person which determine whether or not placement will be successful.

Thus a job might be clerk or storeman while a task might be to type reports from manuscript drafts, or to locate spare parts in a storage area. Activities might be to read

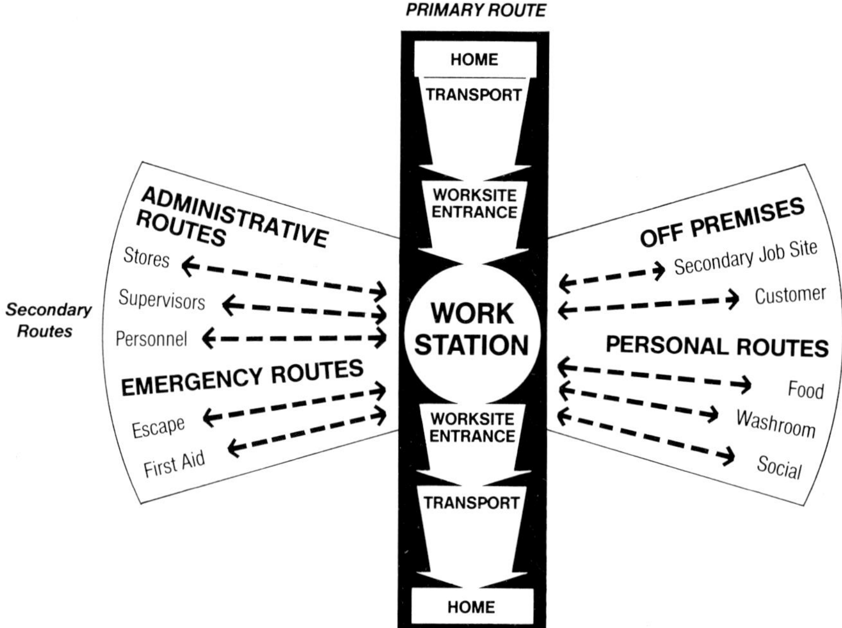

Figure 3: Primary and secondary routes

symbols in a manuscript and to strike successive keys on a typewriter, or to grasp or lift a wrench or a box.

Workspace accessibilty was defined as the ability of an individual to overcome the barriers to movement from the front door of their dwelling to being in their work-station in a position to start work. The *worksite* was defined as the general facility within which the work takes place. The *workstation* was defined as the specific place (or places) in which the client will perform job-related tasks. Worksite accessibility is as-sociated with the *primary route. Worksite mobility* was defined as the ability of an indi-vidual to overcome the barriers to movement within the general area of their work facility associated with performance of necessary personal, social and administrative functions. Worksite mobility is associated with *secondary routes*. The relationship between these route related concepts is shown in Figure 3.

For worksite access, a broad definition of *barriers* was assumed with all factors which may hinder job access or worksite mobility to be considered. For each route and route portion, this included the following aspects: sensory, cognitive, learning, body shape and size, strength, endurance, client attitude, peer attitudes.

Development of the task analysis procedure

A group of four placement specialists (who deal with clients experiencing placement difficulties) were involved in the initial ergonomic development programme. The ob-jective was to familiarize these placement specialists with the application of ergonomics to their work and develop first an ergonomic task analysis procedure and then a route assessment procedure.

The Placement Specialists received five days of ergonomic training spread over a six month period. The training emphasized task description and analysis skills appropriate to their professional needs. In the intervals between training days, each member of the group analysed selected tasks in their current case loads. The final ergonomic task analysis procedure was developed on an interactive basis as the training progressed.

Ergonomic task analysis

Ergonomic analysis has two distinct phases (Drury, 1983). The first phase, task de-scription, provides an objective record of the job and its component tasks. The second phase, task analysis, evaluates selected tasks and their component activities in terms of their human demands such as seeing, hearing, decision making, learning and lifting. In most cases, this evaluation is made so as to cover 90 per cent of a typical industrial population. In the case of rehabilitation, the analysis will be in terms of the capabilities of one particular person who will likely fall outside the usual evaluation range. Therefore, for rehabilitation, a third phase is required: identification of the capabilities of the person concerned.

Of course, these three phases seldom fall into a neat sequence and usually extensive iteration will occur. For example, preliminary information about a disability will guide selection of which jobs to evaluate and on which elements of the job to focus most attention. In turn, task analysis will prompt requests for mere precise information about individual levels disability and probability of improvement.

Analysis procedures

The following sequence of events was assumed for job placement:

(a) Preliminary worker assessment after hospitalization.
(b) Ergonomic description and analysis of potential job(s).
(c) Further evaluation of worker capacity in relation to specific task demands.
(d) Assessment of worker/task compatibility.
(e) Recommendations for task modification and/or worker training.
(f) Placement.
(g) Evaluation of placement.

It was planned to conduct a full ergonomic task analysis for every job evaluation rather than to restrict the analysis to the disabilities of the worker in the case. In this way, the job analysis could be filed and used in the future if the job proved unsuitable in a particular case.

The analysis model

The model shown in Figure 2 was found useful for two reasons. First, any analyst has to keep many things in mind. The person–process–environment model provides an easy to remember conceptual skeleton that emphasizes the interactive nature of each person's relationship with their work. Secondly, the model provides a basis for standardization of investigation and reporting. Inevitably, as an investigation progresses, reports are passed from person to person. It was recognized early during development of these procedures that many people contribute as an injured worker progresses through the steps outlined in Figure 1 and that co-ordination of their contributions is essential.

On-site observation

The sequence of the placement specialist's investigation took the following form and assumed that no more than 4 hours could be devoted to on-site observation. With on-site observations complete, the placement specialist was to be able to describe the job, the tasks it entails, and the demands each task make:

(a) Ask for a breakdown of the main tasks in the job.
(b) Undertake a brief *preliminary* observation. Watch, without interruption or distraction, for at least 15 minutes for a repetitive job with a short work cycle, longer for a more complex one. Observe at least two complete cycles for every task in a job.
(c) Sketch or photograph the layout of the work-site.

(d) Measure and record the physical dimensions (forces, space, repetition rates) and environmental characteristics (heat, noise, lighting, vibration).

(e) Observe each task in more detail. Use a check list. Note observations on an analysis sheet.

(f) Interview the people who do the job now. Make use of the information gained during the previous steps to guide questions. Also question other people able to talk about the work — supervisors, co-workers, maintenance staff. Ask about occasional difficulties (machine breakdown, replenishment of supplies, overtime requirements) and tasks that newcomers take some time to get used to doing.

Once as much on-site information as possible has been gathered, it is put together into a detailed task description but interpretation is not to be attempted until the task description is documented.

Using the task description, the placement specialist then isolates the human factors involved and consults the literature or specialist advisers to assess their general significance as well as their significance for the individual in question. (In many instances it will be necessary to return to the worksite to get additional information.)

Once a suitably compatible job is identified using the ergonomic task analysis procedure, the next step is to identify the client's primary and secondary routes. Routes may also be classified as *access essential* (client mobility unavoidable) and *access optional* (mobility function could be delegated). Examples of the former would be escape and washroom visits and of the latter where materials could be brought to the worksite by someone else.

With primary and secondary routes identified, the person–process–environment model is applied at each point along each route. Measurements along potential routes and client involvement in accessibility trials may also be necessary and barriers identified documented.

The report documents details of the task and route analysis, describes the capacities of the worker in question, (referring to appropriate medical assessment as necessary) and provides answers to the following questions:

(a) Is enough known about the worker and/or the tasks involved to be able to make an informed judgement about worker/task compatibility? If not, what further information is needed?

(b) Is the job suitable for the worker in question as it presently exists? If not, which tasks and which activities are unsuitable and why?

(c) If any task is not suitable, can compatibility be improved by:

    (i) Task or equipment modifications? For example, modification of a seat/workspace configuration to reduce twisting of the spine, more frequent rest breaks, job alteration, enhanced visual presentation of vital task information, ramps for wheelchair access.

    (ii) Worker training or counselling? Examples range from craft training to enhancement of strength or endurance through physiotherapy. Systematic counselling to enhance worker confidence in their capacity to meet the demands of the task is a possible option.

Case studies

Three case studies are described which involved the use of the ergonomic task analysis procedures. Each case was referred to the placement specialists by another department.

Case #1: Tyre inspection

Presenting problem

Immediate re-employment was being offered to an injured worker. The job was tyre inspection and, as an inspection job, this was considered light work and expected to be within the worker's physical capabilities.

The worker had experienced a lumbo-sacral sprain, a left knee sprain and reported ringing in the ears. The worker, familiar with the job being offered, believed he would not be able to cope with its demands and rejected it.

Placement services was asked to assess the appropriateness of the job in relation to what was known about the worker's physical capabilities.

Task description

Looking at tyres and recognizing and marking any of the possible 120 defects. This was done while walking slowly behind the tyre as it was carried through the work area on an overhead carrier. Inspection was carried out partly by sight and partly by touch.

While inspecting the tyre, removing surplus material left by the moulding process with a sharp knife. This required spinning the tyre with one hand while using the knife with the other. Considerable force and acute wrist angles were required.

Lifting and carrying rejected tyres to a conveyor belt which carried them away for re-processing. (See Figure 4.)

The job required standing and walking all day. Lifting occurred approximately once every 3 minutes and the weight of tyres ranged from 15–20 kg. Noteworthy environmental conditions included continuous, high background and impact noise, and a slight but noticeable floor vibration.

Conclusions

The lifting required on the job was assessed as in excess of the worker's capabilities, given his lumbo-sacral injury. In addition, the prolonged standing and walking was likely to cause problems for the knee condition. Noise levels were within regulated levels and in themselves might not have caused rejection of the job for this particular person, but would have impaired speech communication, probably resulting in some distress.

In the absence of very specific knowledge about the worker's lifting capacity and ability to endure walking and standing for long periods, the assessment was cautious.

the procedures and the associated training was as much to make placement specialists sensitive to ergonomic issues and the need for specialist advice as to equip them to give it themselves.

Arising from this is a second limitation, the availability of specialist advice. People experienced and qualified in ergonomics are in short supply and, even if funds were available for employment of suitable specialists, they would be hard to find. A third limitation is the frequent lack of a precise knowledge of the worker's capabilities, existing and potential.

Fourth, in the process of integrating ergonomics into the rehabilitation sequence, it has become clear that the current division of responsibilities between the helping professions poses some difficulties of timing. For example, it is often impossible to provide a sufficiently specific assessment of a worker's abilities until a detailed specification of task demands is available.

An iterative approach is clearly needed and a likely sequence of events is this. First the medical rehabilitation process provides a general indication of the area and level of disability for the person concerned. This, in combination with the placement specialist's own interviews with the worker, will allow some broad screening of job offers. Second, a systematic on-site observation and assessment of task demands of pre-screened jobs will take place. This should uncover more specific questions about worker capabilities. For example, the placement specialist may want to know if the worker can be expected to lift an object with a specified weight for a specified number of repetitions over a certain period.

Currently the answers to such questions are not readily accessible and beyond the placement specialist's own ability. Logically, the next step in the development of the ergonomic task analysis is to provide ready access to such information on a case by case basis.

Provided with this information, the placement specialist could then attempt a more informed assessment of the degree of potential mismatch between worker and job, and the options for workplace modification. In the process of determining the options, access to further specialist advice in industrial ergonomics and industrial engineering would be an advantage.

Taking this view the placement specialist may be seen as a broker integrating information from their medical services, the worker, the potential employer, and technical services about potential job site modifications, training, prosthetic devices, or technical aids.

Future trends

A pyramidal structure is envisaged with general training in ergonomics for all field rehabilitation counsellors (some 200 people). Field counsellors then refer specific cases to a small group of placement specialists trained in ergonomic task analysis and given the job title of work-site analyst. Work-site analysts in turn have access to advice from specialists in different areas of industrial ergonomics.

Expansion of specialist resource staff has commenced and there are now four work-site analysts. All field counsellors receive a two day introductory orientation to ergonomics. This orientation familiarizes them with ergonomics and makes them aware of the availabilty of the more detailed analysis services within the board.

Reference

Drury, C., 1983., Task Analysis Methods in Industry. *Applied Ergonomics*, **14.1**, 19–28.

8:

A physical ability evaluation system used in rehabilitation engineering

M. Rahimi

Safety Science Department
Institute of Safety and Systems Management
University of Southern California
Los Angeles
California 90089–0021
USA

and

D. E. Malzahn

Rehabilitation Engineering Center
Department of Industrial Engineering
Wichita State University
Wichita
Kansas 67208
USA

Abstract

The need for an evaluation system which measures the physical ability of individuals has been expressed. This need was particularly emphasized to comply with federal rehabilitation laws requiring the industrial sector to include research and design of work environments for physically disabled individuals. The Available Motions Inventory (AMI) was introduced as a physical ability evaluation system for upper extremity performance of light manual tasks. The statistically reliable AMI device and its design-oriented scoring system were explained. The profile outputs, obtained from the AMI motion-class scoring system, were used to demonstrate two task design modifications for cerebral palsy individuals. Also, current and future applications of the AMI for disabled employee functional evaluations and placement, injury loss assessment, and synthetic time standard development were discussed.

Introduction

It has been more than ten years since the passage of the Rehabilitation Act of 1973. The Act specified a number of provisions to expand vistas for physically and mentally handicapped individuals. Most notable are Sections 503 and 504 of Title V which

115

specifically prohibit recipients of federal funds from discriminating against handicapped individuals (Silverstein and Kamil, 1979). Moreover, the 1978 amendment to the original legislation mandated that federal funds must be spent to finance research and service delivery systems for the design of independent work and living environments. The Act has become the driving force behind non-discriminating employment of individuals with physical and mental handicaps. At first, non-discrimination issues for the employment of persons with physical handicaps appeared to be easy to resolve. However, U.S. industry is still faced with the problem of objectively assessing job-specific aspects of a physical handicap for job screening and placement purposes.

Functionally, employment of individuals with handicaps has been tied to developments in the field of rehabilitation engineering. Rehabilitation is defined as a process of restoring the handicapped individual to the fullest physical, mental, social, vocational, and economic usefulness (McGown and Porter, 1967). Rehabilitation engineering focuses on applying the process and methods of science and technology to the problems caused by a limitation of function which substantially impede the achievement of goals of individuals. Clearly, human factors specialists can offer a significant contribution to this field by matching individuals' capabilities with requirements imposed by work and living environments. In terms of work, this process concentrates on the design or modification of machines and tasks to optimize human-work interaction.

A major component of the work-related rehabilitation engineering process is an individual's functional assessment. The complex and interdisciplinary content of functional assessment is evident in its definition: the measurement of intentional behaviour, in interaction with the environment, interpreted based on the assessment's intended uses (Halpern and Fuhrer, 1984). Not surprisingly, the literature on functional assessment in rehabilitation is basically scattered, modular, and not comprehensive.

Conventional medical diagnoses of disability are an inadequate basis for developing rehabilitation engineering solutions. Such diagnoses do not delineate the nature of clients' problems with work, severity of those problems, or the engineering interventions that resolve them. Whereas, a comprehensive quantitative functional assessment methodology can lay a foundation for the final engineering solution. Within the area of functional assessment in rehabilitation, Nagi (1969) and Wood (1980) offered conceptual, theoretical, and empirical testing of their disability models. Combining the two models, different stages of human functional assessment contain: (a) pathology or disease, (b) impairment, (c) functional limitation or disability, and (d) handicap. An impairment is the resulting physical limitations from a defect, disease, or injury. A person with an impairment may or may not have a disability, an inability to perform some key life function such as communication, self-care, etc. Only when a disability interacts with the environment to impose impediments to the individual's goals, does the individual have a handicap. Therefore, an impairment can be alleviated through medical restoration. A disability can be remedied through training, orthotic, or prosthetic devices. A handicapping condition, on the other hand, can be remedied through changes in the environment, training of the individual, or both. Also, factors such as age, education, work experience, motivation, or attitudinal barriers play important roles in the alleviation of handicaps.

By definition, rehabilitation engineering involves analysis of functional abilities of disabled individuals. Such analysis must presuppose a sequence of skilled activities to perform tasks. A skill performance is the result of the integration of several more basic abilities (Fleishman, 1979). Ability is defined as the relative capability to perform a fundamental task. Abilities and skills can be differentiated by the relative impact of training on performance, with skills being affected more by training than abilities. Abilities may also be assumed to be more stable over time than skills (Peterson and Bownas, 1982). Consequently, a useful functional assessment system for rehabilitation engineering design should produce a detailed knowledge of individual's *abilities* rather than their disabilities. Such an ability evaluation system can be used to design or modify machines and work environments to improve human–task interaction for general and special populations. Application of results produced by this system should reduce the work environment handicap, increase work efficiency, and increase employability of disabled individuals.

The system described herein is an empirically-derived quantitative physical ability evaluation system called the Available Motions Inventory (AMI). Major advantages of this system are:

(a) Its scoring system can produce comparison scales both between and within individuals.
(b) The scoring system and evaluation results are simple to use by job designers, vocational rehabilitation practitioners, and non-technical individuals involved in rehabilitation counselling.
(c) The results are directly applicable to task design process.
(d) The system is thoroughly tested for its reliability.

The AMI system

The Available Motions Inventory (AMI) was developed to evaluate the upper extremity capability of individuals with neuromuscular impairments such as cerebral palsy. Results of the evaluation are being applied to planning appropriate modifications of a variety of light bench work tasks performed in an industrial setting called the Center Industries Corporation (CIC). The CIC provides employment for individuals with physical disabilities, primarily cerebral palsy. Most of the jobs at the CIC involve light assembly, machine control, and manual operations. The products manufactured include aircraft windows, state auto license tags, a variety of light farm tools, and electrical connectors. The AMI was developed in response to a need expressed by the CIC engineers. The engineers required physical ability information relevant to task design and modification for specific handicapping work conditions.

Generally, in designing job and work stations, the designers must compare and match the task's requirements with worker's abilities. The process is the same when designing a task for a person with a disability, except that the reference data base is not as easily obtained and generalized. Therefore, the need becomes apparent for a system that efficiently defines the residual abilities of individuals. The AMI fulfills a portion of this requirement by evaluating upper extremity abilities.

Background to the AMI

The earliest version of the AMI as an evaluation system for measuring job-related abilities was introduced by Malzahn (1979). The core of the AMI system is an adjustable modular test frame (Figure 1). The test modules have been designed to simulate components of industrial jobs with respect to measures of strength, accuracy, or rate of performance. The system includes 71 separate evaluations for each hand, a total of 142 measures. Table 1 contains a sample of the AMI subtests and its modular positions. The measures are specific functional outputs and the results can be described as functional abilities. It is important that person's abilities be measured as part of a task because the integration of motion is particularly difficult for persons with cerebral palsy and other neuromuscular disorders. These physical abilities are considered to be the precursors to skill development.

Figure 1. *AMI modular test frame and cabinet.*

The AMI scoring and analyses

Data acquisition, scoring, and analyses are prompted and performed by a microcomputer. The entire AMI scoring and analysis system is objective; measuring physical abilities *quantitatively* (Malzahn, 1984; Rahimi and Malzahn, 1984). This is a major advantage of AMI over other vocational evaluation systems (Wygant, 1983). An individual's evaluation can be performed using the following progressing analyses: raw

Table 1.   A sample of available motions inventory sub-tests and positions

Sub-test		Position*			Description
	C L H	S L H	C L V	C U V	S U V	
Swith activation						
Push button	X	X	X	X		Twenty-seven 1·9 cm square push buttons — activate 9
Toggle	X		X	X	X	Twenty-seven 3 position toggle switches — activate 18
Setting						
Finger knob	X		X	X	X	Ten 1·9 cm diameter knobs with pointer — settings at 0·628 rad intervals
Detent knob	X		X	X	X	Ten 1·9 cm diameter knobs with pointer — detent at 0·688 rad intervals
Slide	X	X	X	X		3 horizontal and 3 vertical linear slide switches — settings at 1·3 cm increments
Strength						
Grip						Grip dynamometer
Force vertical		X				Force in frontal plane at 15·2 cm above elbow and 90% of reach
Reach-Reaction						
Lateral reach	X					Time required to move hand from a point in front of body 30·5 cm to the side
Lateral move	X					Time required to move hand from a point 30·5 cm to the side to directly in front of the body
Reaction time	X					Time required to respond to an auditory stimulus

*Positions

CLH (Centre Lower Horizontal) — Horizontal work surface at seated elbow height with centre area 75% of reach

SLH (Slide Lower Horizontal) — Horizontal work surface at seated elbow height with centre 50·8 cm lateral to the CLH position

CLV (Centre Lower Vertical) — Vertical work surface at 90% of reach and 15·5 cm above seated elbow height

CUV (Centre Upper Vertical) — Vertical work surface at 90% of reach and 45·7 cm above seated elbow height

SUV (Side Upper Vertical) — Vertical work surface at 0·785 rad to the frontal plane and 50·8 cm lateral to the CUV position.

scores, ability scores, and motion-class scores. Each analysis uses the results of the previous step.

The first level of the AMI scoring system is the raw scores calculated in units of kilograms for strength tests, and correct actuations per minute for timed tests. The raw scores for only a few sub-tests, primarily those concerning strength, are of direct use in job design and modification. A more comprehensive interpretation for job design, however, requires some means of intra-individual and inter-individual ability comparisons. For example, whether an individual is more capable of using knobs or toggle switches or how his/her capabilities compare with other persons within the same target population.

To provide the required comparison, normalized ability scales are generated. These ability scales are generated by establishing the mean of the dominant-hand scores of a group of able-bodied individuals as the zero point. One standard deviation of that score is the unit of variation. Standards for the dominant hand are used so that the scales generated for an individual's subordinant and dominant hands can be compared directly. The means and standard deviations are established using 80 able-bodied subjects (37 females and 43 males). Females were from 20 to 54 years old and males were from 19 to 55 years old.

The following brief example illustrates how the ability scores can be interpreted. A computer output for ability scores of a disabled individual shows a score of $-0\cdot10$ for handknob setting and $-4\cdot32$ for slide switch. In terms of physical ability, this person is less capable of making fine manual sliding motions than motions involved in setting a knob due to his cerebral palsy impairment (intra-individual comparison). Also, his sliding motion ability appears to be similar to other cerebral palsy individuals (inter-individual comparison). At this level, different device (module) and position analyses can be performed for different populations. Currently, an extensive multivariate data analysis is being performed for comparisons of able-bodied dominant and subordinate hands. The knowledge base obtained from able bodied data will be used to compare and evaluate the disabled population performance.

The third level of the AMI analysis incorporates an ability taxonomy based upon the type of motion and body member involved in performing a sub-task. To classify the AMI motion requirements, an ''order'' of movement has been determined. From fine to gross, the order of movement are related to finger–knuckles, hand–wrist, forearm–elbow, and arm–shoulder (Barnes, 1980; Heyde, 1983). Motions may also be

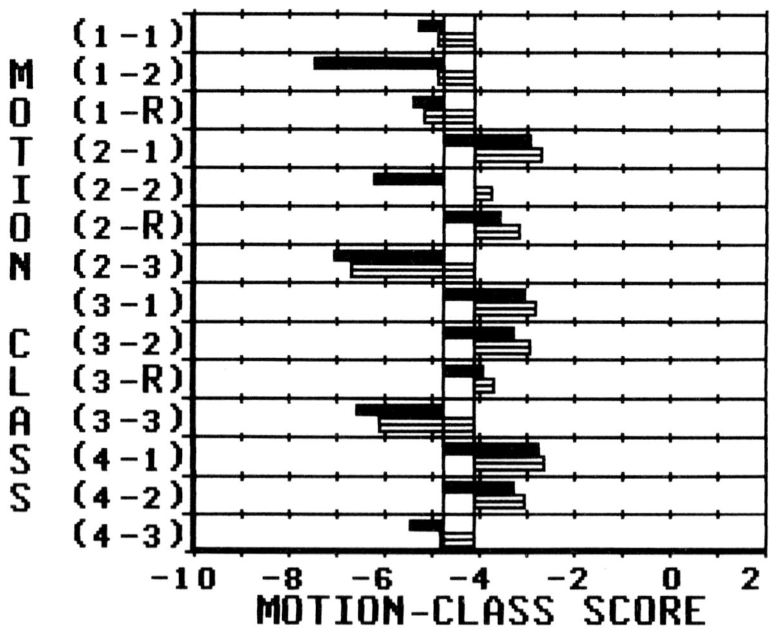

Figure 2.   *An example of a motion-class score profile.*

classified as to the quality of motion control required. The degrees of freedom required to describe a motion can be used to define the complexity of position–feedback–control motion system. One, two, and three degrees of freedom describe one dimensional (linear), two dimensional (surface), and three dimensional (spatial) movements, respectively. A rotational set of motions was also included. By combining the motion–order and degree of freedom classification systems, a total of 14 "motion-classes" have been retained in the final AMI derived motion-class analysis. It is important to note that both human abilities and specific job requirements can be described by thos motion-class profile system.

An individual's 71 AMI ability scores can be transformed into the 14 motion-class scores by a weighting system. Each AMI sub-test has been evaluated as to the degree to which it involves a particular motion-class. The weights are then normalized for each motion-class. These normalized weights are multiplied by the corresponding ability scores for each AMI sub-test and the products summed to yield a motion-class score. Figure 2 is the motion-class score profile of an individual with cerebral palsy. The motion-class score of zero represents able-bodied mean score and a negative score indicates below-standard performance. The profile contains the left hand (dark-shaded), the right hand (light-shaded), and their corresponding mean scores. For each individual, this profile is an indication of relative ability demonstrated by each body member combined with the effects of the AMI sub-task complexity requirements. As an initial evaluation, the Figure 2 profile indicates:

(a) A significantly better score for the right hand (1–2) motion-class. If this person is given a task which requires finger movement in a plane, the task should be assigned to the right hand.

(b) As the complexity of the motion increases, the motion-class scores decrease, e.g., $(2-3)$ less than $(2-1)$, $(3-3)$ less than $(3-1)$, and $(4-3)$ less than $(4-1)$. This person is more able in performing simple motions than complex ones.

(c) Average scores for the more distal body members (fingers) appear to be less than the proximal ones (upper arm). This is a typical scoring profile obtained from individuals with cerebral palsy.

(d) Relative differences between the hand scores appear to be similar to that of able-bodied difference scores. Nevertheless, both hands scored approximately four standard deviations below the able-bodied standard. This person requires an extensive job and workstation modification in order to compete with average able-bodied performance.

Another advantage of the AMI has been its statistical reliability. Statistically significant reliability of the AMI has been computed and documented (Rahimi and Malzahn, 1984).

Examples for task modification

Modification of a task to fit a particular profile of abilities is the major strength of the AMI evaluation system. Three examples are presented describing the application of

AMI evaluation system to task design and modification. Using the AMI, similar strategies can be used to increase productivity of both disabled and able-bodied workers. The modified task illustrated in Figure 3 consists of cutting sections of heat shrink tubing to one inch lengths. The task before modification was performed without a fixture. The employee simply cut tubing to length with a hand-held pair of scissors. Each length of tubing had to include an imprinted identification number. The task was modified for a group of individuals with cerebral palsy. These individuals exhibited two general characteristics on their AMI motion-class score profiles similar to the one shown in Figure 2. First, they indicated lower scores on $(1-1)$, $(1-2)$, and $(1-R)$ compared to upper arm performance scores. Second, their performance scores decreased as the degrees of freedom increased; e.g., $(3-3)$ was lower than $(3-1)$ and $(4-3)$ was lower than $(4-1)$. The principles used in the modification of this task are:

(a)   Employ larger body members.
(b)   Reduce the degrees of freedom for all body members.
(c)   Reduce bi-manual coordination requirements.
(d)   Reduce material handling requirements.
(e)   Allow the task to be performed by either hand.

The modified task illustrated in Figure 3 incorporates a simple lever and fixture to transform a complex hand and finger task into a simple linear whole arm movement task. The requirement for positioning the tubing with the hand and fingers such as $(1-R)$ and $(2-3)$ was changed to a whole arm task $(3-1)$ by providing a guide for the tubing. Other modifications include: making two cuts at a time, allowing either

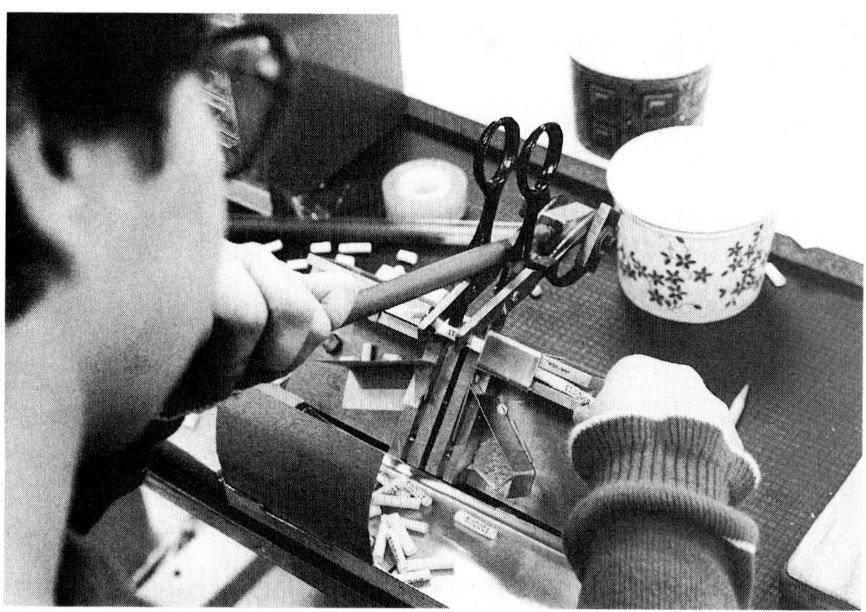

Figure 3.   Modified design for cutting heat tubes.

to perform task components, and providing gravity drops for good parts and scraps. The modified station increased the productivity of the worker with a disability and able-bodied co-workers.

The second example is the device shown in Figure 4 which allows a person with very limited strength (no measurable grip or pinch in either hand) to use correction fluid. An AMI evaluation indicated excellent dexterity with most scores above $-1\cdot00$ in both hands. This acceptable level of dexterity could not be completely used at a secretarial work station because of the difficulty in opening and closing a small bottle of correction fluid. This task requires simultaneous application of pinch, grip, and torque strength and their associated motion requirements. The modification for this task consists of a rotating base to which the bottle is clamped with three threaded aluminium knobs. These knobs also provide spokes so that the bottle can be rotated using simple linear motions of a larger and stronger body member. An inverted ''T'' handle is attached to the bottle cap. The ''T'' handle is stabilized in a slot in the top of the fixture. To remove the cap and brush from the bottle the base is rotated with the knobs until it is disengaged from the threads. The base is then moved forward until the ''T'' handle disengages from the slot on the top of the fixture and the cap removed. The basic principles applied in the modification are:

(a) Increase the size and strength of the body member used.
(b) Allow the task to be performed by one hand.
(c) Increase the effective force applied with levers.
(d) Use sequence of simple linear motions rather than a single complex motion.

Figure 4. *Modified design for opening a bottle cap.*

This modification could also be used for nail polish, paint, glue, cooking ingredients or eye-drop containers. This is an example of a modification that does not significantly increase the performance of able-bodied co-workers but has a variety of applications for individuals with disabilities similar to the above.

The third example is a label separating device as shown in Figure 5. The AMI profile of the individual performing this task indicated acceptable strength in both hands and arms. The dexterity of the right extremity, however, was below $-5\cdot00$ for all motion classes except for $(3-1)$ and $(4-1)$. The dexterity of the left extremity was within acceptable limits except for $(1-2)$ and $(2-3)$. The non-modified task consisted of removing pre-printed gummed labels from a large role by holding a strip of labels in one hand and removing the label with the other. This task required good finger dexterity with the holding hand and excellent finger and hand dexterity with the removing hand. The modification for this task consists of a series of rollers through which the strip of labels are guided by the side members. The rollers are placed so that as the strip is pulled through a gummed label separated from the backing material and positioned at the front of the fixture. Based on the individual's AMI profile, the basic principles applied in this modification are:

(a) Allow the task to be performed with one hand.
(b) Reduce the finger dexterity required for label removal.
(c) Use surface of table to reduce degrees of freedom required for positioning.
(d) Use sequence of simple linear motions rather than a single complex motion.

Each of the modifications described above involved evaluation of abilities demonstrated by the individual which was later compared with the functional requirements of the specific task. The AMI provides necessary information which can be easily used to match individual's abilities with task requirements objectively.

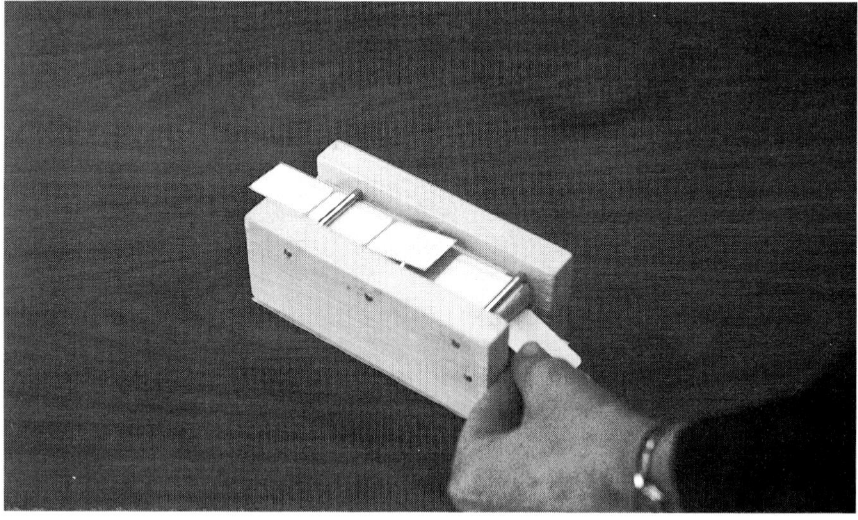

Figure 5.   Modified design for label separating task.

Current and future applications of AMI

The Rehabilitation Engineering Center at Wichita State University is currently involved in several projects extending research applications of the AMI evaluation system. The following is a brief discussion of current and future applications of the AMI system under study.

Placement and task assignment

The motion-class scoring system developed for the AMI can be applied to the documentation of task requirements. After a task has been documented in this fashion, a comparison can be made between an individual's abilities and task requirements. If a data bank of jobs is developed, the most appropriate job for a particular individual can be selected based on his/her AMI results. Or conversely, if a data bank of AMI scores is developed, the most appropriate individual can be selected for any specific job. This is a much more objective approach to placement than the currently used procedures based upon the Dictionary of Occupational Titles (DOT). The DOT describes job requirements in very broad terms and thereby addresses only very general categories of abilities to perform a task. Whereas, the AMI application requires analysis of a specific task based on evaluation of specific abilities. Therefore, the AMI approach has been implemented at the CIC. For several disabled individuals, the most appropriate tasks have been selected from a range of possible task configurations.

Synthetic time standard development

In an attempt to develop individual specific synthetic time standard systems, a study has been performed comparing performance of able-bodied individuals and synthetically developed standards. The synthetic time standard system chosen for analysis was Modular Arranged Predetermined Time Standards (MODAPTS) (Heyde, 1983). This system is similar to other time standard systems currently in use such as MTM-2, MOST and Brief Work-Factor. The MODAPTS system has a unique characteristic in that it emphasizes the body member used in performing a task and the amount of control required. This parallels the scoring system developed for the AMI.

For the able-bodied group of 50 individuals studied, strong correlations were found between the MODAPTS standards and actual performance on the AMI (adjusted R-squared = 0·9927). This high degree of correlation was achieved by adjusting each individual's predicted time on the 71 sub-tests using a unique multiplier for each individual. Each individual's multiplier produced consistently accurate standards for that individual over a wide range of tasks. Currently, research is being conducted to determine the applicability of using a similar approach to develop individual specific synthetic time standards systems for persons with physical disabilities.

Quantification of functional loss

There currently exists no objective, quantitative, and reliable system of determining an individual's functional loss due to an accident. The AMI is being investigated for determining whether a physically impaired individual is capable of returning to the same job. If not, would simple cost effective task modifications make the individual employable again? Also, it is possible to estimate an individual's economic loss due to any impairment-causing accident. If previous production records for an individual are available, a high quality estimate of their likely performance before injury on their AMI can be calculated. This could then be compared to their AMI results after injury. The difference can be easily translated into units of productivity loss using a synthetic time standard system.

Evaluation of disabled employees

After an individual with a disability is placed on a job, it is often difficult to evaluate his/her performance. If the individual's performance is compared directly with able-bodied standards, errors may ne introduced because od special task modifications developed for that individual. Thus, performance evaluation of the individual includes an evaluation of the modified task and its inherent productivity improvement. In order to correct this inconsistency, an individualized task-specific standard can be developed from the AMI performance results. Such a standard can be developed from the AMI performance results. Such a standard would allow the individual to be evaluated bu comparing actual performance to his capability for a specific task. Such a system would also make it practical to determine a realistic level of supplemental payment scales for the disabled employees.

Evaluation of cognitive abilities

Physical abilities are necessary for a successful task performance. However, other non-physical factors may be needed to predict ultimate on-the-job performance. In an attempt to incorporate cognitive abilities in a more comprehensive job evaluation system, the Structure of Intellect (SOI) evaluation system (SOI Institute, 1981) has been selected. The SOI scoring system (normalized scores). Since the units of measure are identical for both the AMI and SOI, the results can be compared directly. Currently data collection from a group of cerebral palsy individuals are being conducted. Future inclusions for this project will consider perceptual and communication evaluation systems.

Conclusion

The functional evaluation system discussed in this paper has demonstrated that if relevant data are available, work environments can be accommodated to individuals with physical impairments. The AMI is one approach to generating this necessary design-oriented data. The AMI evaluation system has applications in many other areas of interest to the rehabilitation community such as job placement, individual specific synthetic time standards, evaluation of functional loss due to injury, and performance evaluation on a modified task.

Acknowledgements

The authors would like to thank Dr John Leslie and the Cerebral Palsy Research Foundation of Kansas for their administrative contribution. Funding was provided by the National Institute for Handicapped Research Grant No. G008005053.

References

Barnes, R.M., 1980, *Motion and time study: Design and measurement of work* (7th edition), (New York: Wiley).

Fleishman, E.A., 1979, Evaluating physical ability required by jobs. *Personnel Administration*, **24**, 89–90.

Halpern, A.S. and Fuhrer, M.J., 1984, Introduction. In A.S. Halpern and M.J. Fuhrer (Eds.), *Functional Assessment in Rehabilitation* (Baltimore, Maryland: Brooks Publishing Co.).

Heyde, G.C., 1983, *MODAPT PLUS*. (Sidney, Australia: Heyde Dynamics Pty).

Malzahn, D.E., 1979, Ability evaluation and job modification for the severely disabled. In *Proceedings of the Spring Annual Conference of the Institute of Industrial Engineers* (Norcross, GA: Institute of Industrial Engineers), pp.44–51.

Malzahn, D.E., 1984, Functional evaluation for task modification using the Available Motions Inventory. In A.S. Halpern and M.J. Fuhrer (Eds.) *Functional assessment in rehabilitation.* (Baltimore, Maryland: Brooks Publishing Co.), pp.131–144.

McGown, J.F. and Porter, T.L., 1967, *An Introduction to the Vocational Rehabilitation Process*. U.S. Department of Health, Education and Welfare, Rehabilitation Services Administration, Washington, D.C.

Nagi, S.Z., 1969, *Disability and Rehabilitation*. (Columbus Ohio: Ohio State University Press).

Peterson, N. and Bownas, D., 1982, Skill, task structure, and performance acquisition. In M. Dunnette and E.A. Fleishman (Eds.) *Human performance and productivity. Vol 1 – Human capability assessment.* (Hillsdale, N.J.: Earlbaum Associates), pp.49–105.

Rahimi, M. and Malzahn, D.E., 1984, Task design and modification based on physical ability measurement. *Human Factors*, **26**, 715–726.

Silverstein, R. and Kamil, B.L., 1979, *Handbook for the implementation of Section 504 of the Rehabilitation Act of 1973*. Department of Health, Education and Welfare, Office for Civil Rights, Washington, D.C.

SOI Institute. 1981, *Technical data manual – SOI learning abilities test*. (El Segundo, California: SOI Institute).

Wood, P.H.N., 1980, Appreciating the consequence of disease: The International Classification of Impairments, Disabilities, and Handicaps. *World Health Organization Chronicles*, **34**, 376–380.

Wygant, R.M., 1983, Job analysis: A review of concepts and methods. In *Proceedings of the Spring Annual Conference of the Institute of Industrial Engineers* (Norcross, GA: Institute of Industrial Engineers), pp.205–213.

9.

Training mentally retarded adults for inspection

R. E. Barnes, C.G. Drury and D. Pomerantz[1]
Department of Industrial Engineering

[1]*Department of Exceptional Children Education*

State University of New York College at Buffalo, Amherst, New York 14260, USA

Abstract

Mentally retarded (MR) adults number about $7 \cdot 2$ million in the U.S.A. and thus comprise a significant sub-population with specific ergonomic needs at home and at work. There have been few ergonomic studies of mentally-retarded adults, despite the fact that problems of human-job fit can be particularly acute for them. This study was aimed at understanding the abilities of MR adults in a cognitive task (inspection) which is usually excluded from their work repertoire. The rationale was that demonstrating abilities in this area would allow a wider range of jobs to come to sheltered workshops, benefiting both the financial stability and job satisfaction of MR adults. An inspection task with decision and search sub-components was devised so that individuals could be trained and performance measured separately on each sub-component. Both Low and High Difficulty conditions were tested using ten subjects at each level of difficulty, with subject groups being matched for age and IQ level. A major finding was that MR adults could be trained to an adequate performance level on all tasks, indeed to a level little different from non-MR subjects in similar tasks. Training times were long by industrial standards but within the capabilities of sheltered workshops. Those subjects not finishing (4 out of 24) were older and had lower IQ scores than those competing. In general, MR subjects were poorer at search but more accurate in decision making than non-MR subjects. Implications of these findings for habilitation of MR adults into the workforce are discussed.

Introduction

As the ergonomics profession has grown from its military origins, so the concept of the human operator has grown from a young, fit male group to both more general populations and other specific sub-populations. The aim of improving the match between operator and system remains identical. The mentally retarded adult as a specific sub-population has received very little attention from ergonomics engineers. Those studies which can be found, e.g., Wade and Gold (1978), Overby *et al.* (1978), stress the need for more research and applications in this field.

Mentally retarded adults form a small percentage of the population, but in total numbers represent a sizeable group. Although definitions of mental retardation differ, causing different estimates of its prevalence, the best estimate is probably by Carter (1975) who estimates 3 per cent of the population, giving 7·2 milion people at the current population total, in the U.S.A. The prevalence of mental retardation is exceeded only by heart disease, cancer, mental illness and arthritis. Given these large numbers, there are impacts both of society on the individual and the individual on society to consider. With the general agreement among workers in the field that productive employment of mentally retarded adults is extremely beneficial to both the employer and employees (and therefore to society), then the impact of employment of these individuals upon industry and the impact of industry upon the individual client need to be assessed.

Perhaps the most telling statistics for mental retardation are those which document the chronic unemployment and under-employment of this group. McInerney and Karan (1981) show that over half a million non-institutionalized mentally retarded adults are economically idle and that 80 per cent of handicapped individuals (of all types of handicap) are unemployed or under-employed within one year of leaving school. If productive work for the mentally retarded helps them live more satisfying lives and yet most remain idle, then perhaps the ergonomics profession can help by maximizing, or at least improving, the 'fit' between the mentally retarded worker and the job. This paper presents a study conducted with the aim of determining whether suitable methods could be found for mentally retarded workers to learn a relatively complex industrial inspection task.

Before ways of improving worker–job fit can be considered, the ergonomics engineer needs data on the worker. Hence a brief review of Mental Retardation is presented here (for more detail, see Barnes, 1984).

Mental retardation

A good definition of mental retardation comes from Carter's (1975) *Handbook of Mental Retardation Syndromes*:

> 'MR is a condition characterized by the faulty development or loss of intelligence, which impairs the individual's ability to learn and adapt to the demands of society at an accepted normal level.'

Although both organic and functional aetiologies can be differentiated, from an employment viewpoint it is the level of functioning which determines the particular needs in terms of workplace design and training. One classification system (function-based) currently in use is shown in Table 1 (Ehlers *et al.*, 1973). It can be seen that th ;

Table 1. A classication of MR subjects (adapted from Ehlers, et al., 1973; Carter, 1975).

Description	Wechsler IQ range	Typical adaptive behaviours	Per cent of all MR subjects
Mild	55–69	Can profit from education programmes Capable of managing own affairs under favourable circumstances Able to work in competitive positions	89·0%
Moderate	40–54	Can be taught to care for themselves Will always need supervision and support Can be taught simple jobs, for example in sheltered workshops Only need residential care if no family/friends available in home	6·0%
Severe	20–35	Limited language, speech and motor capabilities Often have associated physical handicaps Unable to make important life decisions for themselves Can learn some self-care and self-protection skills Can live at home with suitable care	3·5%
Profound	under 20	Need total care and supervision Major impairments in physical and sensory-motor development Need hourly supervision, often in complete custodial care May achieve very minimal self-help skills and some speech development	1·5%

great majority of MR adults will benefit from vocational habilitation, as most have the capability for vocational adjustment.

Many employed MR adults work in sheltered workshops and work activity centres, although a few others hold regular factory jobs. Within a sheltered workshop setting, jobs are accepted from a variety of enterprises by the workshop administrators. Often the perception of what jobs can be performed by the MR workforce is limited to the simplest tasks, despite impressive listings of suitable jobs (Bellamy *et al.*, 1979). Many jobs are rejected (or never offered to the sheltered workshop in the first place) on the grounds that they may contain tasks too difficult for the workers. Other jobs are accepted and considerable effort is spent simplifying the task to aid the worker. However, these policies have two potentially negative consequences. First, they may unnecessarily restrict the jobs which can be accepted, leading to low and highly variable workloads if a steady supply of ''suitable'' jobs cannot be obtained. Second, they may miss the potential to increase both the worker's ability and job satisfaction, hence losing an opportunity for self fulfilment.

There is a large body of literature on the training of MR adults (e.g., Ehlers *et al.*, 1973) which shows that although these individuals learn more slowly than the non-MR worker, they can often achieve comparable levels of ultimate performance within certain boundaries. These boundaries, where the MR worker can achieve comparable performance, vary from task to task and (greatly) from individual to individual. If jobs are broken down into small steps (a typical ergonomics approach) and good training techniques (e.g., queueing, feedback, match-to-sample, delayed match-to-sample, active participation) are used, then these boundaries can be quite wide. Indeed, one school of thought suggests that boundaries merely represent the adequacy or inadequacy of our training techniques. For example, Clarke and Hermelin (1955) used three industrial tasks and showed that with proper training there was no performance difference between MR individuals, even in the Moderate range, and non-MR individuals, although training times of the former group were many times longer. Interestingly, in a study of MR individuals in an information-processing (choice) task, Kirby *et al.* (1982) found that the consistently slower responses obtained could be traced to a S-R compatibility issue, showing up as a response-processing deficit rather than a stimulus-processing one. One other issue of interest is that research findings in job satisfaction and incentives for performance closely parallel those of the non-MR literature (e.g., Talkington and Overbeck, 1975; Martin and Morris, 1980).

After this brief review of the subject population and training interventions, a discussion of the proposed job is required before matching can begin.

Industrial inspection

A number of recent reviews (Sinclair, 1984; Drury, 1984) have shown that inspection jobs for all products have two major components — visual search and decision making. The first component is generally the most time-consuming and consists in itself of two factors — perception within a single glimpse and a strategy for determining where on the item the next glimpse should be located. The former, often referred to as the visual lobe, is a function of the human optical system, the target searched for and the background searched, none of which are likely to be MR-related. In contrast, search strategy is not well understood, even for non-MR operators. Searchers are often modelled as ''random'' or ''systematic'' with little detail as to why the strategy was chosen (Morawski *et al.*, 1980), much less how to train for effective search strategies. There is no literature, to the authors' knowledge, on search strategy for the MR population.

The decision-making component of inspection has been modelled in a variety of ways, of which perhaps the most influential has been the Theory of Signal Detection (TSD). TSD sees decision performance as governed by two independent (with some assumptions) factors — the discriminability and the criterion. Discriminability depends upon the signal strength relative to the noise (visual or neural) in the system, and thus is a measure of the absolute ease or difficulty of the task. The criterion determines the relative weight given to Type 2 errors (misses) and Type 1 errors (false alarms) by the inspector. It is sensitive to the cost structure and probability structure of the task. While few would claim that TSD in its simplest form is a close quantitative model of inspection, all experiments on inspectors have shown qualitative agreement with TSD predictions, suggesting that inspectors do behave as (degraded) optimizers when detecting signals (defects) amongst noise (non-defects).

When analysed into search and decision-making components, inspection performance can be interpreted rather well in terms of individual differences. For example, Drury and Sinclair (1983) were able to determine the strengths and weaknesses of an inspection system and, at the individual level, give feedback to inspectors on their performance on the two sub-tasks. More recently, Drury and Wang (1986) suggested that the decision sub-task was likely to be generalizable across different inspection tasks, whereas search performance was more task-specific.

No studies have been located of MR performance of inspection tasks. Perhaps the closest is an anecdote in Rigby and Swain (1975) which describes the hiring of an MR person to perform the job of locating rare occurrences of debris in soft-drink bottles. With the inspector instructed to report personally to the plant manager any debris found, the outcome was a reduction of customer complaints "practically to zero". The anecdote does provide some encouragement.

Research rationale

One way of expanding the work opportunities available to MR individuals is to determine whether they can perform inspection tasks. As inspection tasks can vary in difficulty, the question of whether MR workers can perform inspection implies finding out which inspection tasks they can perform. But what is "performance"? As all inspection is imperfect (Drury and Fox, 1975), one reasonable measure is to define it as the expected performance of non-MR inspectors. Thus a more precise definition of the research rationale was to determine the envelope of inspection task difficulty inside which performance of MR individuals lies. As was mentioned earlier, all performance requires training so that an alternative wording would be "over what performance envelope can we devise training schemes which will equate MR and non-MR inspection performance?"

For a first study in the area, only two difficulty levels of each sub-task, and two of the overall inspection task, were used. To avoid unwanted carry-over effects between difficulty levels, a between-subjects design was used with separate groups of subjects in the Low and High Difficulty conditions. Because performance in inspection requires performance on both sub-tasks, each group of subjects performed search, decision-making and inspection tasks.

Methodology

Subjects

Participants were mentally retarded individuals vocationally classified as mildly retarded. They were recruited from a typical mentally retarded population working at a work activity centre. A worker was admitted to the study if he/she met the following criteria:

(a) had good eyesight (at least 20/25 corrected),
(b) was no older than 45 years,
(c) had knowledge of simple numbers,
(d) recognized colours (red, yellow and green),
(e) exhibited a good attendance record, and
(f) had no known behavioural problems or adverse reactions to training.

Ten subjects were selected for each group, with the measured characteristics given in Table 2. Some subjects did not finish the research and were replaced, hence Table 2 shows data on subjects completing the research plan. Table 3 compares the twenty subjects completing the experiment with the four who withdrew. Table 2 reveals that there were no significant differences in age or IQ between the High and Low Difficulty groups ($t18 = 0\cdot75$, $P > 0\cdot45$; $t18 = -0\cdot82$, $P > 0\cdot40$ respectively). However, as shown in Table 3, a significant difference in Age between completers and non-completers existed ($t22 = -3\cdot71$, $P < 0\cdot001$; and almost IQ, $t22 = 2\cdot03$, $P = 0\cdot055$, respectively).

Table 2. Subjects completing and research

	Male	Female	Mean age	SD age	Mean IQ	SD IQ
Low Difficulty	5	5	27·5	4·7	62·5	4·8
High Difficulty	7	3	25·8	5·3	64·2	4·5

Table 3. Comparison between subjects completing and not completing the research

	Male	Female	Mean age	SD age	Mean IQ	SD IQ
Completing	12	8	26·6	5·0	63·4	4·6
Not completing	1	3	36·5	3·8	58·5	2·4

Stimulus material

A stimulus material used in previous inspection studies (Czaja, 1980; Gallwey and Drury, 1986; Schwabish and Drury, 1984) was used in this task. It consisted of 35 mm black-on-white slides of computer-generated 60×20 arrays of symbols (Figure 1). Search difficulty could be varied by varying the characteristics of target characters and background density, while decision difficulty could be varied by varying the size difference between target and background characters. Ten background characters (?, @, $, %, !, &, = , − , *, ^) were each drawn so as to subtend 13 minutes of arc vertically and 8 min of arc horizontaly for full-sized characters. Targets were of two types — "Wedge" (>) and "Cross" (+) — and three sizes — Large, 11×11 min; Medium, 8×8 min; and Small, 5×5 min. All had a projected contrast of 75%.

For the training phase of the research, original paper arrays were used rather than the slides to provide a realistic training setting.

Differences between the Low and High Difficulty conditions are summarized in Table 4. The High Difficulty condition had more difficult search (higher number of characters in the array, more target sizes) and more difficult decision making (a third size of target was presented and had to be classified). For both High and Low Difficulty conditions, the "good":"OK":"Bad" ratio of slides was 60:20:20 to maintain a constant fault rate for both conditions.

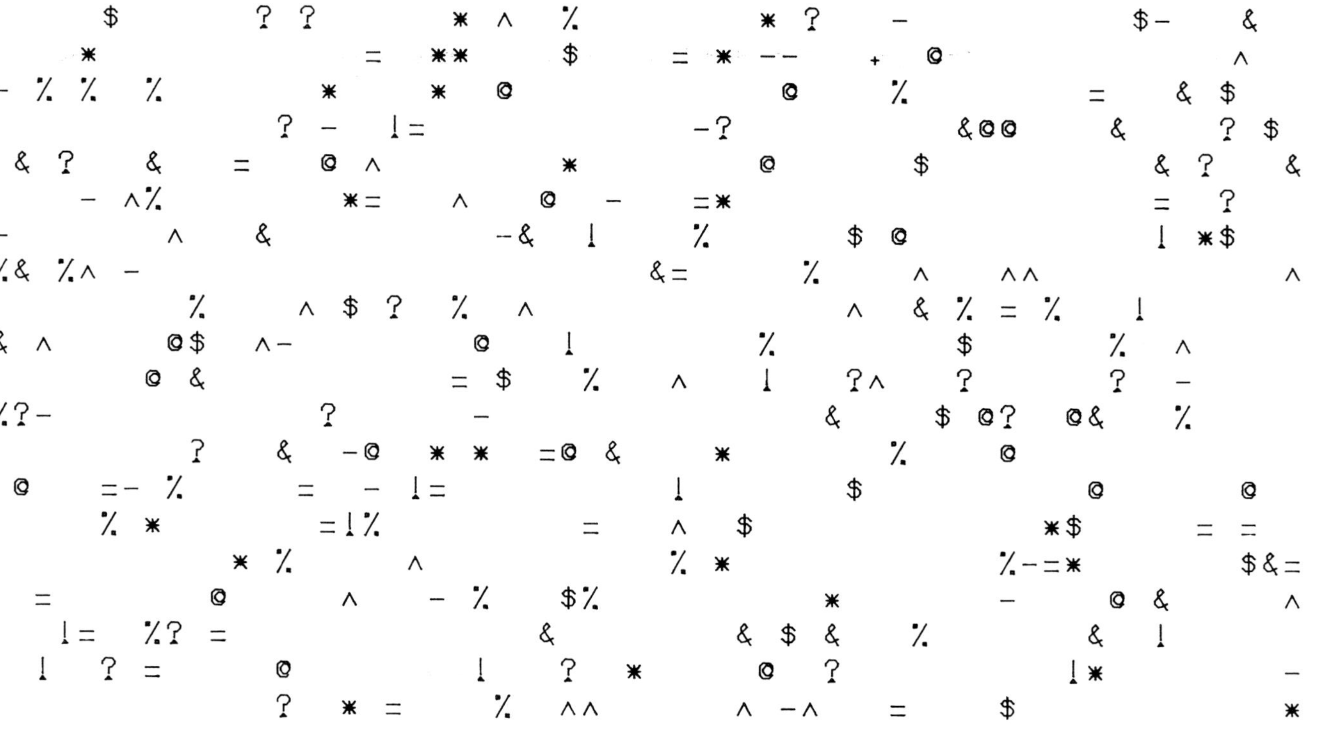

Figure 1. *Simulated product stimulus.*

Table 4. Differences between low and high difficulty conditions

	Low	High
Search related		
Character density	10%	20%
Expected # of characters	120	240
Number of target sizes	2	3
	(small, large)	(small, medium, large)
Decision-making related		
First Quality ("Good")	No target	No target
Second Quality ("OK")	small +	small +
	large >	large >
		medium >
Rejectable ("Bad")	large +	small >
	small >	medium +
		large +

Equipment

Slides were projected onto a 1·22 m square screen located 2·5 m from the seated subject. Two response boxes were constructed for the subject. The first had three rows of three buttons.

Top row: Action taken (Bad, OK, Good)
Middle row: Defect type (Perfect, Wedge, Cross)
Bottom row: Defect size (Small, Medium, Large)

Subjects pressed the Action button when they had come to a final decision on each subject. This provided the response accuracy and response time measures. Next they indicated the type and size of target (defect) found (if required). Areas of the panel not used in any particular phase of training or measurement were covered. When a defect was found, a grid was superimposed onto the stimulus from a second projector, so that subjects could indicate which quadrant of the display contained the defect. The second response box had four switches representing spatially the four quadrants.

Procedure

Subjects took part in four study phases as follows. The numbers are indicative of the quantity of training given at each level (see note below).

Phase 1: Training
        Target identification
            Match-to-sample (8 steps)
            Delayed match-to-sample (8 steps)

        Search (target identification and location)
            Target known (16 steps low, 24 steps high)
            Target unknown (4 steps)

       Size Judgement (6 steps)
       Decision Action (2 steps)
       Inspection (1 step)

Phase 2: Search Sub-task
       Transfer (3 steps)
       Practice (40 trials low, 60 trials high)
       Assessment (40 trials low, 60 trials high)

Phase 3: Decision Sub-Task
       Transfer (1 step)
       Practice (40 trials Low, 60 trials High)
       Assessment (40 trials Low, 60 trials High)

Phase 4: Inspection Task
       (No Transfer required)
       Practice (40 trials Low, 60 trials High)
       Assessment (40 trials Low, 60 trials High)

Note: To accomplish each step, a criterion was enforced, 90% correct in a running total of 10, thus the minimum number of trials per step was nine; one step used a minimum of 18. Failure to meet criterion resulted in the person being excused from the experiment.

Where two numbers are shown in parentheses, these refer to Low/High Difficulty conditions, respectively.

The training programme can best be described as a chained (or a progressive-part) programme with active response and immediate feedback. All of these are re-commended techniques for training non-MR workers in inspection (Czaja and Drury, 1981) and for training MR workers in non-inspection tasks (Bellamy *et al.*, 1979). In each step, subjects were presented with new information to integrate into their know-ledge and skill base. Each trial consisted of presentation of the stimulus material, a response by the subject, and feedback from the experimenter.

The match-to-sample technique used a card with a line of characters, one of which was the target, and another card with just the target. Subjects had to identify the target in the line which matched the symbol on the target card. Delayed match-to-sample used the same technique except that the target card was removed before the card containing the line of characters was presented. Both of these techniques are commonly recommended in the MR training literature (Bellamy *et al.*, 1979).

After the targets were learned, they were searched for and located in arrays of in-creasing size: 21×7, 30×10, 42×14, 60×20, first with the subject prompted for the target by name on each sheet (Target Known) and second without the prompt (Target Unknown). For Size Discrimination, a set of similar cards was used, with each row containing all sizes to be learned in that condition, with the subject naming the size for each card. This was completed without and with the response panel. Decision Action training used the same set of cards, except that the response was now a decision action (OK or Bad) rather than a size judgement. Finally in Inspection, the subjects had to locate a target on an array and then respond with the correct decision action.

Following training on the paper materials, the transfer phase repeated the process using the 35 mm slides and the response panel. Subjects continued to receive feedback after each response.

Practice and Assessment were identical phases except that feedback was only presented at the end of Assessment. In the Low Difficulty condition, there were 40 slides for each sub-task (search, decision making and inspection). The 40 slides contained 24 perfect (no target present) and four each of the four size/target type combinations. Each target appeared once on each of these 16 slides, with all four quadrants of the array having equal numbers of targets. In the High Difficulty conditions, there were 60 slides comprising 36 perfect and four of each size/target type combination.

In the search sub-task, the subject's response was to locate and classify the target using the mid row of the response box. In the decision sub-task, the target was circled to eliminate search, and the subject's response was to give the decision action followed by the defect type and defect size. In the inspection task, targets were not circled and therefore had to be found, and the same responses as the decision sub-task were required.

Measures were taken of response times (to the nearest tenth of a second) and response accuracies, as well as of total time to complete each phase (to the nearest quarter of an hour). The time and accuracy measures reported here are as follows (for more detailed measures, see Barnes, 1984).

Accuracy measures

P (correct) = probability of a correct response during training, transfer, practice and assessment.

P (correct), targets only = probability of correctly responding to a stimulus containing a target. In an inspection task, this is p (hit).

P (false alarm) = probability of responding incorrectly to a perfect item.

Time measures

Search Time = time to locate a target in a search task.
Decision Time = time to identify a target in a decision task.
Inspection Time = time to locate a target in an inspection task.
Stopping Time = time to respond to any item *not* containing a target.

Results

Selected statistics are presented in this paper to demonstrate the overall level of performance, as the emphasis is on determining whether MR individuals can perform inspection tasks, rather than on the internal–consistency details of the tasks themselves. A complete account of the measures is presented in Barnes (1984).

The total time to train MR inspectors is of obvious importance. Table 5 presents the times for each phase of the experiment, with *t*-tests of the differences between Low and High Difficulty conditions. In all phases, the High Difficulty condition took longer

Table 5. Overall times for each phase in hours

Phase	Low Difficulty	High Difficulty	Test statistic	Result
Training	4·42	6·02	$t(18) = 2·69$	$P < 0·02$
Transfer	0·83	1·08	$t(18) = 4·63$	$P < 0·001$
Practice	1·58	2·68	$t(18) = 2·90$	$P < 0·02$
Assessment	1·30	2·35	$t(18) = 3·20$	$P < 0·001$
Total	8·12	12·12	$t(18) = 3·20$	$P < 0·02$

than the Low Difficulty condition. Table 6 shows the overall probabilities of correct response for training, transfer, practice and assessment. It can be seen that peformance level was high throughout, between 80% and 100% correct responses. It is also noticeable that performance was very similar in the Low and High Difficulty conditions. False alarms were low throughout the whole assessment phase, totalling 4 in 1800 slides.

Table 6. P (correct)

	Low	High
1. Training		
Target identification	1·000	0·999
Match-to-sample	1·000	0·999
Delayed-match-to-sample	1·000	1·000
Search		
Target known	0·936	0·918
Target unknown	0·940	0·933
Size judgement	0·984	0·869
Decision action	0·877	0·814
Inspection	0·878	0·897
2. Search Sub-task		
Transfer	1·000	1·000
Practice	0·818	0·870
Assessment	0·848	0·860
3. Decision Sub-task		
Transfer	0·990	0·939
Practice	0·935	0·935
Assessment	0·965	0·972
4. Inspection		
Practice	0·835	0·832
Assessment	0·850	0·853

If performance as measured by P (correct) was high and consistent across conditions, performance times also need to be measured. Figure 2 shows the performance time comparison in the assessment phase of each sub-task. As with non-MR subjects, stopping times were approximately double the search times for search and inspection tasks (Gallwey and Drury, 1986). Times for the decision sub-task were much shorter as no search was involved, with the decision on a perfect item (no defect circled) fastest of all.

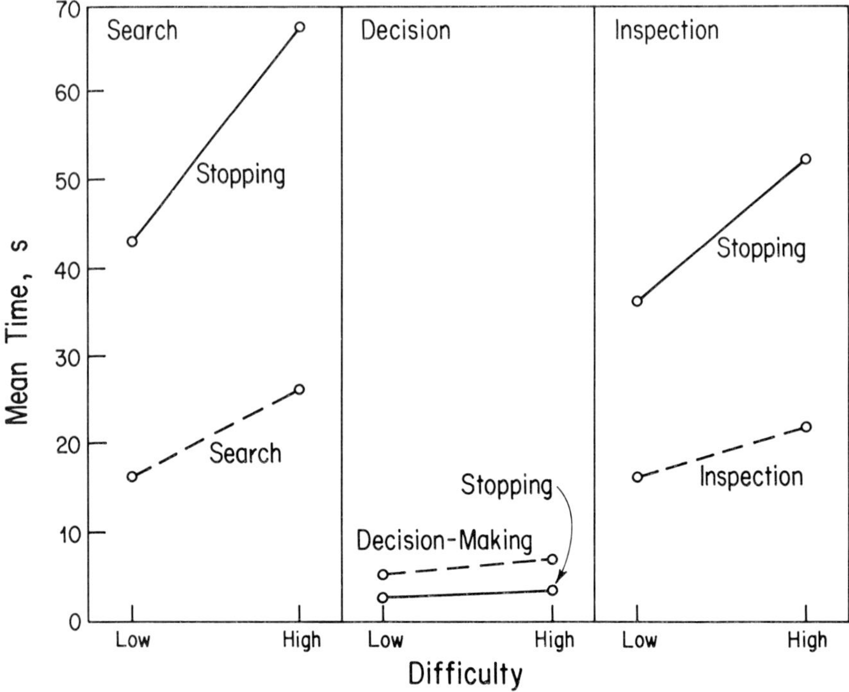

Figure 2. Mean performance times for inspection sub-tasks by diculty level.

Although overall performance was high when measured by overall p (correct), this is a somewhat biased measure. It includes responses to perfect items so that if a subject responded ''good'' to *all* items, he/she would still be correct on 60 per cent of the slides. A more detailed breakdown of the sources of error in the sub-tasks in shown in Table 7. It should be noted that search performance was the lowest, whether it occurred as a separate sub-task or embedded within an inspection task.

Table 7. P (correct) values for each sub-task for slides with targets only.

| | Practice | | Assessment | |
	Low	High	Low	High
Search	0·556	0·679	0·650	0·658
Decision making				
Identification	0·994	1·000	1·000	1·000
Size judgement	1·000	0·858	1·000	0·929
Decision action	0·843	0·976	0·912	1·000
Overall classification	0·838	0·950	0·912	0·983
Inspection				
Search/identification	0·638	0·704	0·675	0·717
Size judgement	1·000	0·858	1·000	0·890
Decision action	0·931	0·979	0·926	1·000
Overall classification	0·594	0·662	0·625	0·679

For an external comparison of these results, perhaps the closest experimental protocol to the one used here is that of Czaja (1980) who used both 'cross' and 'wedge' targets in large, medium and small sizes. She compared three age groups, two pre-training and two training conditions. Table 8 compares the High Difficulty condition of the current experiment to the closest group of Czaja's subjects (young, no pre-training, active training programme group). There was no difference in search time or in size judgment accuracy. MR subjects performed better at decision making and worse at search than non-MR subjects. Although Czaja does not report training times, her subjects all took less than four hours to complete a set of pre-tests, pre-training and training for a more complex task.

Table 8. Comparison between MR subjects (this experiment) and non-MR subjects (Czaja, 1980)

Measure	Test statistic	Result
Mean search time	$t(152) = 1 \cdot 628$	$P > 0 \cdot 05$ ns
Mean stopping time	$t(358) = 8 \cdot 388$	$P < 0 \cdot 001$ *
Mean P (correct)		
Search/Identification	$x^2(1) = 21 \cdot 85$	$P < 0 \cdot 05$ *
Size Judgement	$x^2(1) = 2 \cdot 11$	$P > 0 \cdot 05$ ns
Decision Action	$x^2(1) = 50 \cdot 91$	$P < 0 \cdot 001$ +

ns = no significant difference
* = significant difference, MR worse
+ = significant difference, MR better

Discussion and conclusions

Perhaps the main finding of this experiment was that mentally retarded adults can be trained to perform a relatively complex inspection task, achieving performance levels comparable to non-MR subjects. Putting the emphasis on the experimenter's capability to devise a successful training programme in no way negates the fact that MR clients have special learning and training requirements (Ehlers *et al.*, 1973) and that there are many misconceptions about MR populations which can inhibit learning and training (Olshansky, 1969).

The training programme developed used ergonomics knowledge of the inspection task to break it down into its information processing components (search and decision making). It also used knowledge of MR training to match appropriate instructional techniques to this task (match-to-sample, delayed match-to-sample, simple-to-difficult sequencing and chaining). In addition, non-MR-specific techniques of active learning and immediate, accurate feedback (Czaja, 1980) was used. In combining these principles from different disciplines, one of the primary aims of training, low error rate during training, was achieved (Table 6). Another aim of any training programme was also realized, that of a high completion rate for subjects. Of the 24 who started, 21 completed the training and 20 completed the whole experimental protocol, giving a training completion rate of 87 per cent and an experimental completion rate of 83 per cent. Those subjects who did not complete the whole experiment were significantly older than those who did, and of lower average IQ. The large individual differences between MR workers, even in the same sheltered workshop facility, were again playing a part.

Performance overall was comparable to non-MR clients with deficits in search components being offset to some extent by improved decision performance. It should be noted, however, that the protocols between this experiment and the one chosen for comparison were not identical. Czaja's subjects had four possible target types, each at five possible sizes, although the targets used in the experiment reported here were among the most difficult reported by Czaja. Where the comparison is most useful is in the comparison between MR and non-MR subjects on the component tasks of inspection. Here it confirms evidence from within the current study that search is particularly difficult for MR clients while size judgement and decision action pose no special difficulties beyond the capabilities of the training programme.

Search is both the unreliable component in inspection (Drury and Sinclair, 1983) and difficult to train. Evidence reviewed elsewhere (Gallwey and Drury, 1986) suggests that search strategy is relatively impervious to experimental manipulation. Many studies (e.g., Bloomfield, 1975, for review) report improved search performance with practice, but practice is not training. The experienced searcher may have developed a better understanding of the search field and the relative probabilities of targets appearing at different points, but there is no evidence that the searcher's *strategy* has improved beyond looking where targets are most likely. Indeed, a recent intercorrelational study (Drury and Wang, 1986) showed that search performance on one inspection task did not predict search performance on other tasks. If we do not know what differentiates a good searcher from a poor searcher, how can we teach search? The answer is that at present we cannot; all that we can do is to provide practice-plus-feedback as in the current experiment. However, our results show that for MR workers this is insufficient. Clearly, one result of the study of this special population is to focus on the fact that we need more research into search strategies and search training techniques.

With only two levels of difficulty in each sub-task, it was not expected that a performance envelope would be defined as passing neatly between the two levels chosen. In fact, there was little difference in performance between the two levels, with no significant differences except for stopping times in tasks containing the search component. Even here the drop in performance was little more than could be expected from the fact that twice the characters were being searched in the High Difficulty array. As Drury and Clements (1978) have shown, stopping times bear an approximately linear relationship to total number of characters in the search field. Apart from these obvious differences, the two levels of Difficulty were closely comparable throughout.

A major difference between MR and non-MR subjects, and between the Difficulty conditions, was found in training times. A training programme was developed and it worked, but the cost in training time was high. For all phases up to the end of Practice it took $6 \cdot 8$ hours in the Low Difficulty condition and $9 \cdot 8$ hours in the High condition, considerably longer than Czaja's less-than-four-hours training time for her more difficult task. Furthermore, these times are net times, with fetching subjects, taking breaks, and the general social interaction inherent in any training situation not included. With MR individuals, such social interaction is of utmost importance in maintaining motivation and providing encouragement in what was perceived as a difficult task. The principal author, who performed the training, was made acutely aware of the need for patience and understanding but was rewarded by forming friendships with the subjects and by helping them to achieve new levels of performance. These factors must be acknowledged by any who would work with this particular special population.

This study has shown that when a good training scheme is designed and implemented, mentally retarded adults can achieve industrial levels of performance in a complex task. It is hoped that the results will open up the range of tasks accepted for MR workers, to the benefit of both the worker's job satisfaction and the agency's financial soundness.

Acknowledgements

The authors acknowledge: the subjects for their hard work and perseverance; Allentown Industries, Buffalo, New York for cooperating in this study; the State University of New York at Buffalo Graduate Student Association for their financial support; and the Department of Industrial Engineering, State University of New York at Buffalo.

References

Barnes, R.E., 1984, *Inspection Performance of Mentally Retarded Persons*. Unpublished dissertation, Department of Industrial Engineering, State University of New York at Buffalo, Buffalo, New York.

Bellamy, G.T., Horner, R.H. and Inman, D.P., 1979, *Vocational Habilitation of Severely Retarded Adults — A Direct Service Technology*, (Baltimore, Maryland: University Park Press).

Bloomfield, J.R., 1975, Theoretical approaches to visual search. In Drury, C.G. and Fox, J.G. (Eds.) *Human Reliability in Quality Control*, (London: Taylor and Francis).

Carter, C.C., 1975, *Handbook of Mental Retardation Syndromes*, (Springfield, Il: Charles C. Thomas).

Clarke, A.D.B. and Hermelin, B.F., 1955, Adult imbeciles: their abilities and trainabilities. *Lancet*, **2**, 337–339.

Czaja, S.J., 1980, *Varying Training Techniques Across Age Span for a Visual Inspection Task*. Unpublished dissertation, Department of Industrial Engineering, State University of New York at Buffalo, Buffalo, New York.

Czaja, S.J. and Drury, C.G., 1981, Training programs for inspection. *Human Factors*, **23.4**, 473–484.

Drury, C.G., 1984, Ergonomics in Quality Control. *Proceedings of the 1984 International Conference on Occupational Ergonomics*, Toronto, Ontario, Canada, 56–60.

Drury, C.G. and Clement, M.R., 1978, The effect of area, density and number of background characters on visual search. *Human Factors*, **20**, 597–602.

Drury, C.G. and Fox, J.G., 1975, The inspector in the future: his efficiency and well-being. In Drury, C.G. and Fox, J.G. (Eds.). *Human Reliability in Quality Control*, (London: Taylor and Francis).

Drury, C.G. and Sinclair, M.A., 1983, Human and machine performance in an inspection task. *Human Factors*, **25.4**, 391–399.

Drury, C.G. and Wang, M.-J., 1986, Are research results in inspection task specific? *Proceedings of Human Factors Society 30th Annual Meeting*, Dayton, Ohio, 476–479.

Ehlers, W.H., Krishef, C.H. and Prothero, J.C., 1973, *An Introduction to Mental Retardation: A Programmed Text*, (Columbus, Ohio: Merrill Publishing Company).

Gallwey, T. J. and Drury, C.G., 1986, Task complexity in inspection. *Human Factors*, **28.5**, 595–606.

Kirby, N.H., Nettlebeck, T. and Western, P., 1982, Locating information-processing deficits of mildly mentally retarded young adults. *American Journal of Mental Deficiency*, **87**, **3**, 338–343.

McInerney, M. and Karan, O.C., 1981, Federal legislation and the integration of special education and vocational rehabilitation. *Mental Retardation*, **19**, 21–23.

Martin, A.S. and Morris, J.L., 1980, Training a work ethic in severely mentally retarded workers — providing a context for the maintenance of skill performance. *Mental Retardation*, **18**, 67–71.

Morawski, T., Drury, C.G. and Karwan, M.H., 1980, Predicting search performance for multiple targets. *Human Factors*, **22**, 707–718.

Olshansky, S., 1969, An examination of some assumptions in the vocational rehabilitation of the mentally retarded. *Mental Retardation*, **7**, 51–53.

Overby, C.M., Myer, T.H., Hutchinson, J. and Wiercinski, R., 1978, Some human factors and related socio-technical issues in bringing jobs to disabled persons via computer-telecommunications technology. *Human Factors*, **20**, 349–364.

Rigby, L.V. and Swain, A.D., 1975, Some human-factor applications to quality control in a high technology industry. In Drury, C.G. and Fox, J.G. (Eds.). *Human Reliability in Quality Control*, (London: Taylor and Francis).

Schwabish, S.D. and Drury, C.G., 1984, The influence of the reflective-impulse cognitive style on visual inspection. *Human Factors*, **26.6**, 641–647.

Sinclair, M.A., 1984, Human Factors of Inspection. Workshop materials from 1984 International Conference on Occupational Ergonomics, Toronto, Ontario, Canada.

Talkington, L. W. and Overbeck, D.B., 1975, Job satisfaction and performance with retarded females. *Mental Retardation*, **13**, **3**, 18–19.

Wade, M.G. and Gold, M.W., 1978, Removing some of the limitations of mentally retarded workers by improving job design. *Human Factors*, **20**, 339–348.

PART V
OCCUPATIONAL BIOMECHANICS

10.

Human modelling for rehabilitation

Ronald L. Huston

Mechanical and Industrial Engineering Department
University of Cincinnati
Cincinnati, Ohio 45221–0072, USA

Abstract

This paper presents a methodology for modelling and analysis of the human body and its movements. The objective is to develop a basis for rehabilitation procedures.

The model is based upon a finite-segment division of the human frame, with rigid bodies representing the limbs. If the feet, hands, legs, arms, torso, head, and neck are included, the model can have as many as 17 bodies. Hence, if there are three degrees-of-freedom at each joint, the system can have as many as 51 degrees-of-freedom. As such, the model is a multibody system. Recent advances in multibody systems analysis may therefore be applied with the model. These advances are based upon: Body connected arrays; Euler parameters; and Kane's equations. Each of these is discussed in the paper.

The application of these procedures in rehabilitation research is discussed. This ranges from the design of protective devices, to forces causing injury, to weight training for repair of damaged ligaments and muscles.

Introduction

Recently there has been increased interest in developing rehabilitation procedures for injuries occurring from industrial accidents, from sports, and from long-term repeated forces applied in industrial work stations. This interest is motivated by three factors: (a) new procedures have been developed for muscle stimulation and repair; (b) new rehabilitation devices have been developed; and (c) new modelling procedures based upon computer analysis have become available.

The procedures for muscle stimulation involve direct electrical stimulation of the muscle thus bypassing the normal neural stimulation. This is a means of overcoming paralysis. The rehabilitative devices involve the use of computer equipment, new mechancial designs, and new materials. Finally, the new modelling procedures are based upon recent advances in computer-oriented analyses of multibody systems.

Successful development of muscle stimulation procedures and the successful application of rehabilitation devices can be enhanced by use of the modelling advances. In this paper details of a recently developed computer modelling procedure are presented. This procedure has been successfully applied in the analysis of a variety of human movements [see Huston and co-workers (1973–1986)].

147

The model is based upon a finite-segment representation of the human body. The dynamics is developed by using Kane's equation [see Kane and Levison, 1980, 1985]. This in turn, provides a basis for the development of computer algorithms for the numerical generation of the governing equations.

The paper is divided into four parts with the following section providing a discussion of concepts useful in the sequel. The next two parts discuss the kinetics and dynamics. The final part contains a discussion about applications in rehabilitation and some concluding remarks.

Preliminary considerations

Modelling

Figure 1 depicts a model of the human body. It is a "gross-simulation" model in that it is useful for describing large, major movements of the human body. The model

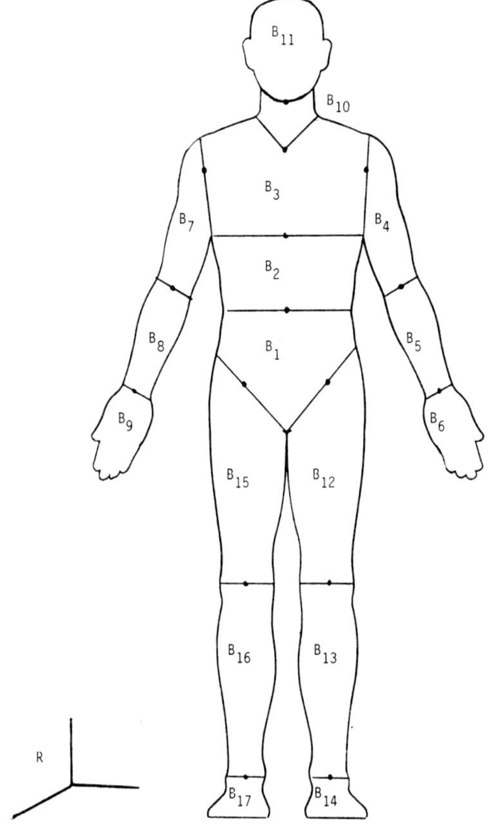

Figure 1.   *A human-body-model*

consists of 17 bodies having the shapes of cones, frustums of elliptical cones, elliptical cylinders, and spheroids. The geometrical and inertia data of the bodies are adjusted to simulate the corresponding data for the human frame.

The bodies are connected by pin, spherical, and sliding joints. The knees and elbows are pins joints. The hips, spine, shoulder, ankles, and neck are spherical joints. The neck also allows for translation between its bodies. The neck thus combines both spherical and sliding joints.

Finally, R represents an inertial reference frame in which the system moves.

Transformation matrices

Consider a typical pair of adjoining bodies of the system such as B_j and B_k as in Figure 2. The orientation of B_k relative to B_j may be defined in terms of the relative orientation of dextral orthogonal unit vectors: n_{ji} and n_{ki} $(i = 1, 2, 3)$, fixed in B_j and B_k as shown.

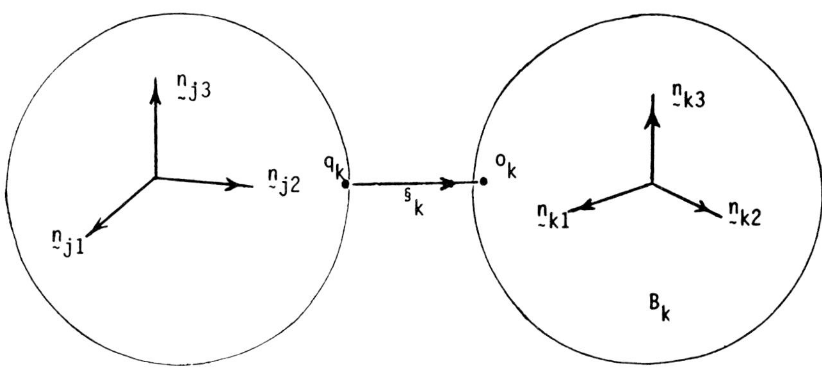

Figure 2.   *Two typical adjoining bodies*

Specifically, the n_{ji} and the n_{ki} are related to each other by the relation

$$n_{ji} = SJK_{im} \, n_{km} \tag{1}$$

where SJK is a 3×3 orthogonal transformation matrix defined as:

$$SJK_{im} = n_{ji} \cdot n_{km} \tag{2}$$

(Regarding notation, the J and K in SJK and the first subscripts on the unit vectors refers to bodies B_j and B_k, and repeated indices, such as the m, in Equation (1) signify a sum over the range of the index).

From Equations (1) and (2), it can be shown that the transformation matices obey the following matrix product chain and identity rules:

$$SJL = SJK \, SKL \tag{3}$$

and

$$SJJ = I = SJK \, SKJ = SJK \, SJK - 1 \tag{4}$$

where I is the identity matrix.

Since these transformation matrices play a central role in the analysis it is necessary to have an algorithm to compute their derivatives. From the definition of Equation (2), and by noting that the $\underset{\sim}{n}_{oi}$ are fixed in R, the derivative of SOK_{ij} may be expressed as:

$$d(SOK_{ij})/dt = \underset{\sim}{n}_{oi} \cdot d^R \underset{\sim}{n}_{kj}/dt \tag{5}$$

where the superscript R indicates that the derivative is computed in reference frame R. Since the $\underset{\sim}{n}_{kj}$ are fixed in body B_k their derivatives are:

$$d^R \underset{\sim}{n}_{kj}/dt = \omega_k \times \underset{\sim}{n}_{kj} \tag{6}$$

where ω_k is the angular velocity of B_k in R. Hence, Equation (5) may be written as:

$$d(SOK_{ij})/dt = - e_{imn}\omega_{kn}SOK_{mj} \tag{7}$$

or the matrix form as:

$$d(SOK)/dt = WOK \, SOK \tag{8}$$

where the ω_{kn} are components of ω_k referred to $\underset{\sim}{n}_{on}$. By comparing Equations (7) and (8) the matrix WOK is defined as:

$$WOK_{im} = - e_{imn}W_{kn} \tag{9}$$

where e_{imn} is the standard permutation symbol. WOK is the matrix whose dual vector ω_k.

Equation (8) shows that the derivative of the transformation matrix may be computed by matrix multiplication — a convenient basis for a computer algorithm.

Kinematics

Degrees of freedom and coordinates

A model with 17 bodies can have as many as 102 (17×6) degrees of freedom. This would correspond to the case where there is full translation and full rotation at each joint. If, however, the model has full translation and rotation only at the neck joint, if it has hinge joints at the elbows and knees, and if it has spherical connectors at the other joints, it will have 52 degrees of freedom.

To define these degrees of freedom, let each body of the model have a rectangular coordinate system whose origin is at its connecting joint with its adjoining lower-numbered body. Let the coordinate systems be oriented so that the axes are parallel to principal inertia axes of the bodies. Let the model be in its "reference configuration" when all the coordinate axes are respectively parallel. Finally, in the reference configuration, let the X, or "1", axis be forward, let the Y or "2", axis be on the left, and let the Z, or "3", axis be up.

Let the displacement of the coordinate origin of the pelvis, O_1, relative to R be defined by the translation variables X_{1i} ($i = 1, 2, 3$). Similarly, let the displacements of the neck and head at the neck joints be defined by the variables X_{10i} and X_{11i} ($i = 1, 2, 3$). Let the rotations at the elbows and knees be defined by the angles β_5, β_8, β_{13}, and β_{16}. Let the rotations at the other joints be defined by the relative angular speeds $\hat{\omega}_{ki}$ ($i = 1, 2, 3$), where k is the index of the higher numbered body at the joint.

Let the degrees-of-freedom be described by *generalized speeds* y_l ($l = ,\ldots,52$) defined as follows:

$$
Y_l =
\begin{cases}
\dot{x}_{1i} & l = 1, 2, 3 \\
\dot{x}_{10i} & l = 4, 5, 6 \\
\dot{x}_{11i} & l = 7, 8, 9 \\
\dot{\beta}_5(= \hat{\omega}_{52}) & l = 10 \\
\dot{\beta}_8(= \hat{\omega}_{82}) & l = 11 \\
\dot{\beta}_{13}(= \hat{\omega}_{13, 2}) & l = 12 \\
\dot{\beta}_{16}(= \hat{\omega}_{16, 2}) & l = 13 \\
\hat{\omega}_{ki} & l = 14, \ldots, 52
\end{cases}
\tag{10}
$$

The generalized speeds have the dimension of coordinate derivatives (either displacement derivatives or angle derivatives). However, the latter $39 y_l$ in Equation (10) are defined as the relative angular velocity components $\hat{\omega}_{ki}$. As such they are not derivatives of any single orientation, or angle, coordinates. Indeed, in general, such coordinates do not exist. Hence they are often called "quasi-coordinates". It is necessary, however, to have variables defining the relative orientations of the bodies. Moreover, it is necessary to be able to define the relative angular velocities in terms of these variables. The Euler parameters are discussed by Whittaker (1937), Kane *et al.*, (1983), and by Nikravesh and Chung (1982).

Angular velocities

From Equations (1) and (10) it is readily seen that the angular velocity of a typical body β_k in the inertial frame R may be written in the form:

$$
\omega_k = \omega_{klm} y_l n_{om}
\tag{11}
$$

where, as before, repeated indices designate a sum over the range of the index. The ω_{klm} are components of the "partial angular velocity vectors" defined by Kane *et al.*

                           Ergonomics in Rehabilitation

(1980, 1985) as $\partial\omega_k/\partial y_l$. From Equations (1) and (10) it is seen[*] that the ω_{klm} may be expressed in terms of the transformation matrices SOK.

Angular accelerations

The angular acceleration α_k of typical body B_k may be obtained by differentiation in Equation (11). Hence, α_k is:

$$\alpha_k = (\dot{\omega}_{klm}\dot{y}_l + \dot{\omega}_{klm}y_l)\,n_{om} \tag{12}$$

where the $\dot{\omega}_{klm}$ may be obtained by employing Equation (8).

Mass centre velocities

In a similar and analogous manner the velocity of the mass centre G_k of body B_k may be expressed in the form:

$$v_k = v_{klm}\dot{y}_l\,n_{om} \tag{13}$$

Where the v_{klm} are components of the "partial velocity vectors" defined by Kane *et al.*, (1980, 1985) as $\partial v_k/\partial y$. From Equations (1) and (10) it is seen that the v_{klm} may be expressed in terms of the ω_{klm} and in terms of components of position vectors locating the mass centres[*].

Mass accelerations

The acceleration a_k of G_k may be obtained by differentiating in Equation (18). Hence, a_k is:

$$a_k = (v_{klm}\ddot{y}_1 + \dot{v}_{klm}\dot{y}_l)\,n_{om} \tag{14}$$

where the $\dot{v}_{klm}$ may be expressed in terms of the ω_{klm} and the derivatives of the transformation matrices[*].

Dynamics

Force systems

Let the model of Figure 1. be subjected to an externally applied force field. On a typical body B_k let the forces be replaced by a force F_k passing through mass centre G_k together with a couple with torque M_k. Then Kane's equations lead to dynamical equations of motion of the form:

$$F_l + F_l^* = 0 \qquad l = 1,\dots,52 \tag{15}$$

---

[*]See Huston *et al.*, (1976b) for additional details.
[*]Ibid.

The F_l are called the "generalized active forces" associated with y_l. The F_l are defined as:

$$F_l = (\partial \underset{\sim}{v}_k/\partial y_l)\,\underset{\sim}{F}_k + (\partial \underset{\sim}{\omega}_k/\partial y_l)\cdot \underset{\sim}{M}_k \tag{16}$$
$$= v_{klm}F_{km} + \omega_{klm}M_{km} \qquad l = 1,\ldots,52$$

where F_{km} and M_{km} are $\underset{\sim}{n}_{om}$ components of $\underset{\sim}{F}_k$ and $\underset{\sim}{M}_k$.

Similarly, in Equations (15) the F_l^* are called the "generalized inertia forces associated with y_l. If the inertia force system on a typical body B_k is replaced by a force $\underset{\sim}{F}_k$ passing through G_k together with a couple with torque $\underset{\sim}{M}_k^*$, then the F_l^* are defined as:

$$F_l^* = (\partial v_k/\partial y_l)\,F_K^* + (\partial \omega_k/\partial y_l)\cdot M_k^* \tag{17}$$
$$= v_{k01m}F_{km}^* + \omega_{klm}M_{km}^*$$

where F_{km}^* and M_{km}^* are $\underset{\sim}{n}_{om}$ components of $\underset{\sim}{F}_k^*$ and $\underset{\sim}{M}_k^*$. Finally, $\underset{\sim}{F}_k^*$ and $\underset{\sim}{M}_k^*$ may be expressed as [see Kane and Levinson (1985)]

$$\underset{\sim}{F}_k^* = -\,M_k\,\underset{\sim}{a}_k \qquad \text{(no sum)} \tag{18}$$

and

$$\underset{\sim}{M}_k^* = -\,\underset{\sim}{I}_k\cdot - \,\underset{\sim}{\alpha}_k - \underset{\sim}{\omega}_k \times (\underset{\sim}{I}_k\cdot \underset{\sim}{\omega}_k) \qquad \text{(no sum)} \tag{19}$$

where m_k is the mass of B_k and $\underset{\sim}{I}_k$ is the central inertia dyadic of B_k.

Joint forces and moments

If there are forces and moments at the joints such as from muscles and ligaments, they may also contribute to the generalized active forces. To determine these contributions, consider again two typical adjoining bodies as in Figure 2. For generality, let there be translation, or separation, between the bodies as in the neck. Let this separation be measured by the vector $\underset{\sim}{\xi}_k$ as shown in Figure 2, where Q_k and O_k are connection and reference points of β_j and β_k. (Note that when $\underset{\sim}{\xi}_k$ is zero, Q_k and O_k coincide).

Let the force system exerted on B_j by B_k at the joint be represented by a force $\underset{\sim}{K}_i$ passing through Q_k together with a couple with torque $\underset{\sim}{T}_j$. Then, by the law of action–reaction and by assuming that $\underset{\sim}{\xi}_k$ is small**, the force system which B_j exerts on B_k is equivalent to a force $\underset{\sim}{K}_k$ passing through O_k together with a couple with torque $\underset{\sim}{T}_k$ such that:

$$\underset{\sim}{K}_k = -\,\underset{\sim}{K}_j \tag{20}$$

and

---

*Ibid.
**See Huston and Kelly (1982).

$$T_k = -T_j \tag{21}$$

Then from Equation (16) the contribution $\hat{F}_l$ from these force systems to the generalized active force F_l is:

$$F_l = (\partial \underline{v}_{ok}/\partial y_l) \cdot K_k + (\partial \omega_k/\partial y_l) \cdot T_k \\ + (\partial \underline{v}_{qk}/\partial y_l) \cdot K_j + (\partial_k/\partial y_l) T_j \tag{22}$$

From Figure 2, it is seen that $\underline{v}_{Ok}$ and $\underline{v}_{Qk}$ are related by expression

$$\underline{v}_{Ok} = \underline{v}_{qk} + \omega_j \times \xi_k + \xi_{km}\, \underline{n}_{jm} \tag{23}$$

Also, ω_k and ω_j are related by the expression:

$$\omega_k = \omega_j + \hat{\omega}_K = \omega_j + \hat{\omega}_{km}\, \underline{n}_{jm} \tag{24}$$

Hence, from Equations (20) to (24) it is seen that the contributions to F_l belong to one of the following three cases:

Case 1: y_l is not equal to either $\hat{\omega}_{km}$ or ξ_{km}

In this case Equations (23) and (24) lead to:

$$\partial \underline{v}_{qm}/\partial y_l = \partial \underline{v}_{Ok}/\partial y_l \text{ and } \partial \omega_k/\partial y_l = \partial \omega_j/\partial y_l \tag{25}$$

Then from Equations (20), (21), and (22), the contribution to F_l is:

$$\hat{F}_l = O \tag{26}$$

Case 2: y_l is equal to one of the ω_{km}.

In this case Equations (23) and (24) lead to:

$$\partial \underline{v}_{Qk}/\partial y_l = \partial \underline{v}_{Ok}/\partial y_l = \partial \omega_j/\partial y_l = 0 \tag{27}$$

and

$$\partial \omega_k/\partial y_l = \underline{n}_{jm} \tag{28}$$

Then from Equation (22), the contribution to F_l is

$$\hat{F}_l = \underline{n}_{jm} \cdot T_k = T_{km} \tag{29}$$

where the T_{km} ($m = 1, 2, 3$) are the joint moment components referred to $\underline{n}_{jm}$.

Case 3: Y_l is equal to one of the ξ_{km}

In this case Equations (23) and (24) lead to:

$$\partial \underline{v}_{Qk}/\partial y_l = \partial \omega_k/\partial y_l = \partial \omega_j/\partial y_l = 0 \qquad (30)$$

and

$$\partial \underline{v}_{Ok}/\partial y_l = \underline{n}_{jm} \qquad (31)$$

Then from Equation (22), the contribution to F_l is

$$\hat{F}_l = \underline{n}_{jm} \cdot \underline{T}_k = T_{km} \qquad (29)$$

where the K_{km} ($m = 1, 2, 3$) are the joint force components referred to $\underline{n}_{jm}$.

From Equations (28) and (32) it is seen that the joint force and movement components each contribute to a *different* generalized active force. This in turn means that in the dynamical equations each joint force and moment component occur in a separate equation — that is, they are *uncoupled*.

Joint moment constraints

The forces and moments generated by the soft tissue, such as muscles and ligaments, at the joints are conveniently represented in the forms [See Tien and Huston (1984)]

$$F = K_f X_D + C_F \dot{X}_D \quad \text{and} \quad M = K_M X_R + C_N \dot{X}_R \qquad (33)$$

where F and M represent the force and moment, X_D and X_R repesent displacement and rotation parameters, and where the coefficients K_F, C_F, K_M and C_M are, in general, functions of X_D and X_R and their derivatives. For example K_F and C_F typically have the forms:

$$K_F = \begin{cases} K_{FO} + K_{F1}(X_D - X_{Dmax}) & \text{for } X_D > X_{Dmax} \\ K_{FO} & \text{for } X_{Dmin} \leqslant XD \leqslant XDmax \\ K_{FO} + K_{F1}(X_D - X_{Dmin}) & \text{for } X_D < X_{Dmin} \end{cases} \qquad (34)$$

and

$$C_F = \begin{cases} C_{FO} & \text{for } X_D \dot{X}_D < O \text{ or} \\ & X_{Dmin} \leqslant X_D \leqslant X_{Dmax} \\ C_{FO} + C_{F1}(X_D - X_{Dmax}) & \text{for } X_D > X_{Dmax}, \dot{X}_D > O \\ C_{FO} + C_{F1}(X_D - X_{Dim}) & \text{for } X_D < X_{Dmin}, \dot{X}_D < O \end{cases} \qquad (35)$$

where K_{FO}, K_{F1}, C_{FO}, and C_{F1} are constants and where X_{Dmax} and X_{Dmin} represent maximum and minimum displacement values.

Similar expressions may be written for K_M and C_M. The values of these parameters are difficult to determine analytically. However, by curve fitting techniques they may be selected to be compatible with experimental data. In te writer's opinion more experimental data needs to be recorded before a comprehensive listing can be obtained for all joints. Preliminary values for the neck may be found in Tien and Huston (1984).

Governing equations

By substituting from Equations (12), (14), (17), and (18) into (15) the equations of motion may be written in the form

$$a_{lp}\dot{y}_p = f_l \qquad l = 1, m, 52 \tag{36}$$

where the arrays a_{lp} and f_l are

$$a_{lp} = m_k v_{klm} v_{kpm} + I_{kmn}\omega_{klm}\omega_{kpm} \tag{37}$$

and

$$f_l = F_l - (m_k v_{klm}\dot{v}_{kpm}y_p + I_{kmn}\omega_{klm}\dot{\omega}_{kpn}y_p \\ + e_{rhm}I_{khn}\omega_{klm}\omega_{kqr}\omega_{kpn}y_q y_p \tag{38}$$

where m_k is the mass of B_k, I_{kmn} is its central inertia dyadic referred to $\underline{n}_{om}$, and where e_{rhm} is the permutation symbol.

Discussion

There are several concluding thoughts which might be helpful in applying the foregoing ideas.

(a) First, a number of computer programmes based upon these concepts have already been written [see for example, Kamman and Huston (1985) and Huston *et al.*, (1982)]. These programmes have been validated against experimental data. They have been applied in crash-victim simulation, analysis of gait, and in the analysis of sporting movements. The crash-victim simulation programmes include both gross-motion (whole-body motion) as well as detailed simulation of the head/neck systems [see Tien and Huston (1984)]. The sports mechanics programmes include simulation of swimming and throwing motions.

The same procedures are being applied in the development of specialized computer programmes for the analysis of rehabilitation devices. For example, in the foregoing analysis it is relatively easy to introduce springs and dampers between the bodies of the human model. These springs and dampers can be used to model rehabilitation devices such as braces and collars. They can be used to model forces between the bodies as would occur from the rehabilitative devices.

A second application is the computer modelling of orthopaedic casts, crutches, and prostheses. Such devices, when employed with the human frame, are equivalent to either the modification of existing limbs, the creation of additional limbs, or the replacing of traditional limbs. The dynamic effects of such devices are still not well understood even though they have been extensively employed. For example, the weight of a walking leg cast can affect the forces in the joints and the soft tissue of the remainder of the frame. Also, the gait pattern can be affected. The long term deleterious effects are not known. The foregoing analysis can be used to study such phenomena. To the writer's knowledge such studies have not yet been performed. Similar studies in the use of crutches and prosthetic devices are also needed.

A third application of the computer programmes is with the development and monitoring of electro-neurological muscle stimulation procedures. There is currently great interest in applying these procedures in paralytic rehabilitation.

To monitor such phenomena it is necessary to develop goniometers to measure joint movement. The development of these devices is ongoing. Carpenter and Huston (1984) have recently presented a description of a typical goniometer. The design objectives are: 1) accurate measurement of joint movement; 2) unrestrained movement; 3) compactness; and 4) ease of attachment and removal.

The output (electrical signal) of the gomiometer is a measure of the joint kinematics. Such data, converted from analogue to digital form, can be used to predict the dynamic effects of the motion — not only on the limbs comprising the joint, but also on the remainder of the frame.

A combination of computer analysis, electro-goniometer measurement, and experimental data are expected to lead to significant advances in this growing field.

(b) Next, the procedures outlined herein are useful for computing the energy in a given motion. Specifically, the kinetic and potential energies can be calculated and used in an energy analysis. Thus motions can be determined which will maximize (or minimize) energy expenditure in a given muscle group.

These analyses are significant since the limbs of the human frame are ''redundant'', That is, they contain more degrees of freedoom than are needed to accomplish a given task. For example, the human arm can be considered to have seven degrees of freedom (three at the shoulder, one at the elbow, and three in the wrist-forearm). However, moving an object with the hand requires at most six degrees of freedom. How the seventh degree of freedom is used determines the energy expenditures, and thus the effort, in the movement of the subject. Such information can be extremely useful in the development of optimal rehabilitative procedures.

(c) Finally, the procedures can also be used for the design of rehabilitation machines such as are used for ''weight-training''. These machines can be designed to isolate muscle groups to provide specific exercises for injury rehabilitation. Arthur Jones (1986), the developer of Nautilus machines, has conducted extensive studies on the effects of these procedures. Again, optimal motions and configurations can be identified. The analysis can thus be used to develop a basis for physical therapy. This in turn could lead to shortened rehabilitation times.

Acknowledgement

The research for this paper was partially supported by the National Science Foundation under Grant MEA8313439.

References

Amirouche, M. and Huston, R.L., 1984, Modelling of human body vibration. In *Trends in Ergonomics: Human Factors I*, edited by A. Mital, (New York: Elsevier Science Publishers [North Holland]) pp.65–70.

Bhattachanga, A., King, T.P., Patel, P.B. and Huston, R.L., 1986, Biodynamic modelling of carpetlayer dynamics. In *Trends in Ergonomics/Human Factors III*, edited by W. Karwowski (New York: Elsevier Science Publishers [North Holland]) pp.759–764.

Carpenter, M.A. and Huston, R.L., 1984, Electrogoniometer measuring system for human gait. In *Trends in Ergonomics/Human Factors I*, edited by A. Mital (New York: Elsevier Science Publishers [North Holland]) pp.77–82

Gallenstein, J. and Huston, R.L., 1973, Analysis of swimming motions. *Human factors* **15**, 91–98.

Huston, R.L., 1986, Useful procedures in multibody dynamics. In *Dynamics of Multibody Systems*, edited by G. Bianchi and W. Schiehlen (Berlin, Heidelberg: Springer) pp.69–77.

Huston, R.L. and Kamman, J.W., 1981, On parachutist dynamics. *Journal of Biomechanics* **14**, 645–652.

Huston, R.L. and Kamman, J.W., 1982, A discussion of constraint equations in multibody dynamics. *Mechanics Research Communication*, **9**, 251–256.

Huston, R.L. and Kelly, F.A., 1982, The development of equations of motions of single-arm robots. *IEEE Transactions on Systems, Man, and Cybernetics*, SMC-12, 259–265.

Huston, R.L. and Perrone, N., 1980, Dynamic response and protection of the human body and skull in impact situation. In *Perspectives in BioMechanics*, edited by H. Reul, D.H. Ghista and G. Rau (New York: Harwood), pp.531–571.

Huston, R.L. and Sears, K., 1981, Effect of protective helmet mass on head-neck dynamics. *Journal of Biomechanical Engineering* **103**, 18–23.

Huston, R.L., Hessel, R.E. and Winget, J.M., 1976a, Dynamics of a crash victim — a finite segment model. *AIAA Journal* **14**, 173–178.

Huston, R.L., Passerello, C.E. and Huston, J.C., 1976b, Numerical prediction of head/neck response to shock impact. In *Measurement and Prediction of Structural and Biodynamic Crash-Impact Response*, edited by K. Saczalski and W. Pilkey, (New York: ASME) pp.137–150.

Huston, J.C., Harlow, N.W. and Huston, R.L., 1978a, A comprehensive, three-dimensional head-neck model for impact and high acceleration studies. *Aviation, Space and Environmental Medicine*, **49**, 205–210.

Huston, R.L., Winget, J.M. and Harlow, M.W., 1978b, A biodynamic model of a parachutist. *Aviation, Space, and Environmental Medicine*, **49**, 178–182.

Huston, R.L., Harlow, M.W. and Gausewitz, N.L., 1982, User's manual for UCIN-EULER a multipurpose, multibody systems dynamics computer program. National Technical Information Service (NTIS) Report No. AD-120403.

Jones, A., 1986, Exercise 1986 the present state of the art, now a science. Natilus Sports/Medical Industries, Inc., Deland, Florida 32721–1783.

Kamman, J. W. and Huston, R.L., 1984, Constrained multibody system dynamics — an automated approach. *Computers and Structures*, **18**, 999–1003.

Kamman, J.W. and Huston, R.L., 1984, Dynamics of constrained multibody systems. *Journal of Applied Mechanics*, **51**, 889–904.

Kamman, J.W. and Huston, R.L., 1985, User's manual for UCIN SYNOCOMBS. National Technical Information Service (NTIS) Report No. PB85–240075/AS.

Kane, T.R. and Levinson, D.A., 1980, Formulation of equation of motion for complex spacecraft, *Journal of Guidance and Control*, **3**, 99–112.

Kane, T.R. and Levinson, D.A., 1985, *Dynamics, Theory and Applications* (New York: McGraw Hill).

Kane, T.R. Linins, P.W. and Levinson, D.A., 1983, Spacecraft Dynamics (New York: McGraw Hill).

King, T.P. and Huston, R.L., 1985, Analysis of simple throwing. In *Trends in Ergonomics/Human Factors II*, edited by Elberts and Elberts, (New York: Elsevier Science Publishers [North Holland]) pp.265–271.

Liu, V. and Huston, R.L., 1984, Analysis of human body dynamics in the frequency domain. In *Trends in Ergonomics: Human Factors I*, edited by A. Mital, (New York: Elsevier Science Publishers [North Holland]) pp.71–75.

Nikravesh, P.E. and Chung, I.S., 1982, Application of Euler parameters to the dynamic analysis of three dimensional constrained systems. *Journal of Mechanical Design*, **104**, 785–791.

Tien, J. and Huston, R.L., 1984, Biodynamic modelling of the head/neck systems. In *Trends in Ergonomics: Human Factors I*, edited by A. Mital, (New York: Elsevier Science Publishers [North Holland]) pp.45–50.

Whittaker, E.T., 1937, *Analytical Dynamics* (London: Cambridge) pp.8, 16.

spinal stresses. If kinematic parameters change, the magnitude of joint stresses will also change.

In previous biomechanical studies of manual lifting activities, only healthy individuals took part and the resulting stress norms were determined from kinematic data collected on these healthy individuals. It is reasonable to expect that load (wrist) kinematics will be different for healthy and disabled individuals, even when both are subjected to the same magnitude of external loading, and job parameters remain unchanged. The slow and deliberate movements of disabled individuals may very well cause these differences. Since the published literature lacks studies comparing the load lifting kinematics of healthy and disabled individuals, this hypothesis remains unconfirmed.

The purpose of this study was to determine if the kinematic parameters of the wrist of disabled individuals (individuals limited in their ability to grasp containers properly) engaged in manual lifting are different from kinematic parameters of the wrist of healthy individuals engaged in an identical activity. It was hypothesized that disabled individuals will have slower and more deliberate movements, and that this would lead to lower peak accelerations of the load (wrist) and, thereby, lower inertial forces.

Methods

In order to verify the proposed hypothesis, an experiment was conducted to determine kinematic parameters of the wrist of disabled and healthy individuals.

Subjects

Eight male volunteers participated in the experiment. All were physically healthy and had no prior history of any physical ailments, and none was on medication. Their age ranged between 25 and 42 years. The weight and stature of the subjects ranged between 51 kg and 84 kg, and 1·68 m and 1·85 m, respectively. During the experiment, subjects wore comfortable work clothes (safety shoes, work shirts and pants).

The disability

The disability was defined as the loss of the use of four fingers in the primary (preferred) hand and the thumb in the non-preferred hand. Under normal circumstances, a box is held in a grip formed by two jaws of the hand; four fingers together form one jaw and the thumb forms the other. Loss of fingers in one hand and thumb in the other means loss of one jaw in each hand. This weakens the coupling and extra effort is required to lift and hold the box. Therefore, for a person suffering from such a disability, the lifting task becomes difficult to perform. If the load kinematics for individuals inflicted with such a severe disability are not found to differ significantly from the load kinematics for healthy individuals, load kinematics for individuals inflicted with less severe disabilities would be even less important. This is the rationale for selecting this disability for study.

Simulation of the disability was accomplished by taping four fingers of the preferred hand and thumb of the non-preferred hand with a nylon tape strip (Figure 1). The taped fingers and thumb did not, in any way, restrict the movement of the fingers or thumb. It was also ensured that the tape, while preventing the movement of taped fingers, was not excessively tight nor restricted the flow of blood.

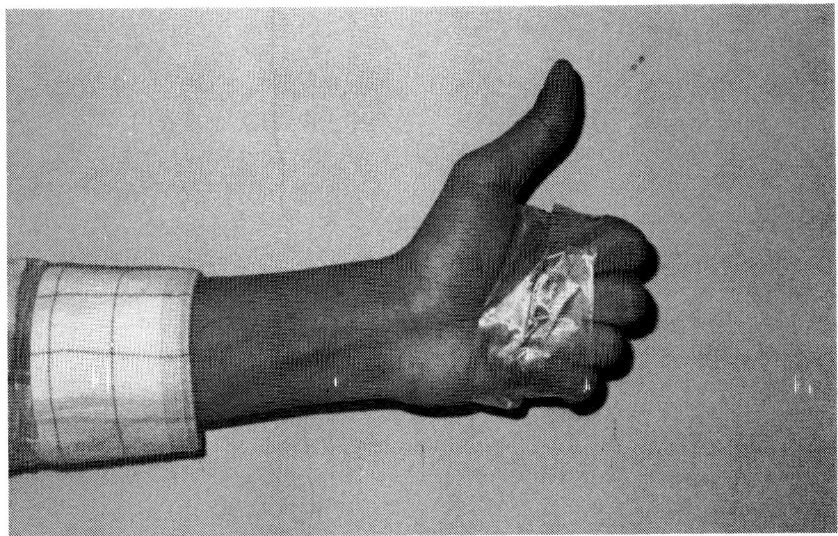

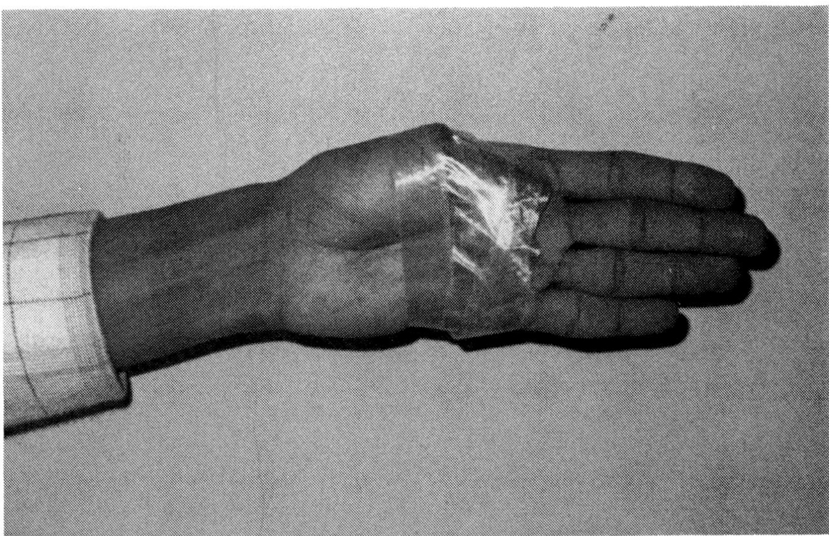

Figure 1.   *Taped fingers of the preferred hand (top), and thumb of the non-preferred hand (bottom).*

Equipment

A Panasonic video camera, mounted on a tripod, was used to film the lifting activity, which was recorded by a video recorder. After recording, the motion was replayed, one frame at a time, on an RCA model FB497W colour video monitor and digitized using a Grafbar Model GP-7 sonic digitizer (Science Accessories Corporation). The compact box lifted by the subjects was wooden and built in-house (0·45 m long — in the sagittal plane, 0·15 m high, and 0·30 m wide — in the frontal plane). A wooden workbench, 0·80 m high, was used for placing the box after it was lifted from the floor by the subject.

Procedure

The experimental procedure required each subject to lift the box from the floor and place it on the workbench. During lifting, the box was held at diagonally opposing corners (preferred hand at the bottom corner closer to the body, and the non-preferred hand at the top corner away from the body). The lifting motion was filmed by a video camera (standard play mode) at a speed of 30 Hz. The distance between the subject and the camera was approximately 3 m and the axis of the camera was perpendicular to the mid-sagittal plane to avoid parallax error. A marker was placed on the subject's wrist joint (the joint facing the camera) to facilitate determination of the wrist's position during lifting.

Each of the eight subjects lifted the box four times. The four lifting treatments were determined by (i) weight in the box (11·35 kg or 22·7 kg) and (ii) presence or absence of the disability (fingers and thumb taped or free). The experimental order of all four treatments [2 weights × 2 lifting conditions (tape or no tape)] was randomized for each subject. The order in which each subject performed the tasks was also randomized. A rest of at least five minutes was provided between treatments to avoid any buildup of static fatigue. Only one subject performed at a time, but all subjects performed on the same day.

Data reduction and differentiation

The recorded motion was replayed, one frame at a time, on a video monitor and, using the sonic digitizer, the absolute position of the wrist joint (X, Y) was acquired and stored on an IBM AT personal computer. Since the raw data were noisy, due to digitization errors, smoothing (reduction) was required prior to determining kinematic parameters (displacement, velocity, and acceleration). The cubic spline function method of data smoothing and differentiation was used for this purpose. The smoothing algorithm consisted of the *FORTRAN* IMSL (International Mathematical and Statistical Libraries) smoothing routine ICSSCU. This cubic spline data smoother accepted raw data and temporal data, necessary for achieving the desired level of smoothing, as input. The raw data consisted of the raw data points $(X, Y$ values), time associated with each data point, the total number of film frames, and the smoothing parameter. The smoothing parameter selected for this study was 0·3.

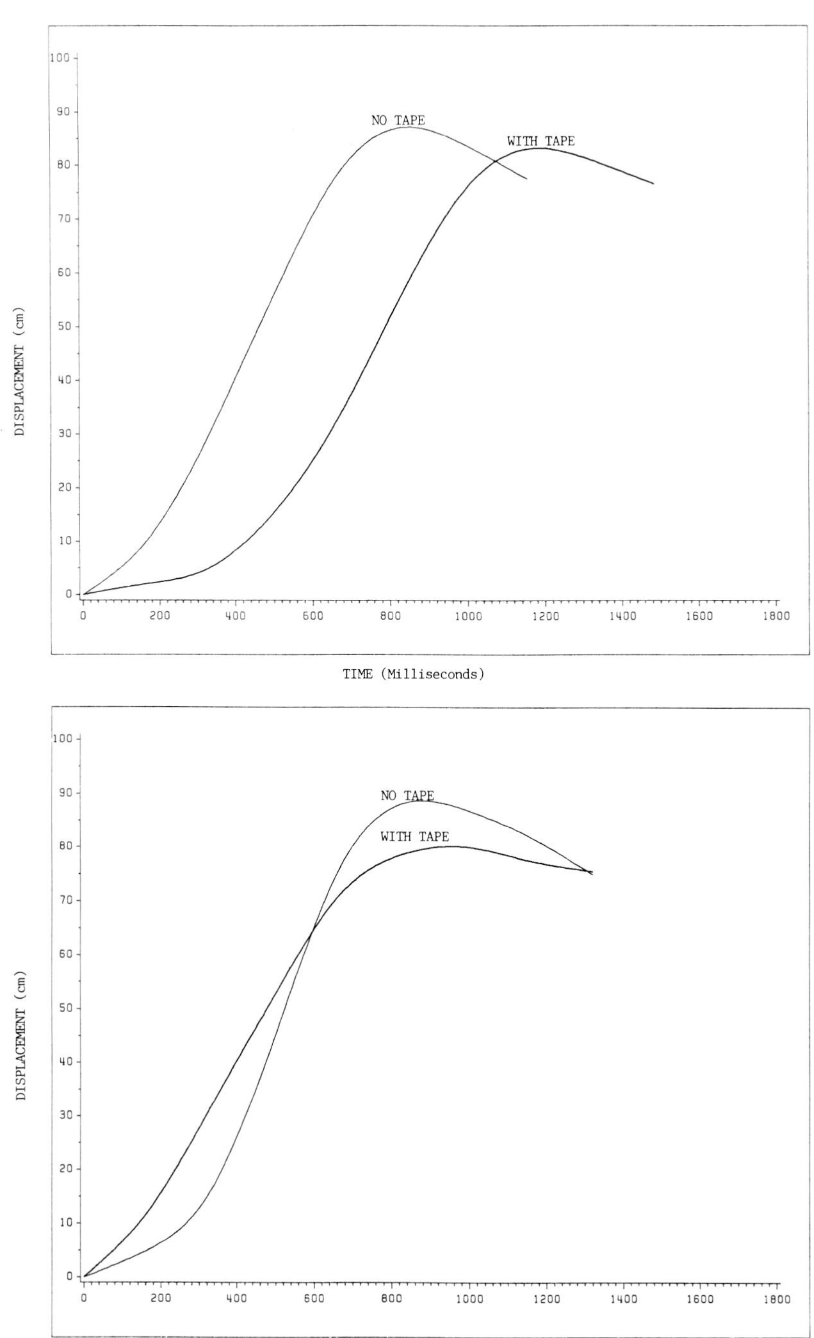

Figure 2. Typical displacement profile of the wrist joint for taped and untaped conditions, for 11 · 35 kg (top) and 22 · 7 kg (bottom) weights.

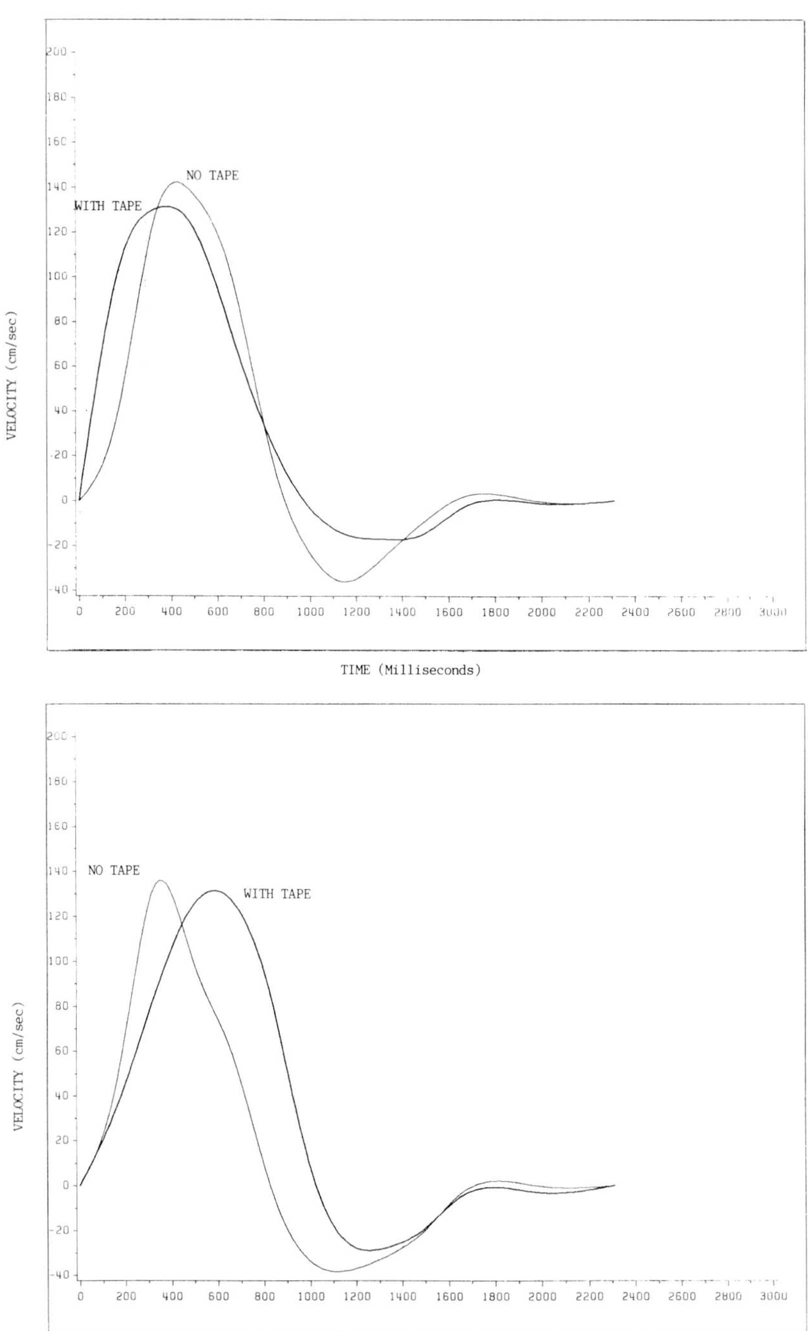

Figure 3.   *Typical velocity profile of the wrist joint for taped and untaped conditions, for 11 · 35 (top) and 22 · 7 kg (bottom) weights.*

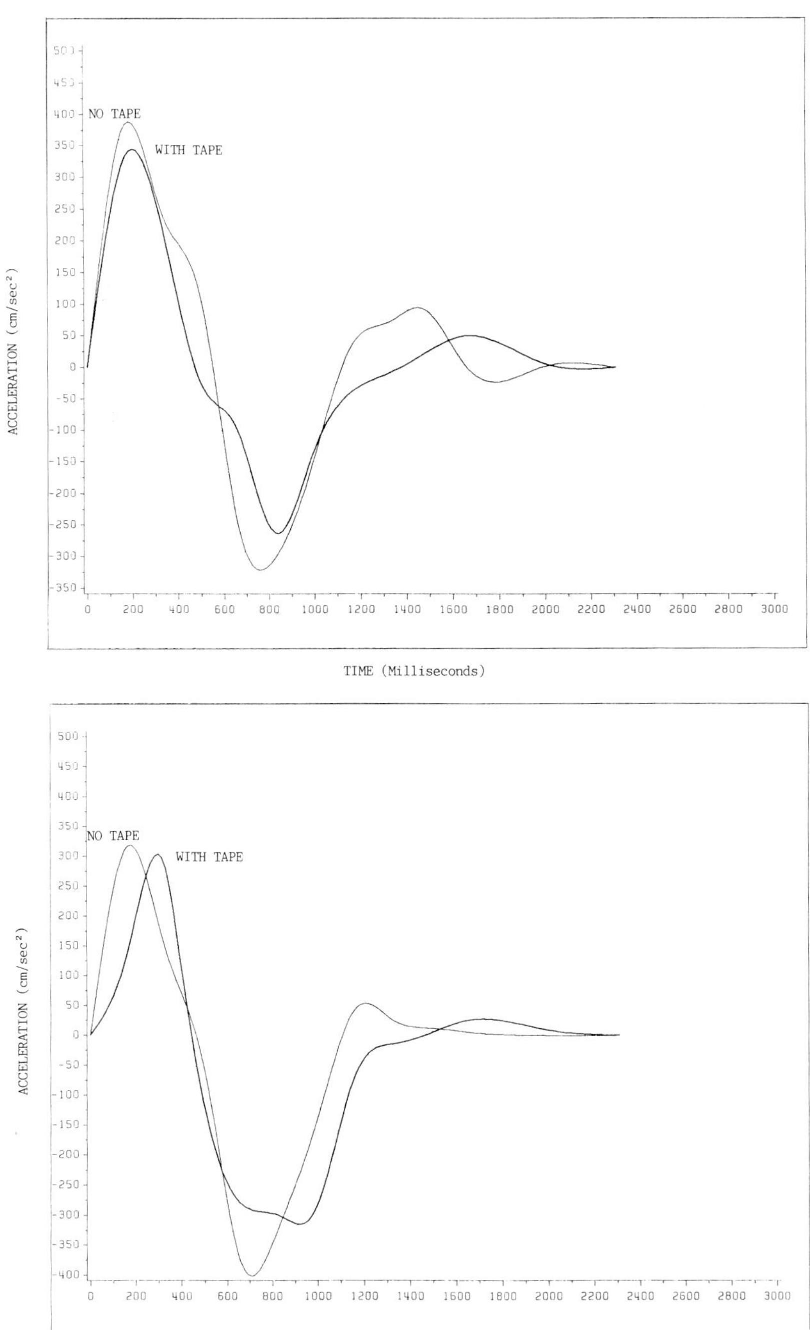

Figure 4. *Typical acceleration profile of the wrist joint for taped and untaped conditions, for 11 · 35 kg (top) and 22 · 7 kg (bottom) weights.*

In order to determine the velocity and acceleration of the wrist joint, the smoothed data were numerically differentiated, The differentiation was performed by another IMSL *FORTRAN* routine, DCSEVU. The velocity and acceleration profiles, thus obtained, were used in subsequent analysis. Figures 2-4 show the displacement, velocity, and acceleration profiles of a typical subject for $11 \cdot 35$ kg and $22 \cdot 7$ kg weight.

Results

In order to test the hypothesis that peak acceleration of subjects with the disability would be lower than the peak acceleration of healthy subjects, peak acceleration data were subjected to an analysis of variance. The results are shown in Table 1. As indicated, the disability (taping of the fingers and thumb) resulted in significantly lower peak accelerations. However, the peak acceleration for both healthy and disabled individuals occurred at approximately the same instant once the lifting activity commenced (Figure 4). This last observation was also found to be true in the case of peak load velocities (Figure 3). The only difference was that, unlike peak load accelerations, peak load velocities with the handicap and without disability were not significantly different from each other $(P > 0 \cdot 10)$.

Table 1.   Analysis of variance with peak acceleration as the dependent variable.

Source	DF	Sum of Squares	F-Value	Prob. $>$ F
Subjects	7	$72991 \cdot 31$	$7 \cdot 76$	$0 \cdot 0001$
Box Weight (W)	1	$1048 \cdot 82$	$0 \cdot 78$	$0 \cdot 3870$
Lifting Condition (L)	1	$8521 \cdot 65$	$6 \cdot 34$	$0 \cdot 0200$
W*L	1	$174 \cdot 84$	$0 \cdot 13$	$0 \cdot 7219$

Note: DF is degrees of freedom
       W*L is weight and lifting condition interaction
       F-Value is the value of F ratio.

Even though the average peak acceleration of 287 cm sec^{-2} obtained with the $22 \cdot 7$ kg weight was lower than the average peak acceleration of 298 cm sec^{-2} obtained with the $11 \cdot 35$ kg weight, the effect of box weight on peak acceleration was not significant $(P > 0 \cdot 10)$. On average, it took subjects 2655 milliseconds to lift the box with the heavier weight, compared with 2590 milliseconds it took them to lift the box with the lighter weight.

When the subjects had their fingers and thumb taped, it slowed them down significantly $(P < 0 \cdot 01)$. The average time for lifting the box with the fingers and thumb taped was 1432 milliseconds. Without the tape, subjects, on average, required 1265 milliseconds to lift the box. As indicated in Figure 2 the trajectory of the box was also different when tape was used; individuals did not overshoot the destination as they did when no tape was used.

Discussion and conclusion

This study was undertaken to determine whether the loss of the ability to grasp containers properly would result in significant changes in peak load (wrist) accelerations during manual lifting. The results support the hypothesis that peak load accelerations during manual lifting are indeed lower for individuals with such a disability.

Lower load accelerations mean a smaller contribution of acceleration in inertial forces (on average 3·7 per cent smaller in this study). This, however, does not mean that overall inertial forces will also be lower for individuals with a disability. Factors such as trunk angle, etc. will determine the ultimate magnitude of inertial forces. It is quite possible, therefore, that spinal stresses for disabled individuals may increase during lifting even when the levels of external loading remain the same. Also, the task becomes somewhat more demanding and awkward for the disabled individual. The lifting cycle time was longer when boxes were lifted with the disability (approximately 13 per cent longer on average) compared with the lifting cycle time when boxes were lifted with no disability. This means that disabled individuals held the load longer. Subjects, at the end of the experiment, also indicated that they had to be more careful during the initial phase of lifting (picking up the box). The results of this study and these subjective opinions, thus, indicate that a disabled person is more cautious during lifting and the additional time taken in initiating the lift is spent in seeking and maintaining a safe grip. This conclusion is also supported by the fact that once the lifting had commenced, peak load accelerations for the disabled and normal lifting conditions were reached at approximately the same instant during the lifting cycle. The same was true for peak load velocities.

It is important to realize that in this study we simulated the disability by taping the fingers and thumb. The actual disability, however, is quite different from the simulated disability. In general, individuals learn to accommodate their disability and there is some learning effect, which could lead to a reduction in lifting cycle time. It is also possible that post-learning load kinematics may not be very different from kinematics for healthy individuals. If so, other measures of job difficulty, such as EMG, may have to be investigated.

Loss of a jaw (four fingers or the thumb), most likely, will also necessitate deviation of the wrist to prevent the box from slipping. Wrist deviation would impose additional stress on the individual and reduce the available grip strength (Drury *et al.*, 1985). In order to maintain the required grip force to prevent the box from slipping, the disabled individual will probably have to recruit muscles of the arm instead of merely the muscles of the hand.

Conclusion

The results of this study suggest that individuals with a disability tend to become more cautious while lifting containers. They take longer to complete the lift and, thus, sustain the external loading longer. The loss of one jaw (fingers and the thumb) may also lead to reduced grip strength and require the disabled individual to recruit arm muscles in addition to the hand muscles. A disabled person, therefore, will be under greater stress during manual lifting than a normal healthy person.

References

Chaffin, D.B., 1969, A computerized biomechanical model development of and use in studying gross body actions. *Journal of Biomechanics*, **2**, 429–441.

Chaffin, D.B. and Baker, W. ., 1970, A biomechanical model for analysis of symmetric sagittal plane lifting. *Transactions of the American Institute of Industrial Engineers*, **2**, 16–27.

Drury, C.G., Begbie, K., Ulate, C. and Deeb, J.M., 1985, Experiments on wrist deviation in manual materials handling. *Ergonomics*, **28**, 577–589.

El-Bassoussi, M.M., 1974, A dynamic biomechanical model for lifting in the sagittal plane. Ph.D dissertation, Texas Tech University, Lubbock, Texas, U.S.A.

Freivalds, A., Chaffin, D.B., Garg, A. and Lee, K.S., 1984, A dynamic biomechanical evaluation of lifting maximum acceptable loads. *Journal of Biomechanics*, **17**, 251–262.

Garg, A. and Chaffin, D.B., 1975, A biomechanical computerized simulation of human strength. *Transaction of the American Institute of Industrial Engineers*, **7**, 1–15.

Kromodihardjo, S. and Mital, A., 1986, Kinetic analysis of manual lifting activities: Part I — Development of a three-dimensional computer model. *International Journal of Industrial Ergonomics*, **1**, 77–90.

Park, K.S. and Chaffin, D.B., 1974. A biomechanical evaluation of two methods of manual load lifting. *Transactions of the American Institute of Industrial Engineers*, **6**, 105–113.

Smith, J.L., Smith, L.A. and McLaughlin, 1982, A biomechanical analysis of industrial manual materials handlers. *Ergonomics*, **25**, 299–308.

12.

Electromyographic methodologies in rehabilitation

T.M. Khalil, S.M. Waly and S.S. Asfour
Department of Industrial Engineering
University of Miami
Coral Gables, Florida 33124
USA

The use of computers to enhance the capabilities of clinical electromyographic laboratories is desirable. In this paper, a methodology based on using digital signal processing techniques for the preliminary diagnosis of the various neuromuscular disorders is presented. The processing techniques, both in the time and frequency domains, are defined. Case studies that provide a comparison between normal and abnormal muscles are given to demonstrate the effectiveness of the proposed methodology. The advantages and limitations of this methodology are discussed.

Introduction

Musculoskeletal disorders are among the most crucial and persistent problems affecting the human race. Muscular injuries also constitute one of the major problems in occupational environments. In 1981 the National Institute of Occupational Safety and Health, reported that these injuries account for over one-quarter of reported occupational injuries in the United States. The study of the electromyographical signals (EMG) is important in the diagnosis, evaluation, and rehabilitation of the musculo-skeletal and neuromuscular disorders. At the present time, most neurologists assess EMG signals by monitoring their pattern as they flash on the screen of an oscilloscope in real time and/or listening to the sound of the amplified EMG signal. This approach relies heavily on the ability of the experienced electromyographer to detect visually and/or acoustically specific characteristics of the various types of neuromuscular pathologies in the EMG pattern. It was the objective of this study to utilize digital signal processing techniques to analyse the complex EMG signal in order to find useful quantitative indices. Such indices can assist the physician in diagnosis and rehabilitation. Case studies are given to demonstrate the use of EMG in both diagnosis and evaluation of the neuromuscular function.

Physiology of Electromyography

Electromyography is the recording of electrical changes which occur in an anatomical muscle by means of electrodes in contact with the skin covering the muscle or by inserting electrodes into the muscle through the skin. The records obtained represent only the potential changes of the muscle fibres, but it is the best tool for discovering minor dysfunctions of the central nervous system and of motor end plates.

The functional entity of a striated muscle is the motor unit which consists of a number of muscle fibres innervated by a single motor neurone. Near the muscular innervation, each motor neurone branches and terminates into end plates stimulating fibres of different lengths. This arrangement results in a time dispersion of the arrival of triggering pulses. Fibres belonging to the same motor unit are intermingled with fibres from other motor units. All fibres of a motor unit are triggered upon reaching a threshold for stimulation. Under normal conditions, an action potential of sufficient magnitude propagating down a motor neurone activates all its branches. When the post-synaptic membrane of a muscle fibre is depolarized, the depolarization propagates in both directions along the fibre causing what is known as an action potential. The summed electrical activity of the muscle fibres belonging to a motor unit is known as the motor unit action potential (MUAP).

In human tissue, the amplitude of the action potential is dependent on the diameter of the muscle fibre, the distance between the active muscle fibre and the recording site, and the filtering properties of the electrode (Buchtal *et al.*, 1957, and Lindstrom, 1970).

The duration of the action potentials is inversely related to the conduction velocity of the muscle fibre, which ranges from 2 to 6 m/sec (Lindstrom, 1970). The relative time of initiation of each action potential is directly proportional to the difference in length of the nerve branches and the distance the depolarizations must propagate along the muscle fibres before they approach the detectable range of the electrode. It is also inversely proportional to the conduction velocities of the nerve branch and the muscle fibre (Basmajian, 1979).

The shapes and frequency spectrum of the action potentials are affected by the tissue between the muscle fibre and the recording site. The presence of a multi-layer structure between the recording site and point of generation of the action potentials creates a low-pass filtering effect. The band width of this filter decreases as the distance increases (Lindstrom, 1970). This filtering effect is much more pronounced for surface electrode recordings than for indwelling electrode recordings because indwelling electrodes are located closer to the active muscle fibres.

If muscle fibres belonging to other motor units in the detectable vicinity of the recording electrode are excited, their motor unit action potentials (MUAPs) will also be detected. However, the shape of each MUAP will generally vary due to the unique geometric arrangement of the fibres of each motor unit with respect to the recording site. MUAPs from different motor units may have similar amplitude and shape when the muscle fibres of each motor unit in the detectable vicinity of the electrode have a similar spatial arrangement.

Normal electromyography

The EMG signal has been defined as a graphic representation of the electrical manifestation of the neuromuscular activation associated with muscle contraction (Basmajian, 1979). It is an exceedingly complicated signal which is affected by the anatomical and physiological properties of muscles, the control scheme of the peripheral nervous system as well as the characteristics of instrumentation that is used to detect and observe it. Considering the various factors that affect the shape of an observed MUAP, it is not surprising to find variations in the amplitude, number of phases and duration of MUAPs recorded with different electrodes. In normal muscles, the peak-to-peak amplitude of MUAP recorded with indwelling electrodes (needle or wire) may range from a few microvolts (μV) to more than 5 mV with a typical value of 500 μV (Basmajian, 1979). According to Buchtal *et al.* (1954), the number of phases of MUAPs recorded with biopolar needle electrodes may range from one to four with the following distribution: 3 per cent biphasic, 37 per cent triphasic and 11 per cent quadriphasic. MUAPs having more than four phases are rare in normal muscle tissue, but do appear in abnormal muscle tissue (Marinacci, 1968). The time duration of MUAPs may also vary greatly, ranging from less than 1 to 12 m/sec (Basmajian and Cross 1971). It should be emphasized that the amplitude and shape of an observed MUAP are a function of the properties of the motor unit, muscle tissue and recording electrode properties. The filtering properties of the electrode and possibly the cable connecting the electrode to the preamplifiers, as well as the preamplifiers themselves can cause the observed MUAPs to have additional phases and/or longer durations.

Surface electromyography in clinical applications

Based on the above discussion, it is obvious that EMG is a valuable tool in clinical neurology and in the evaluation of neuromuscular deficiences. In medical settings, regular routine EMG is usually carried out with needle electrodes. This requires many penetrations in order to sample a large portion of the motor units activated in the muscle. Surface recording of the EMG is a non-invasive technique and is more acceptable by patients. It can be easily used by paramedical professionals. On the other hand, surface EMGs do not provide a high resolution of the signal components due to the sampling of a large population of the activated motor units. This makes it difficult to isolate activities of a single motor unit and it does not reflect the spectral contribution of small motor units which may be of a clinical significance.

Inbar and Noujaim (1984) showed that surface EMG can be successfully used for clinical applications and diagnosis. This is achieved at the cost of higher computational complexity which can be easily performed using the capabilities of microcomputers. Several attempts have characterized the use of surface EMG in clinical applications (Khalil *et al.*, 1982, 1984; Inbar and Noujaim, 1984; Waly and Khalil, 1985). Inbar and Noujaim (1984) used the spectral characterization procedures to extract, from surface EMG, information about the spectral shape of the motor unit action potentials and the firing rate of the motor units. The use of this technique allowed them to classify the recorded abnormal EMG signals into major groups similar to those classified clinically by needle EMG. Only one subject in this study was misclassified.

The EMG is a complex signal which carries more information than can be described by a small number of indices. Therefore, a battery of measures is needed to capture as much information carried by the signal as possible. This could only be achieved by using several digital signal processing techniques that can provide useful quantitative indices. Such indices can assist the physician in diagnosis and rehabilitation.

Computerized EMG digital signal processing

In clinical EMG investigations, the signal is specified in terms of duration, amplitude and number of phases. The use of a computer for the preliminary diagnosis depends upon the ability to describe the characteristics of the EMG signal quantitatively. With the wide use of digital computers several techniques for processing filtered EMG signals have been proposed (Khalil *et al.*, 1984, 1985; Waly *et al.*, 1986). The first step in processing the filtered EMG signals is the conversion of the signal from an analogue form to a discrete signal that can be stored on a disk for further analysis. Therefore, digitizing is required using an analogue-to-digital converter that is capable of sampling at a high rate. The sampling rate has to be at least twice the highest frequency component in the EMG signal. In the present study a sampling rate of 1000 Hz was used. This sampling rate was selected because the spectral study of the surface EMG signals showed that most of the energy contents in the signal are concentrated below 500 Hz. A *FORTRAN IV* software package was developed to compute the indices which identify the EMG signal in both the frequency and time domains. The digital signal processing techniques used in the present study to calculate these indices can be classified as follows:

Amplitude analysis

An increase in loading of the muscle will be accomplished by an increase in electrical activity when symptoms of fatigue are not present. The rise in electrical activity may be a result of firing of a new set of motor units or increasing the firing rate of the active motor units (Basmajian, 1979). The muscular activity may be quantified by different measurements of the amplitudes present in the EMG signal. The amplitude analysis of the EMG is used to obtain parameters or indices that describe the magnitude of the signal. These indices are given below:

(a) Full wave rectified integral (FWRI) is obtained by integrating the rectified EMG waveform using the trapezoidal integration rule. By integrating the EMG signal, the intensity of muscular activity is measured. Also, the relationship between the force output of the muscle and muscular activity is obtained.

(b) Root mean square (RMS) is one of the commonly used methods of reducing the interference pattern seen in the surface EMG. It is defined as the square root of the integration of the squared EMG signal divided by the time period of the integration.

(c) Sum of voltage excursions (SVE) is defined as the summation of the absolute amplitudes which exceed a threshold level. In this study a threshold of 35μ V is used.

(d) Range of amplitudes (R) is calculated as the peak to peak amplitude (maximum amplitude – minimum amplitude). The range of a signal measures the variability of amplitudes. It is not an accurate measure, by itself, because it depends only on two values of the data set.

(e) Amplitude density function is estimated by calculating the random variables for each amplitude and plotting a frequency of occurrence diagram (histogram). The density function is described by different statistical measures such as the mean, variance, and median.

Frequency analysis

One possible approach to extract the required information carried by the EMG signal is to analyse its power spectrum. Since the EMG is a complex waveform, the parameter of frequency could be defined in relation to the number of zero-crossings, reversals of potentials, number of peaks, etc., in the signal. The task is made difficult, however, due to synchronization of motor unit firings as well as noise within the recording system. This type of analysis results in parameters or indices expressing the signal in terms of its frequency contents. These indices are given below:

(i) Frequency spectrum: The frequency spectrum of the EMG can be computed in two different ways. The first way is to form the autocorrelation function which reflects the periodicity of the signal. The zero-order autocorrelation coefficient, [R(0)], provides a measure of the average power in the signal. These values represent a valuable indication of the waveform. Then by an approximate Fourier Transform the power spectrum can be found. The power spectrum of a signal describes how the total average power of the signal is distributed in the frequency domain. The second way of computing a spectrum is to take the sampled EMG and convert a given length of the digitized signal directly into its fourier transform by means of the Fast Fourier Transform algorithm (FFT). In this study the FFT algorithm of base 2 is used. The square of the modulus of this transform then represents the power spectrum. This method is much faster computationally and, by the addition of a second FFT, can be made to provide the autocorrelation function.

(ii) Event markers: The EMG can be considered as a summation of a large number of distorted action potentials. It is useful therefore to signify the occurrence of some events to get an impression of the distribution of these action potentials in time. Each marker does not necessarily represent a unique action potential. Some markers will also be missing because of the merging of the coincident or almost coincident peaks from different sources. The occurrence of an event can be specified by a turning point or a zero-crossing. A turning point is defined as the point at which the EMG signal changes its slope from negative to positive and vice versa. An event marker is placed wherever there is a turning point above an acceptable threshold level. The threshold is chosen to be 35μ V above which there is no contribution due to noise alone. A zero crossing is the point at which the EMG signal changes its magnitude from negative to positive and vice versa. The rate at which zero crossings occur is a simple measure of the frequency content of the signal.

(iii) Inter-Spike Interval Analysis: Inter-spike interval (ISI) is defined as the time between two adjacent spikes that exceed a prescribed threshold (35μ V). In this study the ISI is measured and the frequency of occurrence of these intervals is plotted as a histogram. The mean firing rate (FR) defined as the reciprocal of the average ISI, ISI standard deviation, and the coefficient of variation of the ISI are calculated. Normally a motor unit does not discharge at a constant rate. However, the mean firing rate (FR) represents an average value of the neural discharges that control the rate at which a motor unit fires. This technique provides an approximate quantitative description of the rate at which the action potentials are generated.

Computerized EMG in diagnosis

The examination of electromyographic records plays an important role in the diagnostic evaluation of neuromuscular disorders and the prescription of rehabilitation treatment. There have been two major approaches to the quantification of the EMG signal for diagnosis and clinical applications. The first approach is based on decomposing the EMG signal (recorded by needle electrodes) into its motor unit action potentials (MUAPs). Dorfman and McGill (1984) reported that this approach, used by Basmajian (1975), is based on the Fast Fourier Transforms (FFT) in which the EMG signal is decomposed into its frequency components. According to Basmajian (1975), it is possible to show differences between the fourier transform of myopathic, neuropathic, and normal muscles. It was found that there is a statistical difference between a neuropathic and a normal muscle in those frequencies with maximum energy, however there is no statistical difference between myopathic and normal muscle. At the present time, no single method of quantitative EMG signal processing has gained wide acceptance in clinical applications and rehabilitation settings.

This paper describes a number of experiments which were conducted on a pool of disabled individuals to validate the application of the digital signal processing techniques discussed earlier for the quantification of various types of pathological conditions affecting the neuromuscular system. The EMG signals were recorded by means of Beckman surface electrodes. The electrical changes recorded by the electrodes were amplified, band pass filtered and displayed on the screen of a multi-purpose polygraph. The low pass and high pass filters were adjusted to record a frequency band of $0\cdot01$ to $1000\cdot0$ Hz. A notch filter was used to remove the line frequency signal. The recorded signal was digitized using an analogue-to-digital converter, and stored on a disk using a PDP–11/34 Digital computer.

Case study I

The activity of the orbicularis oris muscle was recorded for a 17-year old female patient. The patient had sustained Bell's palsy of the left side. She had received physical therapy treatment for eight months until there was no further improvement in her condition. For comparison, the activity of both sides of the face were recorded during relaxation, moderate contraction (smiling), and full contraction (blowing). Visual inspection of the EMG signals recorded from both sides were very difficult to interpret. As shown in Figure 1, both sides showed EMG activities indicating return of the lost conduction velocity and disappearance of signs of denervation. However, clinically there was still a difference between the two sides of the face. The results of the processing of the EMG signals from both sides are summarized in Table 1.

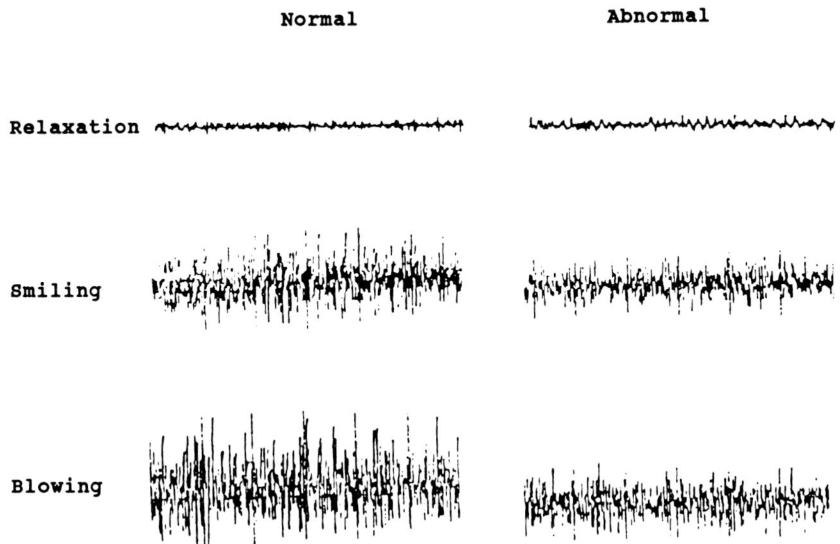

Figure 1.   Recorded EMG for normal and abnormal muscle (Case I)

Table 1   Results of EMG processing for Case I

Parameter	Relaxation		Smiling		Blowing	
	N.	Abn.	N.	Abn.	N.	Abn.
RMS	0·96	1·00	7·77	2·94	8·69	4·22
FWRI	0·80	0·77	6·28	2·38	6·86	3·40
R(0)	46·60	50·68	436·75	170·95	432·50	216·70
NTP	268·00	266·00	301·00	307·00	268·00	266·00
ISI Mean	5·14	5·03	4·33	4·62	5·15	4·96
ISI S.D.	2·87	2·87	2·44	2·62	2·91	2.88

Where: N.   Normal
     Abn. Abnormal

When the patient was asked to contract the muscle by performing acts of smiling (medium contraction) and blowing (strong contraction) the integrated EMG values were observed to increase in both normal and abnormal muscles as shown in Figure 2. However, the values for the decreased muscle were significantly lower than those of the normal side. Similar ratios were obtained for the RMS value of the signal.

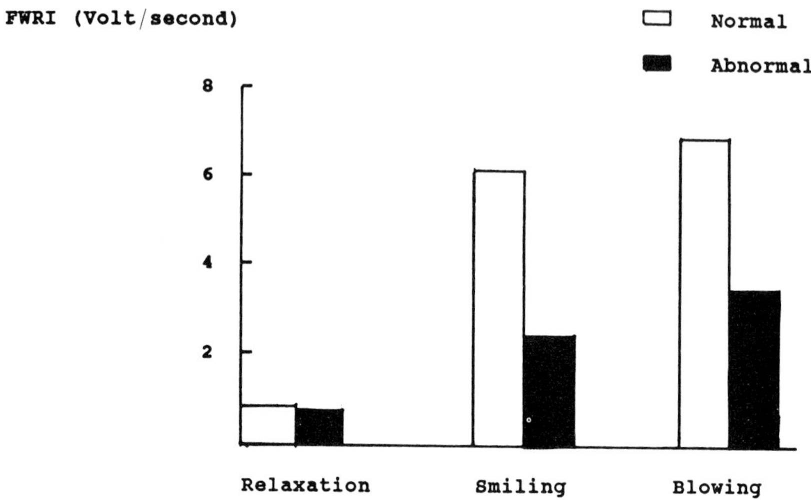

Figure 2. *Changes of FWRI with activity for both normal and abnormal muscle (Case I)*

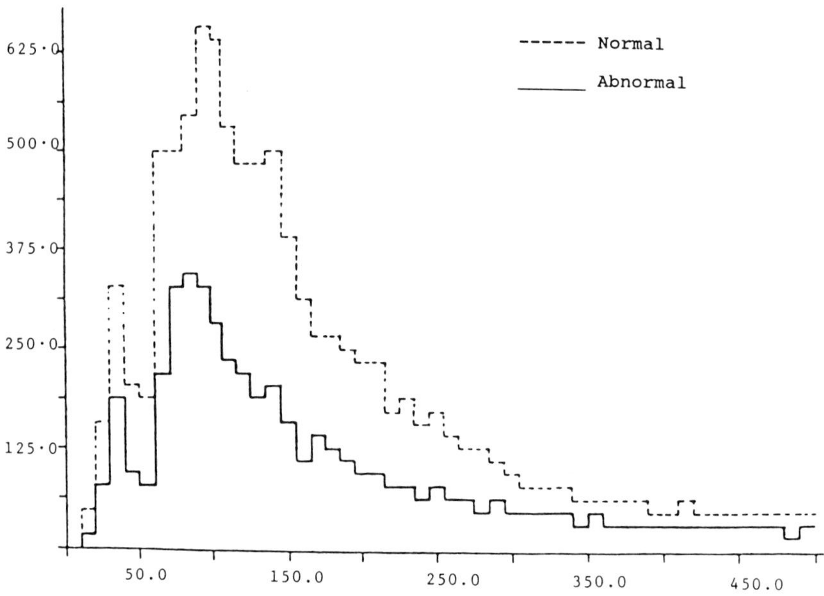

Figure 3. *Frequency profile at blowing for normal and abnormal muscle (Case I)*

The frequency profile during full contraction shows a reduction of average power of about 50 per cent over the entire band as shown in Figure 3. This reduction corresponds to the physician's diagnosis that the patient suffers 40 per cent loss of nerve supply and muscle ability.

This case study shows that quantitative analysis of the EMG signal provides an added dimension to the diagnostic procedure and the determination of the extent of disability. Had the quantitative analysis been conducted before the physical therapy treatment had taken place, the degree of improvement in the patient's condition would have been evaluated.

Case study II

A male patient, aged 60 years, had right hemiplegia (upper motor neurone paralysis) spastic paralysis. Surface electrodes were used to record the activity of both the normal and paralysed biceps muscle. The activity was recorded during relaxation, moderate, and severe contraction. The duration of each record was five seconds. The recorded EMG signals were processed. The averaged results are summarized in Table 2. A comparison between the frequency spectrum of the normal and paralysed muscle at maximum contraction is shown in Figure 4.

Raw EMG signals monitored on the polygraph screen indicated the existence of abnormal activity during relaxation in the paralysed muscle. This indicates that the

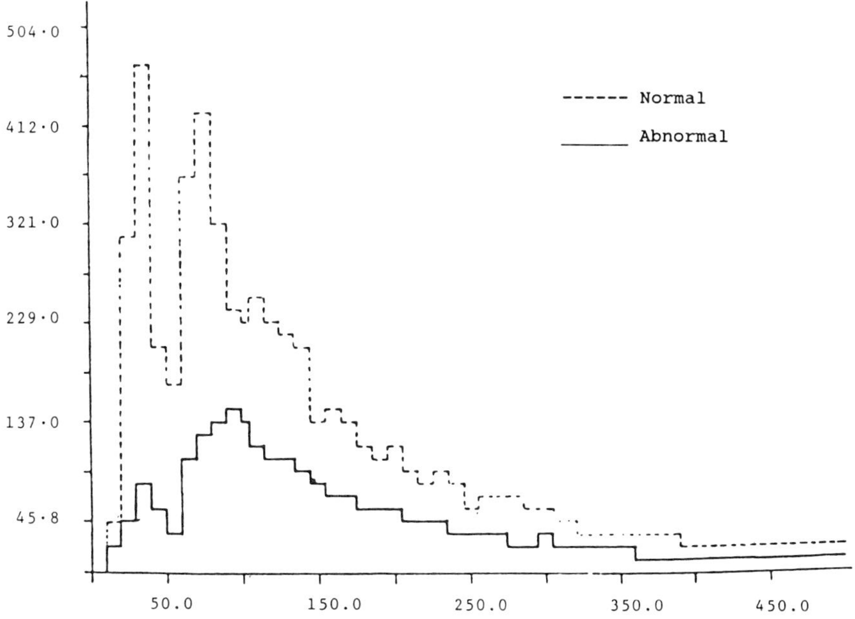

Figure 4. *Frequency profile at full contraction for normal and abnormal muscle (Case II)*

Table 2   Results of EMG processing for Case II.

Parameter	Relaxation		Moderate Contraction		Full Contraction	
	N.	Abn.	N.	Abn.	N.	Abn.
RMS	0·22	0·26	3·40	0·70	5·20	1·89
FWRI	0·17	0·21	2·76	0·56	4·13	1·50
NTP	306·00	275·00	234·00	251·00	241·00	270·00
ISI Mean	4·07	5·39	5·73	5·00	6·00	5·42
ISI S.D.	2·35	3·03	3·20	2·84	3·43	2·92

Where: N.   Normal
        Abn.   Abnormal

muscle is hypertensive. The FWRI and RMS values during relaxation, given in Table 2, reflected the existence of this hypertension. A study of the frequency spectrum in the 50–100 Hz range, where the average power is concentrated, reveals a reduction of about 65% from the normal. This could be attributed to a reduction in the number of active motor units.

Quantitative results were replayed to the physician to assist him in making a determination of the extent of disability in order to make an objective decision related to the disability management and treatment. The data coincided with the physician's clinical diagnosis.

Conclusions

Processed surface EMG records can provide the physican with quantitative information not normally available through visual examination of the EMG signals. Case studies indicated that several of the proposed processing techniques have a good value in describing the nature and extent of the musculoskeletal and neuromuscular abnormality.

Quantitative results, obtained via the computer analysis, in most cases matched the physician's clinical determinations and diagnosis. In fact, the analysis did provide the physician with an added dimension that proved to be helpful in decisions related to management of the disability. Measurement of the degree of patient improvement after a course of treatment is also possible using the described techniques.

References

Basmajian, J.V., 1975, *Computers in Electromyography* (London and Boston: Butterworth).
Basmajian, J.V., 1979, *Muscles Alive*, 4th edition, (Baltimore: The Williams and Wilkins Co.).
Basmajian, J.V. and Cross, G.L., 1971, Duration of Motor Unit Potentials from Fine-wire Electrodes, *American Journal of Physical Medicine*, **50**, 144–148.
Buchtal, F., Guid, C. and Rosenflack, P., 1954, Action Potential Parameters in Normal Human Muscle and their Dependence on Physical Variables, *Acta Physiol. Scandinav.*, **32**, 200–218.

Buchtal, F., Guid, C. and Rosenflack, P., 1957, Multielectrode Electrode Study of the Territory of a Motor Unit, *Acta Physiol. Scandinav.*, **32**, 200–218.

Dorfman, L.J. and McGill, K.C., 1984, Automatic Decomposition of the Clinical Electromyogram, *Technical Report of Rehabilitation Research and Development Center.*

Inbar, G.F. and Nourjaim, A.E., 1984, On Surface EMG Spectral Characterization & Its Applications to Diagnostic Classification, *IEEE Transactions on Biomedical Engineering*, **31**, 597–604.

Khalil, T.M., Banna, M., Taher, M. and Waly, S., 1982, Clinical Applications of Computerized EMG Signal Processing, *Egyptian Journal of Biomedical Engineering*, **3**, 15–28.

Khalil, T.M., Waly, S.M. and Taher, M.F., 1984, Quantitative Evaluation of Muscular Abilities, *Proceedings of the 37th Annual Conference on Engineering in Medicine and Biology.*

Khalil, T.M., Moty, E.A. and Waly, S.M., 1985, Computerized Signal Processing in the Assessment of Muscular Fatigue, *Computers and Industrial Engineering Journal*, **9**, Supplement 1.

Lindstrom, L., 1970, On the frequency spectrum of EMG signals, *Technical Report*, Research lab. of medical elec., Chalmers University of Technology, Sweden.

Marinacci, A.A., 1968, *Applied Electromyography*, (Philadelphia: Lea and Febiger).

National Institute of Occupational Safety and Health, 1981, *Work Practices Guide for Manual Lifting*, Technical Report No. 81–122 U.S. Dept. of Health and Human Services (NIOSH), Cincinnati, Ohio.

Waly, S.M. and Khalil, T.M., 1985, Computerized EMG in Clinical Settings, *Proceedings of the 9th Annual Meeting of the American Society of Biomechanics.*

Waly, S.M., Khalil, T.M. and Asfour, S.S., 1986, Physiological Basis of Muscular Fatigue: An Electromyographical Study. In *Trends in Ergonomics/Human Factors III*, edited by W. Karwowski, (Amsterdam: Elsevier) pp.751–758.

13.

Preventative research – An effective therapy

S. Kumar

Department of Physical Therapy
University of Alberta
Edmonton, Alberta
Canada

Abstract

Electromyographic activity of erectores spinae and external obliques, intra-abdominal pressure and heart rate were monitored during weight lifting. Thirty-two male volunteer subjects (age group 18–33 years, with healthy backs) lifted bar-bell weights from the ground to a table placed in front. The load was increased from an initial 10 kg to a maximum of 55 kg in steps of 5 kg. Temporal translation of traces and numerical extraction of the data were carried out. These values were then subjected to correlation and regression analysis. The myoelectric activity of the erectores spinae (ES), external oblique (EO), and intra-abdominal pressure were strongly correlated with the load of lift ($r = 0 \cdot 9$; $P < 0 \cdot 001$). The intra-abdominal pressure was also strongly and significantly correlated with the electromyographic activities of erectores spinae ($r = 0 \cdot 9$; $P < 0 \cdot 001$) and external oblique ($r = 0 \cdot 83$ to $0 \cdot 93$; $P < 0 \cdot 01$). EMG of ES and EO, and intra-abdominal pressure regressed significantly on load of the lift ($P < 0 \cdot 01$).

Introduction

The idea of prevention may be an old one, but it is certainly not stale. In fact, its significance is increasing for two major reasons:

1. In general, with a progressive change in lifestyle towards more mechanized and efficient methods, we are creating new hazards.
2. The cost of treatment of any given health problem is escalating, and the value of the dollar is steadily declining due to inflation.

In this socio-economic context, a preventative attitude becomes more meaningful than ever.

An inevitability of life is ageing and the normal consequences of ageing; disability and death. However, advances in medical sciences have prolonged life-span and delayed the ageing process. Delaying the disabilities associated with ageing and degeneration is clearly the imminent challenge.

183

Back injury is a major cause of disability both at work and in the home. The 67th annual report of the Workers' Compensation Board of Alberta for the year 1984 shows the operating transaction costs for the years 1984 and 1983 were \$233·5 million and \$250·5 million, respectively. For those years, new back injury claims were 42·72 and 45·42 per cent of total new claims. On the assumption of equal cost per claim, regardless of the body parts, the cost of these claims can be calculated to be \$99·7 million and \$113·7 million for the years 1984 and 1983. It must be noted, however, that back injury claims are generally much higher. In the United States of America, the annual cumulative cost attributed to back pain was calculated to be \$15·87 billion by Holbrook, *et al.*, (1984). Although these injuries are associated with the processes of ageing and degeneration, they are usually precipitated under stress (Magora, 1970; Andersson, 1981: Park and Chaffin, 1974). The process and function of ageing in relation to the intervertebral disc is obvious.

Metabolically, the spine is an inert organ. It begins to degenerate from the early twenties (Armstrong, 1965). The nucleus pulposus, which is fluid and hydrostatic in the beginning, gradually loses its water content, becoming plastic and amorphous and causing various kinds of mechanical derangements (Nachemson, 1960). This, in turn, results in unequal stress being applied to different regions of the spinal unit (Nachemson, 1960).

Mechanically deranged discs are the most liable to sustain injury when exposed to hazardous circumstances (Figure 1). Once the injury is sustained, the course of events may take one of the three possible courses (as illustrated in Figures 2, 3 and 4). However, if sufficient information about preventative procedures, based on sound scientific knowledge is available, many injuries can be prevented or at least postponed (Figure 5).

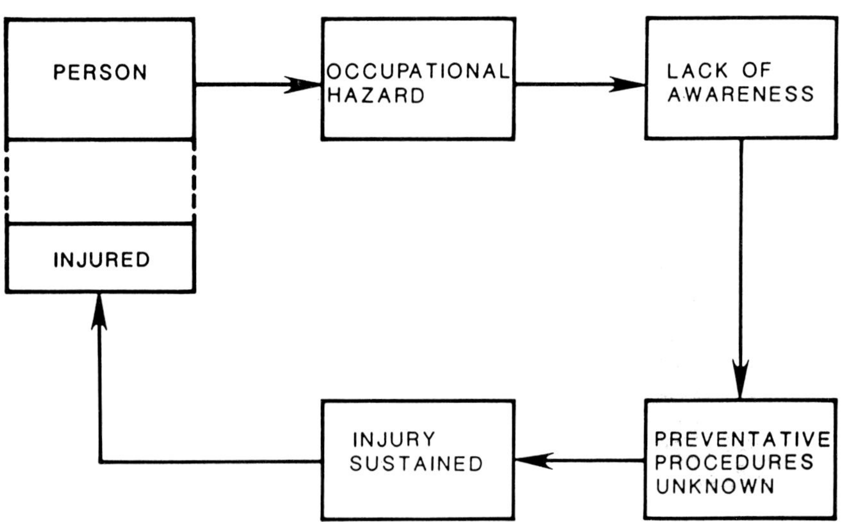

Figure 1.   Accident Loop

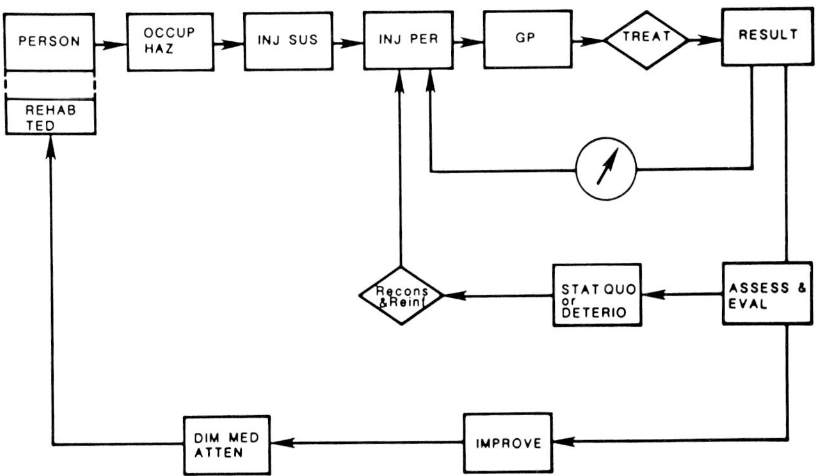

Figure 2.   *Remedial Loop 1*   *Remedy for general care*
            Abbreviations:
            OCCUP HAZ   *Occupational Hazard*
            INJ SUS   *Injury sustained*
            INJ PER   *Injured person*
            G P   *General practitioner*
            ASSES & EVAL   *Assessment and Evaluation*
            STAT QUO or DETERIO   *Status quo or deterioration*
            RECONS & REINF   *Reconsideration and reinforcement*
            DIM MED ATTEN   *Diminishing medical attention*

Since spinal injuries are generally stress precipitated, the larger stresses are more hazardous. Thus, knowledge of the mechanics of stress generation and the pattern of normal physiological response is necessary before any preventative procedure can be established. Major causes of back disorders are weight-lifting and manual material-handling (Hult, 1954; White, 1966; Chaffin and Park, 1973; Magora. 1974; Brown, 1973 and 1975; Kelsey and White, 1980; Andersson, 1981) and it is important to understand the mechanics of stress generation and the physiological responses in these activities among normals.

Despite several studies on weight-lifting (Andersson *et al.*, 1977a; Asmussen *et al.*, 1965; Bartelink, 1957; Davis, 1959; Davis and Stubbs, 1977a, 1977b, and 1978; Floyd and Silver, 1955; Morris *et al.*, 1961), and weight-holding (Asmussen and Poulsen, 1968; Bradford and Spurling, 1941 and 1945; Eie and Wehn, 1962; Morris *et al.*, 1961; and others), a picture of the inter-relationship between the various physiological and biomechanical parameters in these physical activities is still incomplete. No attempt had been made to relate quantitatively the activity of various parameters in a concerted effort until Kumar (1971), Kumar and Davis (1976), Andersson *et al.*, (1977a) and (1977b).

Mechanical stress is accepted as a prevalent precipitating factor of back injuries (Brown, 1973 and 1975; Chaffin and Park, 1973; Farfan, 1973; Nachemson, 1971) and

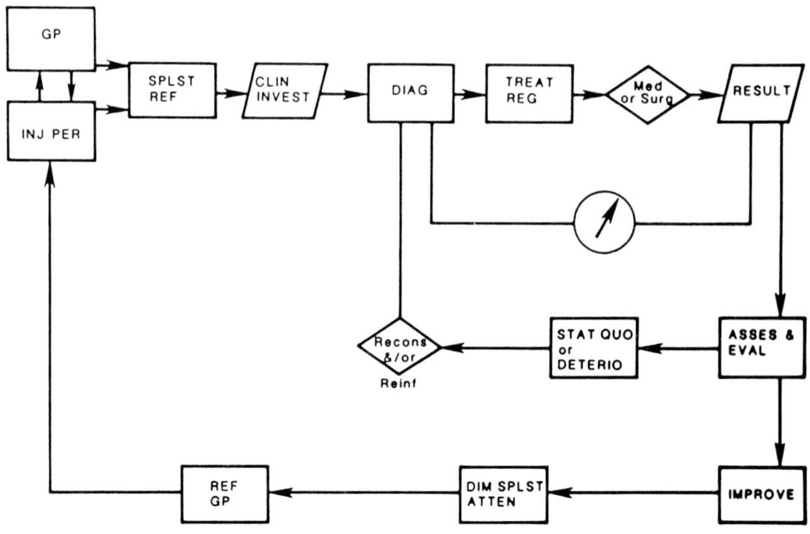

Figure 3	*Remedial Loop 2*	*Remedy for acute care*
	Abbreviations:	
	G P	*General practitioner*
	INJ PER	*Injured person*
	SPL REF	*Specialist referral*
	CLIN INVEST	*Clinical investigation*
	DIAG	*Diagnosis*
	TREAT REG	*Treatment regimes*
	MED OR SURG	*Medical or surgical*
	ASSES & EVAL	*Assessment and evaluation*
	STAT QUO OR DETERIO	*Status quo or deterioration*
	RECONS &/OR REINF	*Reconsideration and/or reinforcement*
	DIM SPLST ATTEN	*Diminishing medical attention*
	REF G P	*Referred back to general practitioner*

compression does elicit mechanical derangement, resulting in back pain or injury. In static postures, the relationship between load and intradiscal pressure, intra-abdominal pressure, and myoelectric activity of back muscles was studied in four subjects by Andersson, (1977b). However, compressive forces of higher magnitude and variable pattern are frequently generated during the process of stoop-lifting. This activity also places a physiological demand. More quantitative information on this dynamic manoeuvre is needed. In this study, myoelectric activity of erectores spinae, external obliques, intra-abdominal pressure and heart rate have been monitored during weight-lifting with successively increasing weights. A proportional quantification of these parameters has been obtained and a statistical analysis performed to synthesize a functional model of lifting activities.

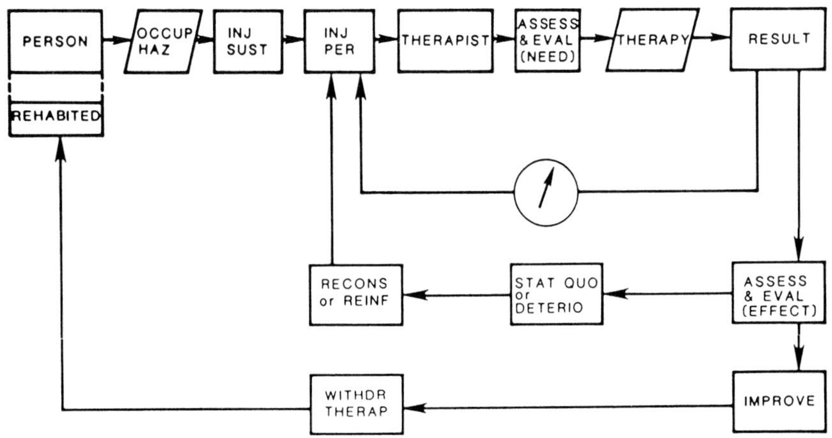

Figure 4   Remedial Loop 3   Remedy by rehabilitation
        Abbreviations:

OCCUP HAZ	*Occupational hazard*
INJ SUS	*Injury sustained*
INJ PER	*Injured person*
ASSESS & EVAL (NEED)	*Assessment and evaluation of need*
ASSESS & EVAL (EFFECT)	*Assessment and evaluation of effectiveness*
STAT QUO OR DETERIO	*Status quo or deterioration*
RECONS OR REINF	*Reconsideration or reinforcement*
WITHDR THERP	*Withdrawal from therapist*
REHABITED	*Rehabilitated*

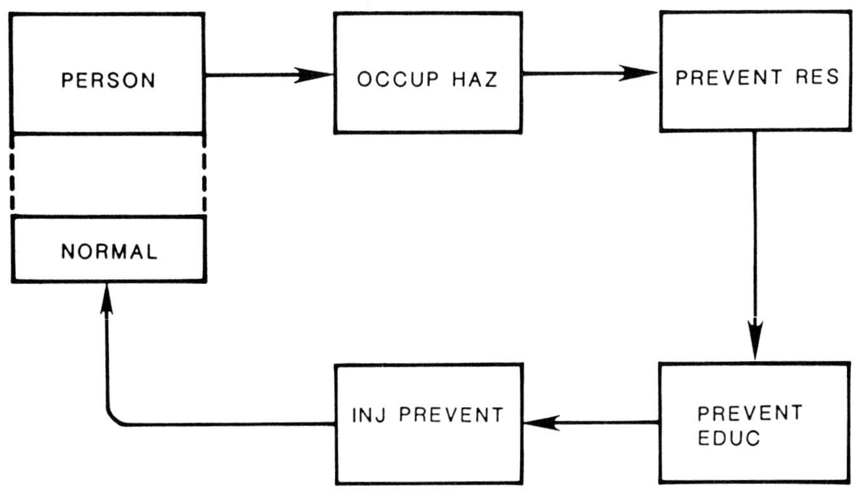

Figure 5   Preventative Loop
        Abbreviations:

OCCUP HAZ	*Occupational hazard*
PREVENT EDUC	*Preventative education*
PREVENT RES	*Preventative research*
INJ PREVENT	*Injury prevented*

Materials and Methods

Thirty-two male volunteers (age group 18 to 33 years) participated, none having a history of either spinal disorder or muscular disease. Bar and bell weights were used and the load was increased from an initial load of 10 kg up to a maximum of 55 kg, in steps of 5 kg.

Task

The cycle of exercise was divided into two phases, "lift-up" and "lift-down". In "lift-up", the subjects stooped down with straight knees and bent back from a standing erect posture. The weight was picked up and placed on a table in front and the original posture resumed. In "lift-down", the weight was lifted from the table and placed on the ground while keeping the knees straight and bending the back. Following weight placement, an erect posture was resumed.

The events of "lift-off" and "touch-down" were registered by a make-and-break circuit. The electromyographic activity of erectores spinae (L_{2-3} levels) and external oblique muscles were recorded by surface electrodes. Heart rate was recorded by electrodes placed on the sternal manubrium and the sixth left intercostal space. The intra-abdominal pressure was monitored using radio-telemetry (Watson *et al.* 1962). The signals were picked up by omnidirectional loop antennae and were received by a radio receiver through an aerial switching unit. A Devices M_4 chart recorder was used to record the traces.

Temporal analysis of the traces obtained was carried out to obtain numerical values for the parameters studied (Kumar, 1971). These values were then subjected to statistical analysis. Simple correlation coefficients were calculated for various parameters in the same phase, and also between "lift-up" and "lift-down" phases. Partial correlation coefficients were also calculated.

Results

Pattern inter-relationship

Electromyography

a) *Erectores spinae:*   Flexion relaxation was maintained in 27 subjects until after the weight had left the ground. In 5 other subjects, a plateau of activity set in before the weight left the ground and further increases occurred during the process of lifting. In both groups, once the weight had left the ground, activity of the muscle started at a low level followed by a sharp increase associated with rapid extension of the lumbar spine. After delivery of the weight to the table, the peak of the erectores spinae suddenly declined and only a low level of activity was left. This was continued in the extension of the trunk from the forward inclined position. In the "lift-down" phase, the "lift-up" was associated with a high peak of activity which quickly declined and settled to a lower value.

b) *External obliques*:   A short burst of activity prior to "lift-up" commonly occurred. This followed the burst of erectores spinae with a time lapse of 0·4 to 1 second. The amplitude of this activity was related to erectores spinae activity. Vigorous activity occurred either coincident with or between 0·2 to 0·4 second before "lift-up". The peak was concomitant with that of erectores spinae. As the moment of greatest mechanical disadvantage passed, the activity dropped to a sustained level. A second peak of activity occurred, which disappeared completely on weight placement, matching exactly forward displacement of weight on the table. Once the stress was withdrawn, the activity disappeared. In the "lift-down", a sharp but short peak occurred during the "lift-up". Otherwise, a low level was maintained until weight placement.

Heart rate

Initial stooping of the trunk was associated with a marked bradycardia in nine subjects. Subsequent rise in the heart rate was delayed until towards or after completion of the operation. In those nine cases, depression of heart rate was of short duration and was followed by a prompt rise. In twenty-three cases, it was spread out with time. In most subjects, a clearer pattern emerged after a few lifts with two peak heart rates becoming evident relating to the two moves of the cycle.

Intra-abdominal pressure

In the standing posture, the intra-abdominal pressure fluctuated cyclically about a mean value, coincident with the respiratory cycle. The maximum peak-to-peak amplitude of this pressure variation was 4 mm Hg. Inspiration caused a trough of low pressure, whereas expiration was associated with a rise. All measurements were made from the mean of these two levels which has been termed "the resting pressure". The cyclic respiratory pressure changes about this resting pressure became more established

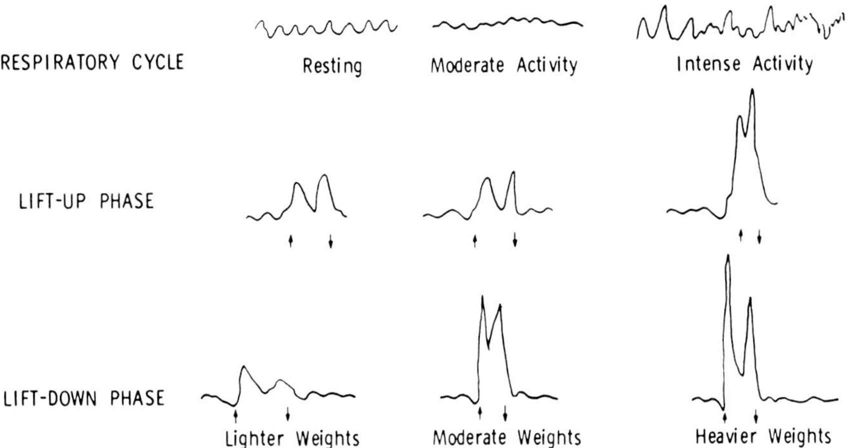

Figure 6   Pattern of intra-abdominal pressure during activities

and pronounced when the subject had been either sitting or standing in a relaxed posture for some time. With intense activity, the cycle was distorted (Figure 6).

"Lift-up" phase

With lower weights, the cyclic respiratory pressure changes continued until just before the start of the lift. However, with moderately heavier weights, a trough set in before the lift but the pressure was restored to the resting pressure just as the lift began. The timing of the onset of this negative pressure was variable. Sometimes it coincided with, and at other times it followed, the outburst of electromyographic activity of erectores spinae. With increase in weights, the magnitude of the negative pressure progressively increased and in seven cases it reached below $-5\,mm\,Hg$. As the time of lift approached, the pressure returned to the resting level, and then rose to exceed this as a small peak or a sustained plateau before "lift-up". The magnitude of this small peak or sustained plateau was similar to the magnitude of the trough.

The intra-abdominal pressure began to rise at the moment of 'lift-up" with smaller weights. With the heavier weights, the onset of rise came from $0\cdot2$ to $0\cdot4$ second earlier than the "lift-up". However, in eighteen lifts of six subjects, the displacement was as much as $0\cdot8$ second. The pattern of time displacement was not found to be characteristic of individuals. The rising intra-abdominal pressure resulted in a peak either coincident with "lift-up" or up to $0\cdot2$ second later.

Following "lift-up", the pressure declined into a sustained plateau until the process of placement of the weight on the table began. Between $0\cdot2$ to $0\cdot4$ second before touchdown on the table, pressure had reached its second peak. With increasing weights, the peaks became steeper and approached the magnitude of the "lift-up" pressure peaks. Within the range of 25 to 40 kg, the second pressure peak markedly exceeded the first one. After touchdown on the table, the pressure peak suddenly declined to the resting level with light weights. It reached a value slightly above the resting pressure after handling heavier weights. In the latter case, the pressure gradually dropped to the resting level and the respiratory cycle was restored.

"Lift-down" phase

At the initiation of the "lift-down" phase, in most cases, the trough of negative pressure was continued for up to $0\cdot2$ second before "lift-up". A peak of intra-abdominal pressure occurred at "lift-up", which increased with increasing weights, and pressure then suddenly dropped to a low sustained level. A final small peak occurred at touchdown, after which the pressure declined steadily to reach the resting level.

Magnitude inter-relationship

The values of the results obtained suggest strongly that there is a close relationship between the magnitude of the weight lifted and that of the other parameters measured.

Table 1 Simple Correlation Coecients:

				Wt	ES	EO	IAP	HR
A.	Peak	Up	Wt	1·000	0·989***	0·955***	0·967***	0·830***
			ES		1·000	0·918***	0·989***	0·832***
			EO			1·000	0·881***	0·760* +
			IAP				1·000	0·849**
			HR					1·000
		Down	Wt	1·000	0·990***	0·943***	0·963***	0·952***
			ES		1·000	0·924***	0·976***	0·969***
			EO			1·000	0·857***	0·822**
			IAP				1·000	0·869***
			HR					1·000
B.	Plateau	Up	Wt	1·000	0·943***	0·933***	0·985***	0·830**
			ES		1·000	0·807**	0·979***	0·844**
			EO			1·000	0·888***	0·709*
			IAP				1·000	0·983***
			HR					1·000
		Down	Wt	1·000	0·987***	0·967***	0·992***	0·952**
			ES		1·000	0·936**	0·989***	0·977**
			EO			1·000	0·939**	0·874*
			IAP				1·000	0·975**
			HR					1·000
C.	Average	Up	Wt	1·000	0·974***	0·921***	0·977***	0·830**
			ES		1·000	0·856**	0·990***	0·835**
			EO			1·000	0·850**	0·679*
			IAP				1·000	0·855**
			HR					1·000
		Down	Wt	1·000	0·991***	0·911***	0·981***	0·952***
			ES		1·000	0·866**	0·992***	0·976***
			EO			1·000	0·830**	0·759* +
			IAP				1·000	0·984***
			HR					1·000

Level of significance ***$P < 0.001$, **$P < 0.01$, *+$P < 0.02$, *$P < 0.05$

Simple and partial correlation coefficients and these variables are presented in Tables 1 and 2, respectively.

All simple regressions calculated (Table 3) for peak, plateau and average activity, during up and down lifts of each relevant variable on the other, were significant ($P < 0.05$).

A comparison of the gradients of the regression lines (*t*-Test) for upward and downward lifts did not show a significant difference. The linear regression lines were, therefore, parallel. The actual values of the various parameters recorded during the experiment, however, show that the values for the downward lift were always smaller than the upward lift, except in the case of change in heart rate. The comparison of the intercepts using *t*-Test revealed an insignificant difference between the up and down

Table 2. Partial Correlation Coefficients:

	Variables	Peak	Plateau	Average
A.	**Between weight and erectores spinae**			
Up	$r_{12 \cdot 3}$	$0 \cdot 96^{***}$	$0 \cdot 89^{**}$	$0 \cdot 93^{***}$
	$r_{12 \cdot 4}$	$0 \cdot 89^{**}$	NS	NS
	$r_{12 \cdot 34}$	$0 \cdot 83^{*+}$	NS	NS
Down	$r_{12 \cdot 3}$	$0 \cdot 94^{***}$	$0 \cdot 93^{***}$	$0 \cdot 98^{***}$
	$r_{12 \cdot 4}$	$0 \cdot 86^{**}$	NS	$0 \cdot 77^{*+}$
	$r_{12 \cdot 34}$	NS	NS	NS
B.	**Between weight and external oblique**			
Up	$r_{13 \cdot 2}$	$0 \cdot 82^{**}$	$0 \cdot 87^{**}$	$0 \cdot 75^{*+}$
	$r_{13 \cdot 4}$	$0 \cdot 85^{**}$	NS	$0 \cdot 82^{**}$
	$r_{13 \cdot 24}$	$0 \cdot 76^{*}$	NS	$0 \cdot 81^{*+}$
Down	$r_{13 \cdot 2}$	NS	$0 \cdot 77^{*+}$	$0 \cdot 79^{*+}$
	$r_{13 \cdot 4}$	$0 \cdot 85^{**}$	$0 \cdot 88^{**}$	$0 \cdot 89^{**}$
	$r_{13 \cdot 24}$	NS	$0 \cdot 83^{**}$	$0 \cdot 81^{*+}$
C.	**Between weight and intra-abdominal pressure**			
Up	$r_{14 \cdot 2}$	NS	$0 \cdot 90^{***}$	NS
	$r_{14 \cdot 3}$	$0 \cdot 89^{***}$	$0 \cdot 94^{***}$	$0 \cdot 96^{***}$
	$r_{14 \cdot 23}$	NS	NS	NS
Down	$r_{14 \cdot 2}$	NS	NS	NS
	$r_{14 \cdot 3}$	$0 \cdot 90^{***}$	$0 \cdot 96^{***}$	$0 \cdot 97^{***}$
	$r_{14 \cdot 23}$	NS	$0 \cdot 77^{*}$	NS
D.	**Between erectores spinae and intra-abdominal pressure**			
Up	$r_{24 \cdot 1}$	$0 \cdot 88^{**}$	$0 \cdot 86^{**}$	$0 \cdot 80^{*}$
	$r_{24 \cdot 3}$	$0 \cdot 96^{***}$	$0 \cdot 96^{***}$	$0 \cdot 96^{***}$
	$r_{24 \cdot 13}$	$0 \cdot 83^{*+}$	$0 \cdot 81^{*+}$	$0 \cdot 74^{*}$
Down	$r_{24 \cdot 1}$	NS	NS	$0 \cdot 78^{*}$
	$r_{24 \cdot 2}$	$0 \cdot 93^{***}$	$0 \cdot 92^{***}$	$0 \cdot 98^{***}$
	$r_{24 \cdot 13}$	NS	NS	NS

Variables   1–Weight, 2–Erectores spinae, 3–External oblique, 4–Intra-abdominal pressure
Significance $^{***}P < 0 \cdot 001$; $^{**}P < 0 \cdot 01$; $^{*+}P < 0 \cdot 02$: $^{*}P < 0 \cdot 05$, NS – not significant

movements for the following variables: external oblique on weight, and external oblique on intra-abdominal pressure. The regression lines, therefore, may be considered coincidental. The intercept of the regression line for the erectores spinae on weight was significantly different in the up and down movements $(P < 0 \cdot 01)$. Thus these lines, though parallel, were significantly different.

Discussion

Pattern inter-relationship

The general pattern of electromyographic activity of the erectores spinae in relation to trunk movement in the present study has been in agreement with previous similar

Table 3. Regression chart: Critical value for f = 6·03

Intercept, gradient and variance ratio

		UP			DOWN		
		a	*b*	*f*	*a*	*b*	*f*
Erectores Spinae on weight	Peak	37·9	0·9	393·2	27·0	1·0	414·7
	Plateau	22·5	0·5	64·8	12·9	0·5	305·0
	Average	29·6	0·7	148·5	19·7	0·7	461·7
External oblique on weight	Peak	−0·3	0·5	82·6	1·1	0·4	64·8
	Plateau	3·1	0·1	53·6	3·0	0·1	116·0
	Average	1·2	0·3	44·9	1·2	0·3	38·9
IAP on weight	Peak	7·4	1·2	114·6	4·2	1·2	102·9
	Plateau	0·6	0·6	259·4	−0·7	0·6	502·6
	Average	3·7	0·9	167·1	1·4	0·9	193·8
Heart rate on weight	Peak	10·2	0·2	17·7	10·3	0·3	78·0
External oblique on erectores spinae	Peak	−18·7	0·5	43·0	−8·8	0·3	46·5
	Plateau	−1·3	0·2	14·9	−0·3	0·2	56·4
	Average	−9·3	0·4	21·5	−5·4	0·3	24·1
IAP on erectores spinae	Peak	−39·8	1.2	361·2	−26·6	1·1	158·6
	Plateau	−25·7	1·2	181·5	−15·7	1·1	389·6
	Average	−33·3	1·2	399·1	−2·5	1·1	483·3
Heart rate on erectores spinae	Peak	3·1	0·1	18·0	1·2	0·3	125·5
	Plateau	2·3	0·3	20·4	1·3	0·7	167·0
	Average	2·9	0·2	18·4	1·3	0·4	163·3
IAP on external oblique	Peak	13·2	1·9	27·6	7·7	2·3	22·1
	Plateau	−7·9	3·8	29·8	−10·6	3·8	59·6
	Average	6·8	2·3	20·8	5·3	2·3	17·7
Heart rate on external oblique	Peak	11·1	0·3	10·9	11·7	0·6	16·7
	Plateau	8·2	1·0	8·3	5·3	2·1	25·8
	Average	11·1	0·4	6·8	12·4	0·8	10·9
IAP on heart rate	Peak	−29·6	4·6	20·6	−28·9	3·3	236·8
	Plateau	−19·9	2·5	24·7	−16·6	1·6	155·2
	Average	−25·0	3·6	21·7	−23·1	2·5	248·8

studies with continuous motion of the trunk (Floyd and Silver, 1951 and 1955; Morris *et al.*, 1961; Morris *et al.*, 1962). In static posture studies (Asmussen *et al.*, 1965; Andersson *et al.*, 1977a; and Andersson *et al.*, 1977b), electromyographic activity of the erectores spinae progressively increased with increased flexion of the trunk. These studies also reported an increased response with higher loads. The pattern of response variation in the current study demonstrated much higher magnitude of activity at greater flexion, with higher weights and, also, with increasing leverage of the weight. The magnitude of the response progressively declined with reduction in the flexion. Thus, the double peak during ''lift-up'' phase was associated with initial extension during lifting and controlled flexion during the placement of weight on the table. During the ''lift-down'' phase of the cycle, the electromyographic activity of erectores spinae was vigorous when picking the weight from the table, due to considerable mechanical disadvantage. It was also large during the delivery phase which involved control of the progressive flexion.

A brief discharge of action potentials from the external obliques, following activity of erectores spinae in most of the subjects during stooping, suggests that the external obliques assist in the control of flexion. However, the massive activity of external obliques, either before or coincident with the event of weight-lifting with a magnitude coinciding and declining with that of the erectores spinae at the moment of maximum stress. The second peak of external obliques during the lift-up phase matched exactly with the carrying of the weight forward to place it on the table. Since the external oblique muscles are both flexors and rotators of the trunk, they may well play a significant synergistic role in these activities. Floyd and Silver (1950) and Bearn (1961) stated that for raising the intra-abdominal pressure, transversus abdominis was the primary muscle.

Such a synergistic role of the external oblique for a large erectores spinae contraction seems to be needed to overcome the inertia of the weight when lifting at a distance from the trunk in a forward inclined posture.

Immediately before the ''lift-up'' phase, with successively increasing weights, an increasing magnitude of negative intra-abdominal pressure was associated with deeper breath-taking and breath-holding in preparation for the lift. These deep breaths seemed to help generate greater intra-abdominal pressure to endure severe spinal compression by providing enhanced stability. The earlier onset of the pressure rise with the heavier weights prior to ''lift-up'' may be associated with a build-up of momentum to accomplish the task. The increasing task of overcoming the inertia of heavier weights and accelerating them required an increasing extensor effect, which was matched by the increasing intra-abdominal pressure. A higher value for the second peak during the ''lift-up'' phase seems to have been associated with the flexor torque exerted by a large weight held in front of the body at nearly full arm's length. This was greater than that exerted by the same weight in the original stooped lifting position. The pattern of intra-abdominal pressure variation during various phases of the activity is shown in Figure 6.

Magnitude inter-relationship

A very high degree of simple correlation between the independent and all the dependent variables suggests that all physiological and biomechanical parameters are strongly affected by the severity of the task. Such a correlation has been reported by previous studies in this and other activities (Kumar, 1971; Andersson *et al.*, 1974; Andersson and Ortengren, 1974; Kumar and Davis, 1976; Andersson *et al.*, 1977a and 1977b). Based on such linear relationship between the trunk stresses and intra-abdominal pressure, Davis and Stubbs (1977a) have determined safe levels of manual forces for young males. The current study also demonstrates a good correlation between the dependent variables (Table 1).

The significant partial correlations between the magnitude of the weight and external oblique activity suggest that the muscles of the abdominal wall are concerned with control of intra-abdominal pressure in this activity and contribute to this. In the "lift-down" phase, the partial correlation is significant for the plateau and average values. Thus, the raised intra-abdominal pressure observed at the beginning of this movement does not receive a proportional contribution from the external obliques, so there is clearly a dissociated activity of the abdominal wall muscles. During the deceleration of the weight in the plateau phase, the external oblique muscles seemed to contribute to abdominal pressure maintenance, and were thus influenced by the magnitude of the weight.

The partial correlation between the weight and intra-abdominal pressure is highly significant. Such a situation is suggestive of a controlling influence of the erectores spinae on the generation of intra-abdominal pressure, and adds support to the hypothesis concerning erectores spinae stretch receptors playing a significant role in control of intra-abdominal pressure (Kumar and Davis, 1973).

Heart rate, however, increases gradually and reaches its peak at the end of the activity, thus the total exercise must be regarded as cumulative, as far as cardiac demand is concerned.

Conclusion

Electromyographic activity of erectores spinae, external obliques, intra-abdominal pressure and heart rate was studied in 32 male volunteers during stop lifting. These variables were highly correlated ($P < 0·01$).

Without exception, all the simple regressions calculated for peak, plateau and average of lift-up and lift-down phases of each variable on the other were significant ($P < 0·01$).

This suggests a strong and dependable relationship between the external load, erectores spinae activity, external oblique and intra-abdominal pressure. Any of these variables may have the potential of being used as an index to determine the stress on the human back. An establishment of a quantitative relationship between these potential indices and the stress, on one hand, and stress parameters and back pain/injury, on the other, will go a long way in assisting the cause of prevention. Prevention clearly has merit for the individual as well as society. A very small fraction of the cumulative annual cost of back injuries in the United States alone could help to bridge the gaps identified.

References

Andersson, G.B.J., 1981, Epidemiological aspects on low-back pain in industry. *Spine*, **6(1)**, 53–60.

Andersson, G.B.J. and Ortengren, R., 1974, Myoelectric back muscle activity during sitting. *Scandinavian Journal of Rehabilitation Medicine*, Supplement 3, 73–90.

Andersson, G.B.J., Johnson, B. and Ortengren, R., 1974, Myoelectric activity in individual lumbar erector spinae muscles during sitting. *Scandinavian Journal of Rehabilitation Medicine*, Supplement 3, 91–108.

Andersson, G.B.J., Ortengen, R. and Herberts, P., 1977a, Quantitative electromyographic studies of back muscle activity related to posture and loading. *Orthopaedic Clinics of North America*, **8**, 85–96.

Andersson, G.B.J., Ortengren, R. and Nachemson, A., 1977b, Intradiskal pressure, intra-abdominal pressure and myoelectric back muscle activity related to posture and loading. *Clinical Orthopedics and Related Research*, **129**, 156–164.

Armstrong, J.R., 1965, *Lumbar Disc Lesions* (Edinburgh, London: E. S. Livingstone Ltd.).

Asmussen, E., Poulsen, E. and Rasmussen, B., 1965, Qualitative evaluation of the activity of the back muscles in lifting. *Danish National Association of Infantile Paralysis*, Communication number 21.

Asmussen, E. and Poulsen, E., 1968, On the role of the intra-abdominal pressure in relieving the back muscles while holding weights in a forward inclined position. *Danish National Association of Infantile Paralysis*, Communication number 28.

Bartelink, D.L., 1957, The role of abdominal pressure in relieving the pressure on the lumbar intervertebtal discs. *Journal of Bone & Joint Surgery*, 718–725.

Bearn, J.G., 1961, The significance of the activity of the abdominal muscles in weight lifting. *Acta Anatomica*, **45**, 83–89.

Bradford, F.K. and Spurling, R.G., 1941 and 1945, *The Intervertebral Disc* (Springfield, Il: C. C. Thomas).

Brown, J.R., 1973, Lifting as an industrial hazard. *American Industrial Hygiene Association Journal*, **34**, 292–297.

Brown, J.R., 1975, Factors contributing to the development of low back pain in industrial workers. *American Industrial Hygiene Association Journal*, **36**, 26–31.

Chaffin, D.B. and Park, K.S., 1973, A longitudinal study of low back pain as associated with occupational weight lifting factors. *American Industrial Hygiene Association Journal*, **34**, 513–525.

Davis, P.R., 1959, The causation of herniae. *Lancet*, **2**, 155–157.

Davis, P.R. and Stubbs, D.A., 1977a, Safe levels of manual forces for young males (1). *Applied Ergonomics*, **8**, 141–150.

Davis, P.R. and Stubbs, D.A., 1977b, Safe levels of manual forces for young males (2). *Applied Ergonomics*, **8**, 219–228.

Davis, P.R. and Stubbs, D.A., 1978, Safe levels of manual forces for young males (3). *Applied Ergonomics*, **9**, 33–37.

Eie, N. and Wehn, P., 1962, Measurement of the intra-abdominal pressure in relation to weight bearing of the lumbosacral spine. *Journal of Oslo City Hospitals*, **12**, 205–217.

Farfan, H.F., 1973, *Mechanical Disorders of the Low Back* (Philadelphia: Lea and Febiger).

Floyd, W.F. and Silver, P.H.S., 1950, Electromyographic study of patterns of activity of the anterior abdominal wall muscles in man. *Journal of Anatomy* (London), **84**, 132–145.

Floyd, W.F. and Silver, P.H.S., 1951, Function of the erectores spinae in flexion of the trunk. *Lancet*, **260**, 133–134.

Floyd, W.F. and Silver, P.H.S., 1955, The function of the erectores spinae muscle in certain movements and postures in man. *Journal of Physiology*, **129**, 184–203.

Holbrook, T.L., Kyle, G., Kelsey, J.L. and Stanfer, R.N., 1984, The frequency of occurrence, impact and cost of selected musculoskeletal conditions in the United States (Chicago, Il: American Academy of Orthopaedic Surgeons).

Hult, L., 1954, Cervical, dorsal and lumbar spinal syndromes. *Acta Orthopaedica Scandinavica*, Supplement 17, p.102.

Kelsey, J.L. and White, A.A. (III), 1980. Epidemiology and impact of low-back pain. *Spine*, **5(2)**, 133–142.

Kumar, S., 1971, Studies of the trunk mechanics during physical activity. Unpublished Ph.D. Thesis, University of Surrey, U.K.

Kumar, S. and Davis, P.R., 1973, The lumbar vertebral innervation and intra-abdominal pressure. *Journal of Anatomy* (London), **114**, 47–53.

Kumar, S. and Davis, P.R., 1976, Interrelationship of physiological and biomechanical parameters during stoop lifting. *The International Congress of Physical Activity Sciences* (Quebec, 1976) p.222.

Magora, A., 1970, Investigation of the relation between low back pain and occupation. *Industrial Medicine and Surgery*, **39(12)**, 504–510.

Magora, A., 1974, Investigation of the relation between low back pain and occupation — 6. Medical history and symptoms. *Scandinavian Journal of Rehabilitation Medicine*, **6(2)**, 81–88.

Morris, J.M., Lucas, D.B. and Bresler, B., 1961, Role of trunk in stability of the spine. *Journal of Bone & Joint Surgery*, **42A**, 327–351.

Morris, J.M., Benner, G. and Lucas, D.B., 1962, An electromyographic study of the intrinsic muscles of the back in man. *Journal of Anatomy* (London), **96**, 509–520.

Nachemson, A., 1960, Lumbar intradiscal pressure. Experimental studies on post-mortem material. *Acta Orthopaedica Scandinavica*, Supplement 43.

Nachemson, A.L., 1971, Low back pain, its etiology and treatment. *Clinical Medicine*, **78**, 18–24.

Park, K.S. and Chaffin, D.B., 1974, A biomechanical evaluation of two methods of manual load lifting. *AIIE Transactions*, **6**, 105–113.

Watson, B.W., Ross, B. and Kay, A.W., 1962, Telemetering from within the body using a pressure-sensitive radio pill. *Gut*, **3**, 181–186.

White, A. W., 1966, Low back pain in men receiving workmen's compensation. *Canadian Medical Association Journal*, **95(2)**, 50–56.

Workers' Compensation Board — Alberta, 1985, 67th Annual Report.

14.

Body height changes as a measure of spinal loads and properties of the spine

Jörgen A.E. Eklund

Department of Industrial Ergonomics
University of Linköping
S-581 83 Linköping
Sweden

Abstract

The shrinkage method has recently been developed for assessment of spinal loads and properties of the spine. Using precise measurements of body height changes, the viscous disc compression, or disc creep, can be assessed. In this paper, the method, the equipment and the procedure used are described.

It was shown that the contribution to the body height reduction from structures other than the spine is negligible in relation to the contribution from the spine. It has also been shown that the body height shrinkage is well correlated with spinal loads, and that the method is sensitive enough to differentiate between spinal loads of 100 N difference. Other studies have demonstrated that the shrinkage is also related to the perception of discomfort.

The method can be used not only for evaluation of spinal loads or effects of spinal loads, arising from different activities, postures, and workplace equipment, but also for evaluation of treatment methods. Finally, it can also be used to assess different individual properties, with respect to shrinkage when subjected to standardized loadings.

Introduction

The intervertebral discs have viscoelastic properties. The rate of the viscous height loss (creep) depends on the magnitude of the load on the disc, the time period under load, the age of the person, the vertebral level, the state of degeneration of the disc, and the temporal pattern of preloads or vibratory loads (Virgin, 1951; Hirsch and Nachemson, 1954; Kazarian, 1972 and 1975; Markolf and Morris, 1974). A review of the physical properties of the discs can be found in Eklund (1986). The creep behaviour of single discs has been modelled with simple Kelvin units by Burns and Kaleps (1980). Many Kelvin units can be connected together in series or in parallel to model different characteristics. However, the number of relaxation periods involved in the response of

199

the disc is uncertain (Burns and Kaleps, 1980; Kazarian and Kaleps, 1979). The following simple one-Kelvin unit model can, for example, describe the disc height as a function of time through the equation:

$$H(t) = A + Be^{-Kt}$$

where $H(t)$ is disc height as a function of time, and A, B and K are constants.

A young disc can bind approximately 90 per cent water, but with an increase in age and increased degeneration, this value can decrease to about 65 per cent (Krämer, 1973; Holm, 1980). Increased load on the disc results in an outflow of fluid and thereby a decrease in disc height (Virgin, 1951; Armstrong, 1958; Krämer, 1973). The opposite occurs when the load is released, i.e. an influx of fluid to the disc and an increase in disc height. This mechanism for aiding the nutritional supply of the discs has been referred to as the pump mechanism. Kraemer *et al.* (1985) showed that a disc after a long period of loading and water loss demonstrates a higher concentration of electrolytes. This increases its osmotic absorption force and also aids retention of the remaining water. After the disc is unloaded, water is absorbed and the disc regains its height. It has been shown after injection of saline in a disc that its height increases by up to $2 \cdot 5$ mm (Markolf and Morris, 1974). A height increase of more than 1 mm in a disc specimen after an injection of chymopapain was demonstrated by Brinckmann and Horst (1986). The specimen was thereafter held under 300 N static load, which subsequently caused a creep of over 2 mm in 8 hours.

Body height changes

The height of the human body changes throughout the day. Forssberg (1899) mentioned that it is well known that body height decreases during the day and increases during the night. This has also been shown by Beneke (1897) and Backman (1924). From his sample of over 1200 people, De Puky (1935) found that people were, on average, 1 per cent shorter by the evening than in the morning. Corresponding figures were 2 per cent for children and $0 \cdot 5$ per cent for 70–80 year-old people. Forssberg (1899) also showed that cavalrymen who rode energetically during the day, decreased body height more than when they rode casually and did not expose themselves to any vigorous activity. Fitzgerald (1972) was concerned that load on the shoulders from safety belts used in aircraft could increase the stiffness of the discs and thereby increase the risk of injury from an ejection. He therefore applied loads on the shoulders of his subjects and showed that the greater the load, the greater was the decrease in body height. These results were confirmed by Gritz (1975) and Krämer and Gritz (1980). The body height increased when the load on the spine was partially removed by letting the subjects lie down. This effect was also recognized in astronauts following space flights, who had spent days or weeks in weightlessness. On returning to the Earth, they experienced an increase in height of up to 5 cm (Jayson, 1981).

Spinal traction has also been shown to lead to increased disc height and increased body height (Worden and Humphrey, 1964). Nachemson and Elfström (1970) measured the disc pressure in the L3 disc during traction and found the pressure decreased.

However, Andersson *et al.* (1983) showed that increases in the disc pressure could also appear in some subjects during the exertion of the traction force. On the whole, traction had little effects on the disc pressure. During autotraction, when the subjects exerted traction forces up to a magnitude of 500 N, increases in the disc pressure were always recorded.

The question has been raised as to what extent body height shrinkage originates from structures other than the spine. It is reasonable to assume that compression of the foot, knee, and hip joints, soft tissues under the feet, and flattening of the foot arch contribute to the body height decreases recorded during the day. However, little is known about the magnitude of these influences. Forssberg (1899) measured the height decrease during the day in 78 subjects. All the subjects were measured in both the sitting and the standing positions. His results showed that their sitting height decreased 17 mm and their standing height decreased 18 mm. He concluded that the 1 mm difference was a result of compression of the lower extremities. Markolf and Morris (1974) attributed most of the body height decrease to a decrease in disc height. Also, the results from experiments of disc creep based on single disc specimen and using loadings in the range of 500–1000 N show that height decreases for single discs in the magnitude of 1 mm are realistic (Kazarian, 1975; Markolf and Morris, 1974). All the above-mentioned results support the assumption that the dominating part of the body height change occurs in the discs.

The shrinkage method

It was proposed that because of the viscous properties of the discs, the body height shrinkage could be used as a measure of the effect of loads on the spine (Eklund and Corlett, 1984). Equipment for precise measurements of body height was therefore developed for that purpose. The first equipment, built in Nottingham, has been modified in many respects, and other designs have also been developed (Reilly *et al.*, 1984). The version currently used in Linköping is presented below.

The Measurement Rig

The measurement rig (Figure 1) consists of a stiff rectangular platform, 90 × 60 cm, reinforced by a framework of beams. A 210 cm high tube, 10 cm in diameter, is attached to the frame at right angles. It can easily be removed from the frame for transport. Aluminium has been chosen for its light weight. The entire rig can be tilted backwards 0–20 degrees by adjusting the length of its two front legs. Backward tilt of the rig permits improved muscular relaxation, and causes practically no change in compressive load on the spine due to gravity. Too much tilt is perceived to be uncomfortable and has also proved to be more strenuous for getting in and out of the measurement rig. Pilot experiments have indicated that around 15 degrees backward tilt is suitable, and this value has therefore been chosen.

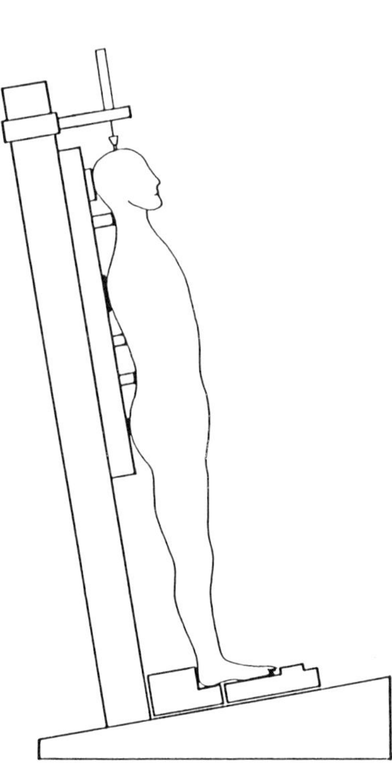

Figure 1.   The principle of body height measurements (left), and the "Linköping" version of the body height measurement equipment (right).

A rectangular plate is placed on the platform. Two V-shaped profiles are mounted on the plate at an angle of 20 degrees to each other, for exact positioning of the participant's feet and heels. The heels are thereby positioned 4 cm apart. The soles are supported on a weighing scale with its top surface on the same level as the surface for the heels. This arrangement allows accurately repeated foot positions and measurement of the weight distribution between heels and soles.

Six supports are mounted along the tube. They control the positions of selected points along the back, namely the sacrum, the mid-lumbar spine, the lower thoracic spine, the mid-thoracic spine, the mid-cervical spine, and the head. All of the supports are adjustable in height, depth, and to the sides, in order to accommodate anthropometric variations of more than 95 per cent of the population. Scales mark the positions of the supports for quick readjustment to an individual's previously used values.

The sacrum support consists of a 9 cm high, 5 cm wide and 1·3 cm thick plate attached to a 20 cm high and 40 cm wide plate, both of wood. The larger one increases the stability around a vertical axis when standing against the supports. It also makes it easier for the subjects to step in and out of the rig. The four back supports are 2 × 5 cm, except for the mid-thoracic support, which is 2 × 10 cm. All four back supports are rounded in the sagittal plane. A thin, flexible brass sheet is mounted on each of them with a small air gap. These metal sheets are in part electrically insulated and designed so that a very small force is needed to create contact between the support and the metal. Hence, they function as micro-switches. When all four switches are in contact at the same time, a light bulb visible to the experimenter is illuminated. This arrangement ensures a more reproducible posture, even though the pressure against the supports cannot be controlled.

The head support consists of two wooden plates, 12×10 cm. These are mounted vertically in a 'V' at right angles, to increase the precision of the head positioning. The subjects wear a spectacle frame with markers for collimation, allowing them to adjust the angle of the head themselves by looking to the front in a mirror. Both these arrangements position the head with an accuracy of a few millimeters in the horizontal plane.

The body height is measured by lowering a measurement head on to the top of the subject's head. It is connected to a linear transducer, which has an accuracy better than $0 \cdot 05$ mm. Its measurement range is 14 cm, and it can be positioned on the tube at intervals of 10 cm. The measurement head consists of a 90 g weight, with five parallel cylindrical pins attached to the underside. These are 1 mm in diameter and 17 mm long. Four pins make up the corners of a $5 \cdot 5$ mm square, with the fifth pin placed in the centre. This construction penetrates thick hair and does not cause feelings of discomfort or pressure upon the subject's head.

The measurement procedure

The following procedure was adopted when the body height measurements were performed: The subject stepped into the rig, positioned the feet and leaned back against the sacrum support. The subject then folded the arms over the chest, inhaled, straightened the back and made contact with the supports. The head angle was adjusted by looking in the mirror, and then the subject exhaled to a relaxed level, relaxed tensed muscles, and gave a signal when ready for measurement. The measurement head was lowered onto the subject's head for approximately one second, and after another second the height was recorded. The whole procedure was repeated five times, which allowed the determination of a mean value, giving a more correct approximation of the body height. It also enabled the standard deviation to be calculated. After each measurement, the subject stepped off and back onto the plate for positioning of the feet, in order to avoid systematic errors due to the feet position. The wooden plate and the scale could be moved slightly forwards or backwards, if a subject experienced knee instability or discomfort. By using this action, the knees were locked in an extended position without causing excessive movements on the knees.

The experimenter noted the height readings and the weight on the scale. A 5 kg variation of the weight on the soles was allowed, because a larger variation could possibly increase the variability of the height measurements. If a subject noticed that a measurement felt strange or was different, if there was no signal confirming contact with all back supports, or if the weight on the scale was outside the approved range, a new measurement was taken to replace the incorrect one.

Two additional measurements were always taken before each set of five measurements, in order to train the subjects in the routine and to check that everything worked according to plan. Small final adjustments often had to be carried out before the start of the experimental session. The subjects kept their clothes on during the measurements, but took off their shoes.

Every subject was given training and instruction for about 20–60 minutes. This was done a few days before the first experimental session. They were taught to peform the procedures without commands from the experimenter. On the first training session, the positions of the supports were noted for future sessions.

Normally, the measurements started 75–90 minutes after the subjects got up in the morning. They were instructed to keep their sleeping hours, morning activities, and travel to the laboratory or workplace consistently and close to their normal pattern. Usually, an experimental session incorporated 45 minutes of work activity. The height measurements were taken immediately before and then immediately after the work period.

Shrinkage and spinal load

Several studies have been performed in order to establish the relation between shrinkage and spinal load. In one laboratory experiment, four subjects were shown to undergo an average reduction in body height of $3 \cdot 2$ mm when standing for 1 hour with a 14 kg shoulder load. The same subjects shrank $1 \cdot 4$ mm on average when standing for 1 hour on a different day without shoulder load.

In another experiment with four different subjects, the shrinkage was $1 \cdot 6$ mm when standing for 1 hour. During the following 30 minutes when standing and carrying a 14 kg weight in one arm, the shrinkage was $2 \cdot 8$ mm (Eklund and Corlett, 1984).

Eight subjects stood for 45 minutes with shoulder loads of 0, 10, 20, and 25 kg on different days. The starting time, the sleeping time and the spinal loading preceding the experiments were controlled and kept constant. The shrinkage amounted to $1 \cdot 8$, $2 \cdot 5$, $3 \cdot 4$ and $4 \cdot 2$ mm respectively (Eklund, 1986).

The body height change for four subjects was measured in the late afternoon with the subjects lying on a horizontal surface. The body height increased $5 \cdot 6$ mm during 1 hour (Eklund and Corlett, 1984).

The biomechanical load on the L3 disc was estimated for the studies presented above, and plotted against the rate of shrinkage in Figure 2.

Despite the fact that the studies were not controlled for the time of the day, experimental time or subjects, a relationship between shrinkage and spinal load is clear.

The method has also been used to evaluate spinal loadings in different activities and work tasks. Dynamic activities such as repetitive lifting, circuit weight-training and running have been investigated by Leatt *et al.* (1986) and Wilby *et al.* (1987).

In another experiment three subjects sat for $1 \cdot 5$ hours in three different chairs and performed sedentary work. One chair was an easy chair with a full-size backrest inclined at 110 degrees and with a 4 cm deep lumbar support. The next chair was an

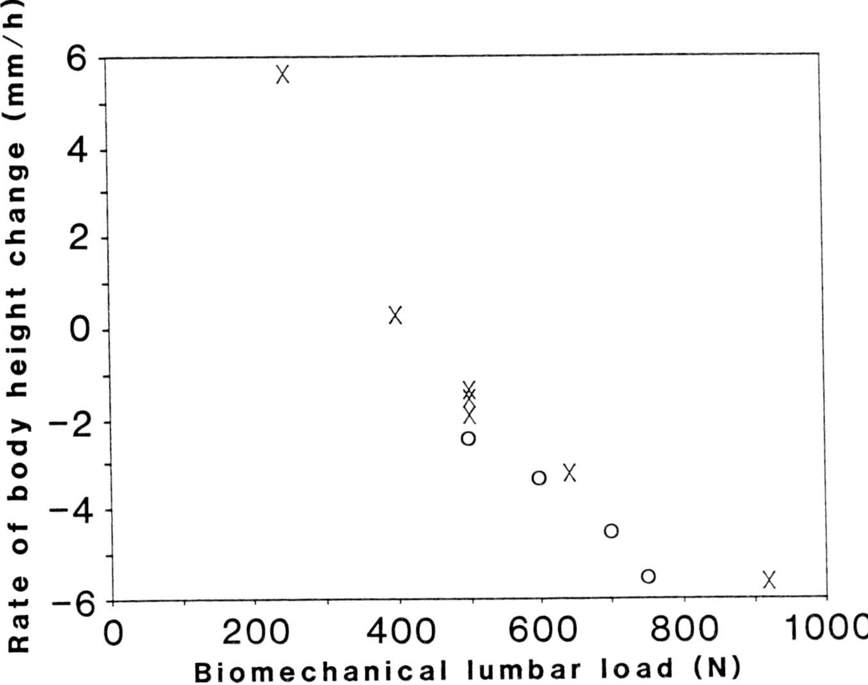

Figure 2. Average rate of body height change from loading experiments, plotted as a function of estimated load on the L3 disc. The following loads were used: lying 250 N, sitting in an easy chair inclined at 110 degrees with a 4 cm deep lumbar support 400 N, sitting on an office chair 500 N, standing 500 N, standing + 10 kg load on the shoulders 600 N, standing + 14 kg shoulder load 640 N, standing + 20 kg shoulder load 700 N, standing + 25 kg shoulder load 750 N, standing + carrying 14 kg in one hand 920 N. Studies in which the loads preceding the experiments were uncontrolled (x), and studies which were controlled (0).

office chair with a lumbar support, and the third one was a stool. The resulting body height changes were an increase of 0·4 mm with the easy chair, a decrease of 0·1 mm with the office chair, and a decrease of 4·6 mm with the stool (Eklund and Corlett, 1984).

In a laboratory study of industrial seat design, a seat with a low backrest was compared to a seat with a high backrest in a task which required exertion of 25 N forwards with the hands. The task was performed for 45 minutes on separate days, and eight subjects participated. The subjects shrank 1·4 mm when using the low backrest and 0·7 mm when using the high one. Eight industrial grinders compared the same seats in a grinding task demanding force exertion and materials handling, performing the experimental sessions for 45 minutes on separate days. The grinders shrank 2·8 mm when using the low backrest and 2·5 mm when using the high one.

In another laboratory study, a sit-stand seat was compared to a conventional seat with a low lumbar support, involving an assembly task with restricted knee-room. The task was performed on separate days by eight subjects. They shrank 0·9 mm when using the sit-stand seat and 2·4 mm when using the conventional seat. The same two seats were also evaluated in punch press work in a factory. The task demanded a fair amount of materials handling and the knee-room was restricted. Eight operators participated in sessions for 45 minutes on separate days. They had worked several hours with heavier work tasks before the experiments started, so the conditions were not comparable with the other experiments reported. The operators increased their body height 0·5 mm when using the sit-stand seat, and they increased their height 0·2 mm when using the conventional seat. It should be noted that the sit-stand seat caused less discomfort for the back, but substantially more discomfort for the buttocks.

In another laboratory study, a seat with a low backrest was compared to a seat with a high backrest in a task demanding vision to the side. The task was performed for 45 minutes on separate days by eight subjects. They shrank 0·9 mm when using the low backrest and 1·4 mm when using the high one (Eklund, 1986).

Shrinkage and subjective experiences

Ten subjects took part in a study, in which they lay supine on a padded board. This board was inclined at 50, 70, and 90 degrees to the horizontal, with the head downwards. This was performed for 30 minutes, followed by 20 minutes of standing. In addition to body height changes, the perception of comfort was recorded. It was found that ratings of discomfort were related to failure to achieve height gains when inverted. In another study, traction in standing position was compared to standing with and without shoulder loads, while height changes and discomfort were recorded. Increased shoulder loads caused increased shrinkage and increased discomfort. The traction situation also caused increased discomfort and increased shrinkage compared to standing without shoulder loads (Troup *et al.*, 1985). There was also an agreement between shrinkage and the perception of discomfort in the industrial seat studies reported above.

A study of the influence on shrinkage from the lower extremities

The purpose of the study was to obtain further quantitative results concerning influence on shrinkage from the lower extremities. Therefore an experiment was set up, in which two female and four male subjects participated. The mean age of the subjects was 28 years (range 19–35 years), their mean height was 171 cm (range 152–191 cm), and their mean weight was 65 kg (range 53–75 kg).

Four colour marks were made on the skin. One mark was made in the mid-cervical region, the second at the right iliac crest, the third at the right knee joint (at mid-patella height), and the fourth at the right lateral malleolus. The heights of these four colour marks were measured with a tape measure. This was done at the same time as the body height was measured in the measurement rig described above. The subjects were measured in the morning, and the measurements were repeated five times. Another set of five measurements was taken in the late afternoon. About ten minutes of bed rest preceded the morning measurements. At least ten minutes of spine loading activities, such as lifting boxes or sitting on a low stool, preceded afternoon measurements.

As shown in Table 1, the body height decreased on average 8 mm over the working day. The height of the skin mark at the neck changed approximately as much, but there was no significant height change of any of the three other marks. The higher standard deviation of the mark at the iliac crest, compared to the knee and foot marks may partly have been due to movements of the skin above the bony structures. Increased difficulties in making a precise height measurement due to the larger measurement distance may also have contributed.

It can be concluded that the influence from the compression of the structures in the lower extremities was very small in relation to the body height change. The majority of the height change occurring over a day took place between the pelvis and the head.

Table 1. The height changes during the day, expressed as mean, and standard deviation within the sets of five measurements and based on the 12 sets.

	Mean (mm)	S.D. (mm)
Body height change	$-8\cdot0$	1.0
Skin mark height change:		
Neck	$-8\cdot7$	$1\cdot2$
Iliac crest	$-0\cdot6$	$1\cdot8$
Knee	$-0\cdot3$	$1\cdot0$
Foot	$-0\cdot4$	$0\cdot7$

Discussion

The concept of the method is that load on the discs cause creep, which results in body height changes. The majority of the body height change originates from the spine, which enables the use of body height changes as a measure of spinal loading. The equipment and procedure for the body height measurements has now been developed so that determination of body height can presently be made with a standard deviation below 1 mm. This value refers to a set of five consecutive height measurements, and it is based on several thousand measurements of almost 100 people. Their ages varied from 18 years to 61 years, and their occupational backgrounds were students, office workers, and industrial workers. They all learned the procedure in less than one hour. It is clear that some performed the height measurements more consistently than others, although the reasons for this are not known.

A body height decrease of 17 mm in a day is approximately 1 per cent of the body height. It can be assumed that about one third of the vertebral column length consists of discs, and that the vertebral column length is about 35 per cent of the body height. In that case a body height shrinkage of 1 per cent means around 8 per cent compression of the discs.

The disc height is an interesting factor to consider in the discussion of possible causes of back pain. A decreased disc height means geometrical changes and changed physical properties of the spine. It reflects increased disc bulging, decreased room for the nerve roots, increased load on the apophyseal joints, increased tension in the collagen fibres of the annulus fibrosus, increased stiffness of the disc, and nutritional supply of the disc is affected. There are also reports that the stability of the spinal joints is affected (Koeller *et al.*, 1984). Since shrinkage is affected by both the spinal load and its temporal pattern, it is more directly linked to the above-mentioned factors, and can be a more relevant predictor for the risk of back pain than a measure of the load alone.

The method has so far been used to evaluate spinal loads or effects of spinal loads arising from different activities, postures or workplace equipment. The influence from the temporal pattern of loadings has also been evaluated to some extent. Shrinkage has been shown to be well correlated with spinal load. The method is sensitive enough to differentiate between spinal loads of 100 N difference. The diurnal variation in body height is in agreement with the pattern of spinal loading arising from the erect postures adopted during day time and the supine postures adopted during night time, and can be explained in that way.

Further, shrinkage has been shown to be related to subjective experiences of discomfort. This finding emphasizes the importance of comfort, and the method might possibly be used in some situations as an objective measure of comfort/discomfort in the future.

A further possibility for the use of this method is as a measure of individual properties. These have also been measured by letting a number of subjects go through standardized loading regimes, and using the responses as measures of individual properties. Considerable individual differences in shrinkage ability, and also differences due to age, have been seen.

Treatment methods such as spinal traction have also been evaluated with the method. The result from Andersson *et al.* (1983), showing that the disc pressure increased during autotraction, has been supported from shrinkage measurements (Pope and Klingenstierna, 1986). It was also demonstrated in a traction situation that comfort is important for the possibility of achieving increased disc height (Eklund, 1986).

Significant individual and age differences have been shown regarding shrinkage for standardized loadings (Eklund, 1986). These differences emphasize the necessity of using the subjects as their own controls in comparative studies. The control of the loads preceding the experiments is also very important, considering the body height change during the day and the change of physical properties of the discs after being subjected to load.

The quick recovery when unloading the spine implies that the temporal pattern of work loads is important. Several short periods of rest for unloading the spine, or in other words more dynamic loading, would consequently cause less disc creep than fewer and longer periods of rest interspersed with prolonged static loading.

An advantage of this method is that it can assess work loads on the discs *in vivo* without any equipment attached to the subject while working. Since the method is non-invasive, it is suitable for field application. It also enables the assessment of loadings from passive structures. The equipment is not expensive, and it does not require extensive education or instruction for use.

A limitation of the method is that there is no indication of whether certain parts of the spine have been subjected to a higher load than other parts, or whether certain discs have caused more height decrease than others. Another disadvantage with the method is that all the control measures which have to be taken are time-consuming.

References

Andersson, B.J.G., Schultz, A.B. and Nachemson, A.L., 1983, Intervertebral disc pressures during traction. *Scandinavian Journal of Rehabilitation Medicine*, Suppl. 9, 88–91.

Armstrong, J.R., 1958, *Lumbar disc lesions. Pathogenesis and treatment of low back pain and sciatica*, (Edinburgh: E & S Livingstone Ltd).

Backman, G., 1924, *Körperlänge und Tageszeit*. (Uppsala: Uppsala Läkareförenings förhandlingar 28), pp.255–282.

Beneke, R., 1897, *Zur Lehre von der Spondylitis deformans* (Braunschweig: Beitrag zur Wissenschaftlicher Medizin), pp.109–131.

Brinckmann, P. and Horst, M., 1986, Short term biomechanical effects of chymopapain injection; an in vitro investigation of human lumbar motion segments. *Clinical Biomechanics*, 1, 14–19.

Burns, M.L. and Kaleps, I., 1980, Analysis of load-deflection behavior of intervertebral discs under axial compression using exact parametric solutions of Kelvin-solid models. *Journal of Biomechanics*, 13, 959–964.

De Puky, P., 1935, The physiological oscillation of the length of the body. *Acta Orthopaedica Scandinavica*, 6, 338–347.

Eklund, J.A.E., 1986, *Industrial seating and spinal loading*. Ph.D. Thesis, University of Nottingham, U.K.

Eklund, J.A.E. and Corlett, E.N., 1984, Shrinkage as a measure of the effect of load on the spine. *Spine*, 9(2), 189–194.

Fitzgerald, J.G., 1972, *Changes in spinal stature following brief periods of static shoulder loading*, IAM report No 514, (Farnborough, Hampshire: Royal Air Force Institute of Aviation Medicine).

Forssberg, E., 1899, Om vexlingar i kroppslängden hos kavallerirekryter. *Militär Hälsovård*, 24, 19–28.

Gritz, H.A., 1975, *Die physiologischen Längenänderungen der menschlichen Wirbelsäule im Verlaufe eines Tages sowie der Einfluss von Be- und Entlastung auf den Intervertebralabschnitt*. Dissertation. (Düsseldorf: Ortopädischen Klinik der Universität Düsseldorf).

Hirsch, C. and Nachemson, A., 1954, New observations on the mechanical behaviour of lumbar discs. *Acta Orthopaedica Scandinavica*, 23(4), 254–283.

Holm, S., 1980, *Nutrition of the intervertebral disc: Transport and metabolism*. Thesis. (Department of Orthopaedic Surgery I, University of Göteborg).

Jayson, M.I. V., 1981, *Back pain. The facts*. (Oxford: Oxford University Press).

Kazarian, L., 1972, Dynamic response characteristics of the human vertebral column. *Acta Orthopaedica Scandinavica*, Suppl. 146.

Kazarian, L., 1975, Creep characteristics of the human spinal column. *Orthopedic Clinics of North America*, 6(1), 3–18.

Kazarian, L.E. and Kaleps, I., 1979, *Mechanical and physical properties of the human intervertebral joint*, Report AMRL-TR-79-3, (Ohio: Wright-Patterson Air Force Base).

Koeller, W., Funke, F. and Hartmann, F., 1984, Biomechanical behavior of human intervertebral discs subjected to long lasting axial loading. *Biorheology*, 21(5), 675–686.

Kraemer, J., Kolditz, D. and Gowin, R., 1985, Water and electrolyte content of human intervertebral discs under variable load. *Spine*, 10(1), 69–71.

Krämer, J., 1973, *Biomechanische Veränderungen im lumbalen Bewegungssegment* (Stuttgart: Hippo-krates-Verlag).

Krämer, J. and Gritz, A., 1980, Körperlängenänderungen durch druckabhängige Flüssigkeits-verschiebungen im Zwischenwirbelabschnitt. *Zeitschrift für Orthopädie und ihre Grenzgebiete*, **118**, 161–164.

Leatt, P., Reilly, T. and Troup, J.D.G., 1986, Spinal loading during circuit weight-training and running. *British Journal of Sports Medicine*, **20(3)**, 119–124.

Markolf, K.L. and Morris, J.M., 1974, The structural components of the intervertebral disc. *Journal of Bone and Joint Surgery*, **56-A(4)**, 675–687.

Nachemson, A. and Elfström, G., 1970, Intravital dynamic pressure measurements in lumbar discs. *Scandinavian Journal of Rehabilitation Medicine*, Suppl. 1.

Pope, M.H. and Klingenstierna, U., 1986, Height changes due to autotraction. *Clinical Biomechanics*, **1**, 191–195.

Reilly, T., Tyrrell, A. and Troup, J.D.G., 1984, Circadian variation in human stature. *Chronobiology International*, **1(2)**, 121–126.

Troup, J.D.G., Reilly, T., Eklund, J.A.E. and Leatt, P., 1985, Changes in stature with spinal loading and their relation to the perception of exertion or discomfort. *Stress Medicine*, **1**, 303–307.

Virgin, W.J., 1951, Experimental investigations into the physical properties of the intervertebral disc. *Journal of Bone and Joint Surgery*, **33-B(4)**, 607–611.

Wilby, J., Linge, K., Reilly, T. and Troup, J.D.G., 1987, Spinal shrinkage in females: Circadian variation and the effects of circuit weight training. *Ergonomics*, **30(1)**, 47–54.

Worden, R.E. and Humphrey, T.L., 1964, Effect of spinal traction on the length of the body. *Archives of Physical Medicine and Rehabilitation*, **45**, 318–320.

15.

Mechanical and physiological gait factors of above knee amputees and of paraplegics activated by functional electrical stimulation

J. Mizrahi[1], E. Isakov[2], Z. Susak[2] and E. Becker[2]
[1]*Department of Biomedical Engineering
and the Julius Silver Institute of Biomedical Engineering
Technion-IIT, Haifa 32000, Israel.*
[2]*Loewenstein Rehabilitation Hospital
Raanana 43100, Israel*

Abstract

In this work, the mechanics and energetics of walking in two severe, yet different, groups of locomotor disabilities were studied. The first group consisted of above knee (AK) amputees, and was divided into two sub-groups related to the cause of amputation: vascular and traumatic. The second group consisted of paraplegics whose leg muscles were activated for standing and reciprocate walking by functional electrical stimulation (FES). Biomechanical and physiological parameters, including gait, weight bearing on the legs, heart rate, oxygen consumption and lactic acid were recorded and studied. The results related to the AK amputee group are discussed in the light of the effect of existence of a knee mechanism (open or locked) on gait performance and metabolic energy cost. In the FES activated group of paraplegics, the oxygen consumption during various activities studied is discussed. While in the sitting position FES of the muscles seemed beneficial in terms of physical activation, the standing and reciprocal walking activities involved exhaustive efforts, especially due to the intensive involvement of the upper body limbs.

Introduction

An important goal in the rehabilitation process is to restore ambulation ability of the handicapped in order to make them more independent in daily life. Evaluations of the physical effort of individuals with different locomotor disabilities during ambulation is, therefore, of major interest. In doing so, it is relevant and convenient to use the gait of normal and healthy individuals for reference. The evaluated parameters include walking speed and energy consumption expressed in terms of oxygen consumption per

211

unit of body weight, per unit of distance travelled (Waters et al., 1978). Alternatively, energy consumption is also defined as the oxygen consumption per unit of body weight, per unit of time. However, the first definition is found preferable as it is more closely related to efficiency of walking.

Concerning the walking speed of normal individuals, Finley and Cody (1970) observed the gait of 1006 pedestrians during ambulation along sidewalks in urban areas, under natural conditions. The average walking speed was reported to be 82 m/min in men and 74 m/min in women. In another study, Corcoran and Brengelmann (1970) found that in a group of 32 men and women with an average age of 38 years, the mean walking speed in unrestrained gait was 83 m/min. They also reported the oxygen consumption (0·167 ml/kg/m).

According to Saunders *et al.* (1953), minimization of the metabolic energy cost during ambulation is dependent on six biomechanical determinants, aimed at limiting and smoothing out the displacement path at the centre of gravity of the body. These determinants are listed as follows: (a) transverse rotation of the pelvis; (b) downward tilt of the pelvis on the unsupported side; (c) knee flexion in the stance phase; (d) and (e) coordination between foot, ankle and knee flexions to reduce jerky motion of the centre of gravity; and (f) the adducted position of the shaft of the femur, which allows transmission of the body weight over the supporting leg without the need of large lateral shifts of the centre of gravity. Lack of one or more of these determinants, as in the case of a locomotor disability, results in reduced mechanical efficiency of gait and, hence, increased energy consumption. An extreme example is the gait of paraplegic patients in which the lower limbs participate minimally, or do not participate at all, in weight-bearing and walking unless supported by long braces mounted on the patient. The main effort required is thus accomplished by the upper extremities and trunk muscles above the level of the lesion. This kind of gait is very costly and inefficient in terms of metabolic energy expenditure.

While reviewing works on physical effort during ambulation in the various locomotor disabilities, it is convenient to refer to the following main groups: patients with orthopaedic deficiencies, those using walking aids, amputees, hemiplegics and paraplegics.

Fisher and Patterson (1981) and Patterson and Fisher (1981) tested young volunteers with an average age of 26 years, during 2–point ambulation assisted by auxilliary and forearm crutches and non-weight-bearing on the affected leg. The walking speed of their patients was almost similar to that of normal subjects, 79 m/min. The oxygen consumption was 0·306 ml/kg/m.

The influence of different casts on the requested effort during ambulation was reported by Waters *et al.* (1982). The patients in this study were on the average of 27 years old and, while walking at their comfortable speeds, wore three kinds of casts: long, cylindrical and short. Two walking tests were carried out: one with full-weight-bearing (FWB) and the other with non-weight-bearing (NWB), while using crutches. In FWB with the long cast, speed of ambulation was 55 m/min and oxygen consumption was 0·240 ml/kg/m. In NWB, speed of ambulation was 51 m/min and oxygen consumption was 0·400 ml/kgm. With the short cast in FWB, speed of ambulation was 70 m/min and the oxygen consumption was 0·190 ml/kg/m. In

NWB, while using crutches, the speed of ambulation was 60 m/min and the oxygen consumption was 0·360 ml/kg/m. These results clearly indicate that immobilization of the knee joint and use of crutches increases the metabolic energy consumption during gait.

McBeath *et al.* (1980) studied the gait of 60 elderly patients four years after unilateral total hip replacements. The average free speed of ambulation was 64 m/min and the oxygen consumption was 0·174 ml/kg/m. In another group of 17 patients in whom both hips were replaced, walking speed was higher, 72 m/min. The resulting oxygen consumption was 0·212 ml/kg/m. In both these groups, gait ability improved considerably after the operation as compared to the pre-operative state of patients.

Amputees are also a non-homogeneous group, and are generally divided into subgroups depending on the severity and level of amputation. Most of the patients in this group are also elderly and suffer from diabetes mellitus or from peripheral vascular diseases, the main causes of amputation. As a matter of fact, only a small proportion of amputations are for traumatic reasons. The effect of amputation level on walking speed and oxygen consumption was investigated by Waters *et al.* (1976). In elderly above-knee (AK) vascular amputees, over 60 years of age, the speed of comfortable walking with prostheses was found to be 36 m/min and oxygen consumption was 0·350 ml/kg/m. In below-knee (BK) amputees from a comparable age group, the speed of walking was 45 m/min and oxygen consumption was 0·260 ml/kg/m. At a lower level of amputation, specifically Syme's amputation, walking speed increased further to 54 m/min and oxygen consumption was 0·213 ml/kg/m.

For comparison with the results of the above mentioned vascular amputation groups, traumatic amputees, much younger in age (average 30 years) were also evaluated in the study of Waters *et al.* (1976). In the AK traumatic amputees, walking speed was 52 m/min and oxygen consumption was 0·250 ml/kg/m. In the BK traumatic amputees, walking speed was 71 m/min and oxygen consumption was 0·200 ml/kg/m.

Another important factor influencing gait ability and performance in amputees is the stump length. Gonzalez *et al.* (1974) studied two groups of BK amputees, differing from one another by their stump lengths. With the shorter stump, approximately 9·5 cm, the walking speed was 61 m/min and metabolic energy consumption was 0·186 ml/kg/m. With the longer stump, approximately 16·9 cm, the walking speed was 72 m/min and oxygen consumption was 0·234 ml/kg/m. Similar results have also been reported by Pagliarulo *et al.* (1979).

The highest degree of impairment among amputees exists in the bilateral AK amputees. In these patients, a particularly high effort is required during ambulation (Huang *et al.*, 1979). In Huang *et al.*'s study, walking speed was as low as 23 m/min and the oxygen consumption was 0·578 ml/kg/m.

A completely different group of patients is the hemiplegic group, in whom the walking difficulties are due to muscular weakness, spasticity and partial control ability on voluntary movements. Corcoran *et al.* (1970) studied the gait of 15 hemiplegics with an average age of 45 years. Walking speed was reported to be 41 m/min and oxygen consumption was 0·311 ml/kg/m. Of special interest in these patients was the effect of a foot-ankle brace (for the prevention of foot-drop) on gait. It was observed that ground clear-off was indeed improved, preventing the circumduction movement normally present in these patients. The reported walking speed increased to 49 m/min and the metabolic energy consumption decreased to 0·282 ml/kg/m.

In paraplegia, mobility is possible only with the help of crutches, long-leg-braces or wheelchairs. Fisher and Gullickson (1978) have shown the existence of a relationship between metabolic energy cost during ambulation and the level of injury. For instance, in paraplegics with a high thoracic lesion and walking with crutches, the metabolic energy cost was 0·675 ml/kg/m, compared to 0·508 ml/kg/m in lower level lesion paraplegics. This difference is attributed to increased difficulties in controlling the pelvis in patients with a higher spinal level lesion.

In comparison to crutches, ambulation of paraplegics by wheelchair was found to be much less energy consuming. Smith *et al.* (1983) studied seven paraplegics with an average age of 24 years, moving at the speed of 50 m/min. The resulting oxygen consumption was 0·203 ml/kg/m. Similar results were also reported by Waters *et al.* (1978), who found that the comfortable speed during pushing of a wheelchair was 82 m/min. The resulting oxygen consumption was 0·191 ml/kg/m.

It should be noted that in considering the results related to the paraplegics, this group of patients is very non-uniform in its pathology, due to both the level and severity of the spinal injury. For this reason, and the fact that most studies on paraplegics involve only a small number of patients, variability of the reported findings is relatively high. All these results are combined in Table 1, for comparison.

The focus of this paper was on mechanical and physiological gait factors of two severe, yet different, groups of motor disabilities: above-knee (AK) amputees walking with a prosthesis and paraplegic patients activated by functional electrical stimulation (FES). The parameters, including kinematics of the stride, weight bearing on the legs, heart rate, oxygen consumption and blood lactic acid, were recorded and studied to evaluate ambulation performance of the groups studied. Among the gait parameters, walking speed is of special importance. Andriacchi *et al.* (1977) found good correlation between this and many other kinematical parameters and therefore suggested using walking speed as a basis for normal and abnormal gait measurements. The significance of walking speed for expressing gait improvement during rehabilitation was reported by Mizrahi *et al.* (1982a), who studied the time–distance parameters of the stride of hemiplegic patients, as related to the clinical improvement of these patients. Of the physiological parameters studied, metabolic energy consumptions clearly reflect the physical effort invested during ambulation. Walking speed and oxygen consumption were thus selected to be the main parameters to evaluate the ambulation of the two groups of patients studied.

In the AK amputees group, the influence of open or locked position of the knee mechanisms of the prosthesis on gait performance were studied. This effect was found significant in patients who were over 50 years of age and were suffering from diseases such as diabetes mellitus, hypertension or other diseases related to the cardio-vascular system, and for whom minimization of the metabolic energy consumption during ambulation is crucial.

In the group of paraplegics activated by FES, metabolic energy consumption of these patients was studied along with other physiological and gait factors during reciprocal

Table 1   *Combined results of speed of ambulation and oxygen consumption for various patient types, collated from several studies.*

Disability	Number of subjects	Age of subjects	Sex	Speed of Ambulation (m/min)	O_2 consumption ml/kg/m	Study
None	1006		M	82		Finley & Cody (1970)
			F	74		
None	32	38 (average)	M/F	83	0·167	Corcoran & Brengelmann (1970)
2-point ambulation assisted by auxiliary and forearm crutches		26 (average)		79	0·306	Fisher & Patterson (1981) Patterson & Fisher (1981)
FWB & NWB; Long, cylindrical and short casts		27 (average)				Waters *et al.* (1982)
FWB-Long				55	0·240	
NWB-Long				51	0·400	
FWB-Short				70	0·190	
NWB-Short				60	0·360	
4 years after unilateral hip replacement	60	Elderly	M/F	64	0·174	McBeath *et al.* (1980)
Both hips replaced	17			72	0·212	
Vascular amputation & prosthesis		Elderly				Waters *et al.* (1976)
Above knee (AK)		> 60yrs		36	0·350	
Below knee (BK)				45	0·260	
Syme's amputation				54	0·213	
Traumatic amputation		30				Waters *et al.* (1976)
AK				52	0·250	
BK				71	0·200	
Stump length (BK)						Gonzalez *et al.* (1974)
9·5cm				61	0·186	
16·9cm				72	0·234	
Bilateral AK amputees				23	0·578	Huang *et al.* (1979)
Hemiplegic	15	45		41	0·311	Corcoran *et al.* (1970)
(with foot brace)		(average)		49	0·282	
Paraplegic (crutches)						Fisher & Gullickson (1978)
With high thoracic lesion					0·675	
With lower level lesion					0·508	
Paraplegic (wheelchair)	7	24		50	0·203	Smith *et al.*, (1983)
				82	0·191	Waters *et al.* (1978)

supported gait. The difficulties involved in ambulation of paraplegic patients by means of FES alone are discussed in the light of the resulting high metabolic energy cost, requiring the support of the anaerobic energy sources.

Materials, patients and procedures

Evaluation tests and instrumentation

The physiological and biomechanical evaluation tests made in this study were as follows:

(a) *Evaluation of cardiac response.* During the test the patient was connected to a telemetric ECG monitor (*Mobile Cardio/Sentinel* with *Telemetry, M.G.* Model 656/F) and heart rate was measured at each stage of the test. The electro-cardiogram was displayed on an oscilloscope and recorded from chest electrodes in V5 placement, enabling the continuous detection of arrhythmias or possible S–T changes.

(b) *Metabolic energy cost evaluation.* Oxygen consumption (OC) was evaluated by measuring the expired air, which was collected in a 100 litre Douglas bag. The sequence of air sampling from the patient is detailed later. The percentage of oxygen in the expired air (OP) was measured by a wet gasometer (*Meterfabrik Dordrecht, Type 5*). Oxygen consumption was calculated from the following formula (Saksena, 1979):

$$OC = A \times EX\ (0 \cdot 265 - 1 \cdot 265 \times OP - 0 \cdot 265 \times CP)$$

where:

OC	= oxygen consumption (l/min);
EX	= expiratory volume of air (l/min), measured at ATPS;
OP, CP	= fractional concentrations of expired oxygen and carbon dioxide, respectively; and
A	= the adjustment factor of volumes from ambient to STPD conditions.

(c) *Metabolic evaluation.* Venous blood, for lactic acid measurements, was obtained from an antecubital superficial vein with venous constriction. Blood lactate was calculated from enzymatic techniques and compared to the rest values. It was reported (Astrand and Rodahl, 1986) that blood lactate exceeding 50 mg% indicates intensive physical efforts requiring anaerobic energy sources.

(d) *Evaluation of weight-bearing on the legs during standing.* Weight-bearing on the feet was measured by *Kistler* force platforms (Kistler Instruments AG, Winterthur, Switzerland) on which the patient was required to stand; his walking aid was supported outside the platform area. Percentage of body weight borne by the feet was calculated after time-integration of the force curves over a standard period of 5 minutes.

(e) *Evaluation of gait parameters.* Gait of the patients was evaluated, as previously reported (Mizrahi *et al.*, 1985), on a walkway which included a 5 m long electrical contact system for monitoring the time–distance parameters of the stride. This method is suitable for monitoring of supported gait, such as exhibited by the patients tested in this study. The parameters measured included stance time, swing time, stride length, walking speed and step symmetry. The contact system, on-line and connected to an *IBM-PC* computer, has been described by Mizrahi *et al.* (1982 a,b).

Patients

AK amputees

Seventeen above-knee amputees (14 vascular amputees and 3 traumatic amputees) as detailed in Table 2, were included in this study. In the vascular group, the age ranged from 50 to 75 years (average age of 60·5 years). Most of these patients were under medication due to various health problems including diabetes mellitus, hypertension, heart failure, and myocardial infarction. They were fitted with an *Otto-Bock* modular above-knee prosthesis (Otto-Bock, Orthopadische Industrie KG, Duderstadt, W. Germany). For open-knee ambulation in the 3R15-type 'safety knee' was mounted; for locked-knee ambulation, the 3R17-type was used instead. These two types are easily interchangeable and do not require major re-adjustments of the prosthesis. A fibre-reinforced plastic quadrilateral socket, with pelvic belt suspension was used. The walking test in the vascular group of patients was conducted about five weeks after fitting of the prosthesis and gait training.

Table 2. *Details of AK amputees taking part in this study. (V = vascular amputee, T = traumatic amputee).*

Patient	Age	Sex	Side of amputation	Associated diseases[*]
V1	57	M	LT	BP
V2	65	M	LT	H,BP
V3	70	M	LT	T, H, RF
V4	50	M	RT	DM
V5	58	M	LT	DM
V6	59	M	LT	H
V7	60	M	RT	DM, CH
V8	50	M	RT	DM, MI
V9	75	F	RT	T, CH
V10	60	M	LT	DM, BP
V11	59	M	RT	DM, BP
V12	60	M	LT	DM, BP
V13	60	M	RT	MI
V14	55	F	LT	DM
T1	48	M	RT	–
T2	26	M	LT	–
T3	32	M	LT	–

[*]Legend of associated diseases: BP = bypass for vascular disease; CH = congestive heart disease; RF = renal failure; DM = diabetes mellitus; H = hypertension; MI = sp myocardial infarction; T = tumour in lower limb.

The group of traumatic AK amputees consisted of three patients aged 26, 32 and 48 years. All three had been amputated for a period exceeding five years. They were in an excellent state of health, and had adapted well to their prostheses, achieving good walking ability. The prosthesis was fitted with a *Mauch S-N-S* knee hydraulic mechanism (Mauch Company, Dayton, OH, USA) which can be easily changed from the open-knee to a locked-knee configuration. The socket was a fibre-reinforced plastic quadrilateral total contact socket, with suction suspension.

FES – activated paraplegics

Seven paraplegic patients as detailed in Table 3 were included in this part of the study. All of the patients participating had to satisfy the following prerequisites:

(a) Spastic paralysis of the lower limbs due to spinal trauma and complete motor neurone lesion.

(b) Complete anaesthesia of the lower extremities.

(c) Full range of passive motion in the joints of the lower limbs.

(d) No pressure sores.

(e) High motivation, readiness for cooperation and good general physical condition.

All patients underwent a physical check-up, blood and urine laboratory tests, rest ECG and X-rays of the lower limbs and spine. Detailed explanations about FES training and about the different testing procedures were given to all of the patients.

Table 3. Details of paraplegic patients taking part in this study.

Patient	Age	Sex	Height (cm)	Mass (kg)	Level of injury	Years of injury	Cause of injury[*]
P1	30	M	165	64	D8–9	7	WA
P2	31	M	185	75	D6–7	10	GSW
P3	37	M	162	69	D6	3	WA
P4	19	M	173	60	D10–11	1	RA
P5	40	M	182	62	D5–6	8	RA
P6	48	M	160	70	D5	30	RA
P7	22	F	157	60	D5	2	RA

[*]Legend of cause of injury: GSW = gun shot wound; RA = road accident; WA = work accident.

Procedure

AK amputees

All walking tests were performed after the patients had rested for half an hour in a quiet room, and at least two hours after their last meal. The vascular amputees walked with the support of a cane and the open-knee walking test was carried out first. The locked-knee walking test was carried out on the following day. The traumatic amputees walked freely without any support. Here too, the first test was with the open-knee; however, the locked-knee test, in this case, followed after one-half hour of rest.

The patients were asked to walk at their own unrestrained walking pace for 5 minutes. Within this period of time heart rate is believed to reach a steady-state value (Astrand and Rodahl, 1986). Heart rate was measured at rest and at the end of the 5 minute walk. Time and distance parameters of the stride were also measured. Oxygen consumption was measured for the traumatic amputees only; the vascular amputees being unable to perform the test due to their unstable gait and slight discomfort imposed by the measuring system.

FES - *activated paraplegics*

In FES-activated paraplegics, the initial period is always devoted to training, aimed at strengthening the paralysed muscles, towards standing and reciprocate walking by means of FES. The detailed stimulus characteristics, stimulating apparatus used and training procedure are described elsewhere (Braun *et al.*, 1985; Isakov *et al.*, 1985, 1986). The procedure for the tests of the present study were as follows.

Prior to all tests and after a rest period of one-half hour, ECG electrodes, in V5 placement, were attached to the patient and blood pressure, body temperature and heart rate were measured. Expired air was collected into a Douglas bag for oxygen consumption evaluation, as described above. At this stage, while the patient was seated in his wheelchair, FES was externally applied to the quadriceps muscles and the peroneal nerve of each leg for a period of 20 minutes. Minimum intensities required for knee extension and ankle dorsiflexion were used; however, these intensities had to be readjusted from time to time due to muscle fatigue. The other stimulus parameters used were those reported by Mizrahi *et al.* (1986) and by Susak *et al.* (1986) for reducing fatiguability of the muscles, namely frequency of 20 Hz and pulse width of 0·3 ms.

Following 20 minutes of exercising, the heart rate was measured and expired air was collected into a Douglas bag for 2 minutes for determining oxygen consumption.

At this point each patient was given a rest of 30 minutes, after which he was re-activated by FES and was asked to stand between parallel bars on the *Kistler* platform for at least 5 minutes. During this time, weight bearing on the legs was measured as described earlier. In this mode of activation, FES was given to the quadriceps muscles continuously. During the last 2 minutes of standing, heart rate was measured and expired air collected for estimating oxygen consumption.

The patients were again given a rest of 30 minutes. In the first 4 minutes of the rest period, a blood sample was taken for lactic acid determination. Following rest, the patient was reactivated by FES for reciprocal walking, while being supported by parallel bars, or by a walker for at least 5 minutes. At this stage, time distance para-meters of the stride were measured from the walkway. During the last 2 minutes of walking, heart rate was measured and expired air-collection was made. After 4 minutes of rest, a second blood sample for lactic acid was taken.

The procedure described above is shown schematically in Figure 1.

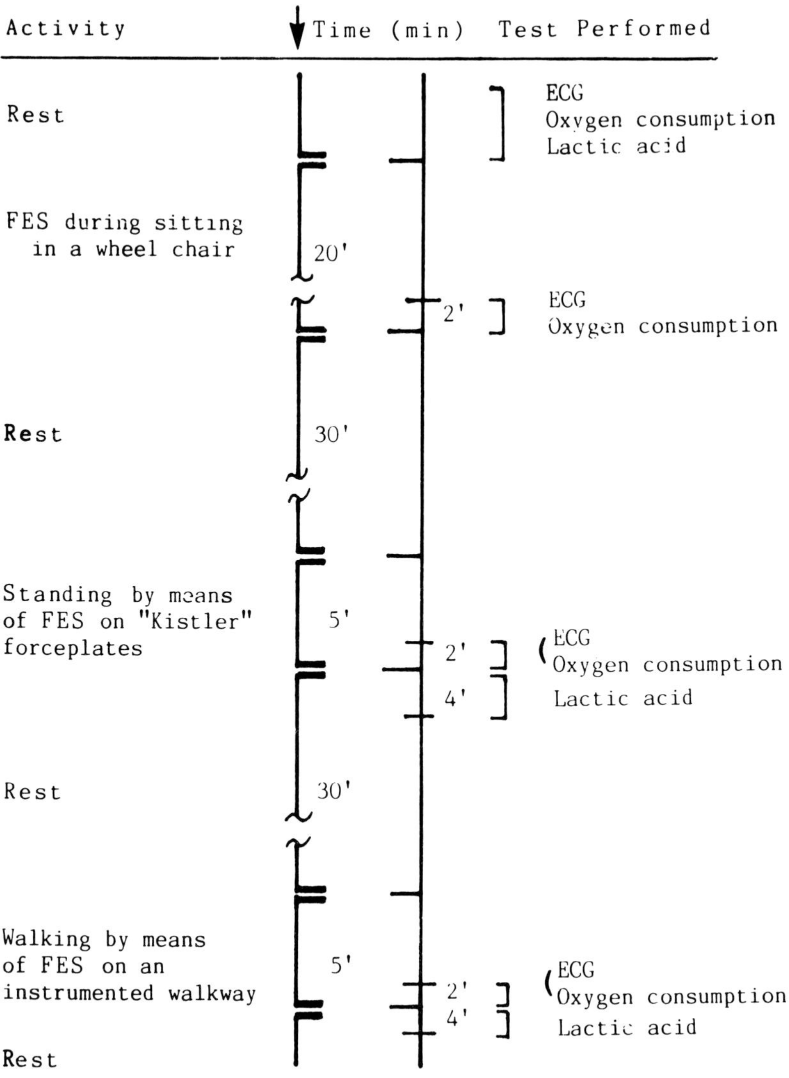

Figure 1: Schematic diagram of procedure carried out for the paraplegic patients during biomechanical and physiological evaluations.

Results

AK amputees

Figures 2 and 3 show the results obtained for the vascular and traumatic amputees, respectively. In Figure 2, the open circles represent the walking results of all the vascular amputees with the open knee mechanism. The black circles present the walking results obtained with the locked knee mechanism. A straight line connecting the walking results with the two knee mechanism tests was drawn for every patient. It is clearly seen that for all the vascular patients, the black circles lie both below and to the right of the open circles, indicating that in these patients the locked knee mechanism enabled both a higher walking speed and a lower HR increase, as compared to the open knee mechanism.

In Figure 3, the results of the HR increase versus walking speed for the traumatic amputees in the open and locked configurations of the knee mechanism are similarly presented (solid lines). The increase in oxygen consumption due to walking is also displayed (dotted lines). It is clearly seen that in the traumatic amputees, unlike the vascular amputees, oxygen consumption during walking, compared to the resting position, is higher with the knee mechanism in the open position (open circles) as compared to the locked position (black circles). The open knee mechanism, however, allows a higher walking speed of these patients as compared to the locked knee mechanism.

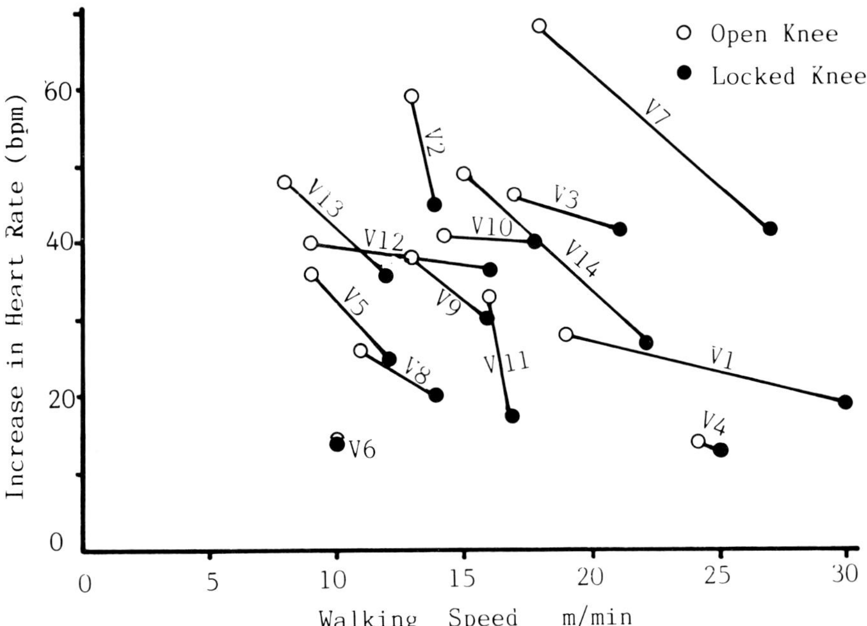

Figure 2: Increase in heart rate versus walking speed for all vascular AK amputees, with open and locked knees.

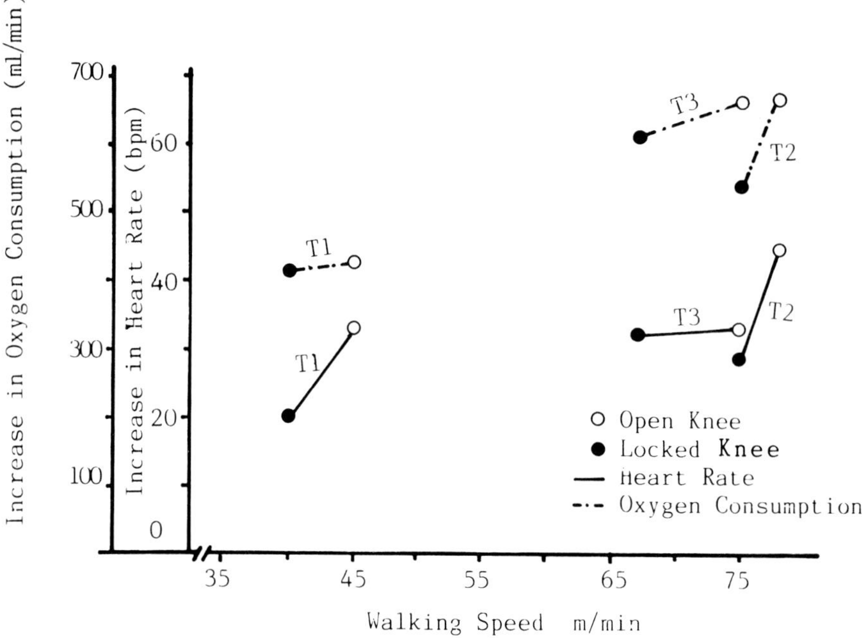

Figure 3: Increase in heart rate and oxygen consumption versus walking speed for all traumatic AK amputees with open and locked knees.

FES-activated paraplegic patients

Cardiac response. Results of heart rate (HR) measured at rest and during FES-activation in the sitting position while exercising, supported standing and supported walking positions are shown in Table 4. Values of HR are also expressed in terms of percentage increase from the resting position.

Following 20 minutes of training of the lower limbs in the sitting position, increase in HR ranged between 10 to 19 bpm (3–24 per cent increase). Standing between parallel bars for 5 minutes, while activating the extensors increased the HR by 25–69 bpm (34–103 per cent increase). Reciprocal gait between the parallel bars, or with a walker, increased the HR by 70–102 bpm (or 79–162 per cent).

Throughout these mentioned activities, subjects were on-line connected to an ECG monitor, showing neither arrhythmias nor S-T changes.

Oxygen consumption. Mean results of oxygen consumption obtained at rest and during FES in the sitting position, supported standing and walking are given in Table 5. Oxygen consumption per unit time is expressed in ml/min and the percentage increase is from the resting position.

In the sitting positions, activation of the quadriceps and dorsi-flexors by means of FES raised the oxygen consumption by 28–238 ml/min (11–101 per cent increase). Standing between the parallel bars for 5 minutes increased the oxygen consumption, compared to the oxygen consumption at rest, between 148–468 ml/min (a 56–208 per

Table 4. Heart rate (bpm) values at rest and during the last 2 minutes of FES in the sitting position, supported standing and reciprocal walking. Increase in % relates to the resting heart rate value. Patients P1, P2 and P4 used a walker as support and patients P2 and P5 used parallel bars (values in brackets are the standard deviation of the mean, SD).

Patient	n =	Rest		Functional Electrical Stimulation (FES)					
				Sitting		Standing		Walking	
	(Observations)	Mean	(SD)	Mean (SD)	Increase(%)	Mean(SD)	Increase(%)	Mean(SD)	Increase(%)
P1	3	62	(4·0)	77 (10·1)	24	92 (7·7)	48	155 (2·6)	150
P2	3	63	(1·5)	70 (7·5)	24	88 (8·1)	40	165 (7·5)	162
P3	3	82	(1·7)	101 (8·1)	23	110 (6·1)	34	152 (5·1)	85
P4	2	82	(1·0)	96 (2·0)	17	120 (4·0)	40	170 (1.0)	107
P5	2	68	(2·0)	70 (1·0)	3	96 (4·0)	41	147 (2·0)	116
P6	4	67	(3·5)	92 (13·6)	22	136 (9·9)	103	– –	–
P7	1	80	(–)	90 (–)	12	– –	–	– –	–

Table 5. *Oxygen consumption (ml/min) values at rest and during the last 2 minutes of FES in the sitting position, supported standing and reciprocal walking. Increase in % relates to oxygen consumption at rest (values in brackets are the standard deviation of the mean, SD).*

| Patient | n = | Rest | | Sitting | | Functional Electrical Stimulation (FES) Standing | | Walking | |
	(Observations)	Mean	(SD)	Mean (SD)	Increase(%)	Mean(SD)	Increase(%)	Mean(SD)	Increase(%)
P1	3	164	(2·0)	330 (70·0)	101	469 (1·5)	186	812 (29·6)	395
P2	3	241	(34·4)	479 (74·5)	99	702 (51·01)	191	1381 (29·6)	473
P3	3	227	(21·3)	423 (71·1)	86	613 (33·0)	170	1095 (35·4)	382
P4	2	262	(4·5)	290 (7·5)	11	410 (12·5)	59	823 (–)	214
P5	2	304	(37·2)	353 (63·5)	16	501 (86·0)	65	840 (–)	176
P6	4	204	(15·8)	391 (67·7)	92	628 (68·5)	208	– (–)	–
P7	1	218	(–)	334 (–)	53	(–) (–)	–	– (–)	–

Table 6. *Summary of results of weight bearing on the legs, stride distance, walking speed, oxygen consumption expressed per unit time (termed oxygen cost) and per unit distance walked (termed efficiency) and lactic acid (values in brackets are the standard deviation of the mean).*

Patient	Standing weight bearing (% W)	Assistive devices	Ambulation Parameters Stride distance (m)	Walking speed (m/min)	Efficiency ($\dot{V}O2 = ml/kg/m$)	Oxygen cost ($\dot{V}O2 = ml/kg/m$)	Lactic acid (mg%)
P1	65	Walker	0·46 (0·02)	8·31 (0·6)	1·52 (0·7)	12·68 (0·4)	68·3
P2	92	Parallel bars	0·67 (0·05)	11·65 (0·9)	1·58 (0·1)	18·41 (0·3)	51·1
P3	98	Walker	0·32 (0·09)	5·40 (0·7)	3·36 (0·9)	18·15 (0·7)	70·1
P4	54	Walker	0·61 (0·08)	8·03 (0·6)	1·70 (–)	13·71 (–)	–
P5	75	Parallel bars	0·51 (–)	4·2 (–)	3·22 (–)	13·54 (–)	–
P6	69	Parallel bars	–	–	–	8·97 (0·9)	74·5
P7	–	Wheelchair	–	–	–	5·56 (0·7)	–

cent increase). Reciprocate walking between parallel bars, or with the support of a walker, while activated by FES, increased the oxygen consumption by 536–1140 ml/min (a 176–473 per cent increase).

Lactic acid. Lactic acid was measured during reciprocal walking in patients P1, P2 and P3 and during standing in patient P5. The concentrations, ranging between 51·1–74·5 mg%, are shown in Table 6.

Evaluation of weight-bearing on the legs in supported standing. Weight bearing on the activated legs during standing was expressed in terms of body weight percentage. The weight borne ranged between 54–98 per cent body weight, depending mostly on the actual posture of every patient. For example, patient P3 was able to bear his body weight almost entirely on his legs while maintaining his postural equilibrium by minimal involvement of his hands. It should be noted that although the weight bearing measurements were done over a period of 5 minutes, the patients were able to maintain the standing position for much longer periods of time, especially in advanced stages of training (Braun *et al.*, 1985).

Reciprocal walking. All patients who achieved walking ability could use parallel bars as a support, however, patients P1, P3 and P4 could use a walker as well. In these patients the average stride distance was 0·46, 0·32 and 0·61 m, respectively. Walking speed was 8·31, 5·40 and 8·03 m/min, respectively (Table 5).

In subjects P2 and P5, supported by parallel bars, stride distance was 0·67 and 0·51 m, respectively. Walking speed was 11·65 and 4·20 m/min, respectively. As training progressed, walking ability improved both in quality, i.e. stride length and speed, and in terms of the total walking distance ability (Mizrahi *et al.*, 1985; Braun *et al.*, 1985).

Evaluation of ambulation efficiency. Knowing the oxygen consumption in ml/min and the walking speed in m/min, the walking efficiency, defined as the ratio between these two values, can be calculated and expressed as oxygen consumption per unit distance travelled (normalized to body weight). The walking efficiencies given in Table 6 ranged from 1·52 to 3·36 ml/kgm (average of 2·27 ml/kgm). No correlation was found between walking efficiency and the walking aid used.

Discussion

In the present work, we studied the mechanics and energetics of walking in two different groups of patients with severe locomotor disabilities: AK amputees and FES-activated paraplegics. In order to evaluate the results, it is important to compare them with those obtained for other groups of patients suffering from locomotor deficits. Figure 4 shows the oxygen consumption from various studies, including the present study. In considering the results found in the literature, the following points should be made: (a) existence of large variability within the individual studies, and, (b) existence of differences between the studies due to differences in populations and methodologies. Therefore, to allow comparison, all the results were standardized in terms of oxygen consumption per unit of body weight, per unit of distance travelled.

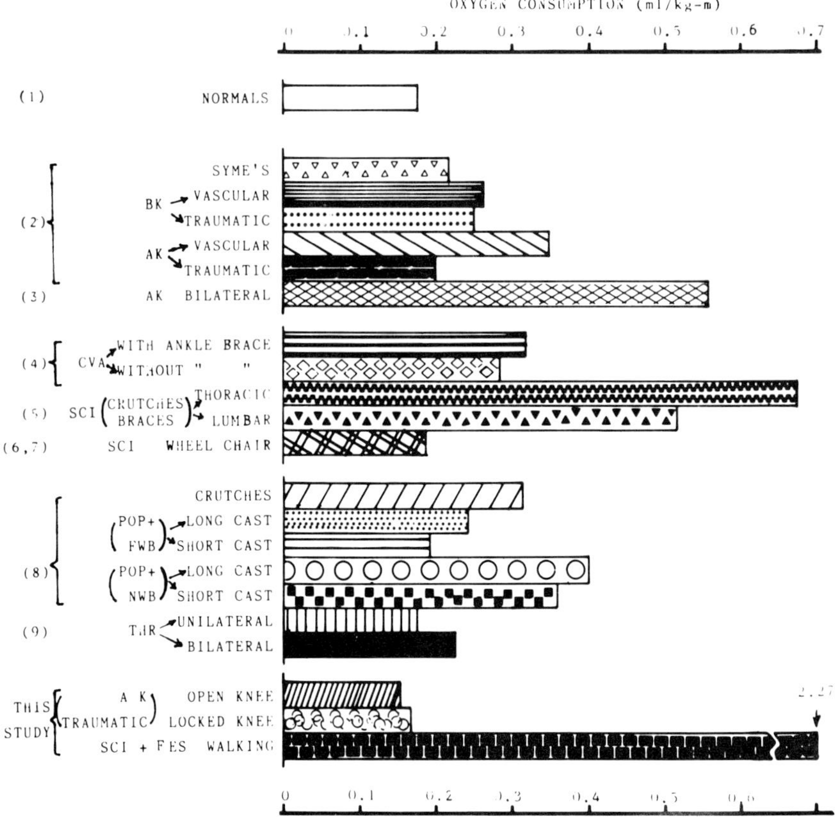

Figure 4: Oxygen consumption per unit body weight per unit distance travelled (ml/kg/m) for various groups of patients, as reported in the literature and from this study. The data presented are grouped from top to bottom in the following order: normals, amputees, hemiplegics (CVA) and paraplegics (SCI), patients with orthopaedic deficiencies and finally the results from this study, including AK amputees and paraplegics (SCI).
Numbers in bracket on the left hand side of the Figure correspond to the following references:
(1) Corcoran and Brengelmann (1970); (2) Waters et al. (1976); (3) Huang et al. (1979); (4) Corcoran et al. (1970); (5) Fisher and Gullickson (1978); (6) Smith et al. (1983); (7) Waters et al. (1978); (8) Waters et al. (1982) and; (9) McBeath et al. (1980).

In evaluating our results for the AK amputees, it should also be remembered that the vast majority of lower limb amputations performed today are due to peripheral vascular diseases in elderly people; these people generally suffer from various associated diseases. In the group of AK amputees, it is accepted that preservation of the knee joint is advantageous, both from the biomechanical and from the kinesiological points of view (Waters *et al.*, 1978; Huang, *et al.*, 1979). In the advanced stages of peripheral vascular disease, AK amputations may be unavoidable, implying that the state of health of AK vascular amputees is normally very low. On the other hand, the group of traumatic AK amputee's average age is much younger. These patients are otherwise in good health and are physically active, making intensive use of their prostheses.

Our results show that the knee joint mechanism of a prosthesis for AK amputees has a different effect in the amputee groups studied. In practice, neither of these groups realized the biomechanical advantages expected in an anatomical and normal knee joint. Among the aged vascular amputees, it was the locked knee mechanism which proved to be most advantageous, as it enabled a higher walking speed and a lower heart rate increase (Figure 2). The patients themselves, however, preferred the locked knee mechanism, with which they felt more secure; and indeed this mechanism was eventually incorporated in their prostheses. In the young traumatic amputees, the open knee enabled faster walking. However, this was at the expense of higher heart rate and metabolic energy cost. This open-knee configuration was also more acceptable to patients. Oxygen consumption of these patients, with the knee joint in the open or locked position, was comparable to that of normal individuals; walking speed, though, was considerably slower.

The second set of results related to the FES-activated paraplegic patients. Biomechanical and physiological evaluations of the activated patients are of great importance due to the increasing interest in applying FES on paraplegics for the restoration of standing and walking abilities by the activation of certain muscle groups in the lower limbs. In this study, evaluation of the activated patients took place in three positions: sitting, supported standing and reciprocal walking. The main parameters studied were weight bearing, gait parameters, heart rate, oxygen consumption and lactic acid. Secondary goals included looking at the effects of FES on the respiratory muscles, spasticity, bladder activity, muscle fibres and blood circulation in the lower limbs.

Activation in the sitting position was based on the application of minimal stimulus intensity required to fully extend the knee joints, while the patient himself remained passive. The principle of minimum stimulus intensity was also applied in other activities, namely standing and reciprocal walking. It, however, necessitated frequent readjustments of the intensities to meet the increasing requirements from the muscle due to fatigue.

Mortimer and Peckham (1973) reported the effect of FES on muscle fibres. They found that the influence on contractility of the muscle was due to transformation of fast-twitch fibres to slow-twitch fibres. It is known that postural muscles are rich in slow-twitch fibres (Astrand and Rodahl, 1986), which are characterized by high endurance ability. The existence of such transformation is therefore very positive and it enables standing and walking with higher endurance levels. However, as has been suggested by Dimitrevic and Nathan (1970), the muscle fatigue observed may also be partly due to habituation of the nerve conduction mechanism to FES.

The increase in oxygen consumption while activating the lower limbs by FES in the sitting position was moderate. The increase in heart and respiratory rates stabilized after a period of 2–3 minutes, corresponding to the steady state stages determined by Astrand and Rodahl (1986). During moderate and medium effort, the oxygen consumption is less than half of the maximal aerobic capacity, hence the amount of oxygen transported to the activated muscles is sufficient for the metabolic reactions taking place, and the muscular contractions are on the whole beneficial for the activation not only of the otherwise inactive lower limbs but also of the whole body. Other outcomes of FES training were increases in the activated muscle's volume and strength.

The importance of standing for paraplegic patients is well recognized, being related, among other things, to blood pressure regulation, calcium metabolism in the lower limbs, osteoporosis, and bladder and intestinal function. Standing via FES activation is achieved by stimulating the quadriceps and gluteus muscles, while the subject is supported by parallel bars or by a walker. Weight-bearing on the legs, which was measured for all the standing patients (Table 6), showed a large variability. Unlike passive standing of a paraplegic, such as on a tilt board, in which no significant increase in heart rate or blood pressure is noted, standing by means of FES was accompanied by a significant increase in heart rate (Table 4) and oxygen consumption (Table 5). In patient P6, the oldest, these increases were extreme. In fact, efforts during standing in this patient reached almost maximal levels as demonstrated also by the lactic acid levels (Table 6).

In the more advanced training stage, reciprocal gait was achieved, especially by younger patients. From the initial standing position, swinging of the leg was produced by stimuli given to 'trigger points' on the anterior aspect of the shank, which activated the withdrawal reflex of the entire limb.

Involvement of the supporting upper limbs is more pronounced in reciprocal gait than in standing. Reference should thus be made to upper limb ergometry in paraplegics, if energetics of reciprocal walking is to be understood. Coutts *et al.* (1983) reported that maximal aerobic capacity in paraplegics with lower thoracic spinal injury, in their thirties, corresponded to a heart rate of 185 bpm and oxygen consumption of 1520 ml/min. Maximal aerobic capacity for lower limbs caused the same increase in heart rate but an oxygen consumption of 3500 ml/min (Astrand and Rodahl, 1986). This striking difference is attributable to the histological and volumetric differences existing between upper and lower limb muscles.

In the five subjects who were able to walk, heart rate increased to 147 bpm (Table 4) and the oxygen consumption exceeded 812 ml/min (Table 5). Lactic acid levels in patients P1, P2, and P3 were 68·3, 51·1 and 70·1 mg%, respectively, (Table 6), the oxygen consumption of a healthy normal subject of 70 kg is 938 ml/min on the average (Astrand and Rodahl, 1986). The extensive usage of the upper trunk and limbs during reciprocal gait of FES-activated paraplegics is thus the main cause for extremely high energy expenditure in these patients.

Astrand and Rodahl (1986) also reported that physical efforts requiring over 50% of the maximal aerobic capacity result in a rise in the blood lactic acid. For a normal subject in unrestrained ambulation, the physical effort corresponds to about 27% of the maximal aerobic capacity of the lower limbs; correspondingly, no increase in the lactic acid level is present. In contrast, FES-activated paraplegics require a physical effort during reciprocal gait of over 50% of the maximal aerobic capacity of the upper limbs; also confirmed by the high levels of lactic acid measured. Such an exhaustive effort can be made only for a very limited period of time. In young patients, this time is of the order of 15 minutes. It is thus clear that optimization procedures aimed at reducing the energy costs of FES, especially in those activities involving the upper limbs, such as supported standing and walking, should be developed in order to safely allow FES application for ambulation of paralysed patients.

Acknowledgement

This study was supported in part by Technion VPR — W. Levenson, Biomedical Eng. R. Fund and by a grant from the Israel Ministry of Defence.

References

Andriacchi, T.P., Ogle, J.A. and Galante, J.O., 1977, Walking speed as a basis for normal and abnormal gait measurement. *Journal of Biomechanics*, **10**, 261–268.

Astrand, P.A. and Rodahl, K., 1986, *Textbook of Work Physiology* (New York: McGraw–Hill).

Braun, Z., Mizrahi, J., Najenson, T. and Graupe, D., 1985, Activation of paraplegics by functional electrical stimulation: training and biomechanical evaluation. *Scandinavian Journal of Rehabilitation Medicine*, Supplement, **12**, 93–101.

Corcoran, P.J. and Brengelmann, G.L., 1970, Oxygen uptake in normal and handicapped subjects, in relation to speed of walking beside velocity-controlled cart. *Archives of Physical Medicine and Rehabilitation,* **51**, 78–87.

Corcoran, P.J., Jebsen, R.H., Brengelmann, G.L. and Simons, B.C., 1970, Effects of plastic and metal leg braces on speed and energy cost of hemiparetic ambulation. *Archives of Physical Medicine and Rehabilitation*, **51**, 69–77.

Coutts, K.D., Rhodes, E.C. and McKenzie, D.C., 1983, Maximal exercise response of tetraplegics and paraplegics. *Journal of the American Physical Society*, **55**, 479–482.

Dimitrevic, M.R. and Nathan, P.W., 1970, Studies of spasticity in man. Changes in flexion reflex with repetitive cutaneous stimulation in spinal man. *Brain*, **93**, 743–768.

Finley, F.R. and Cody, K.A., 1970, Locomotive characteristics of urban pedestrians. *Archives of Physical Medicine and Rehabilitation*, **51**, 423–426.

Fisher, S.V. and Gullickson, G., 1978, Energy cost of ambulation in health and disability: a literature review. *Archives of Physical Medicine and Rehabilitation* **59**, 124–133.

Fisher, S.V. and Patterson, R.P., 1981, Energy cost of ambulation with crutches. *Archives of Physical Medicine and Rehabilitation*, **62**, 250–256.

Gonzalez, E.G., Corcoran, P.J. and Reyes, R.L., 1974, Energy expenditure in below-knee amputees: correlation with stump length. *Archives of Physical Medicine and Rehabilitation*, **55**, 111–119.

Huang, C.T., Jackson, J.R., Moore, N.B., Fine, P.R., Kuhlemeier, K.V., Traugh, G.H. and Saunders, P.T., 1979, Ambulation: Energy cost of ambulation. *Archives of Physical Medicine and Rehabilitation*, **60**, 18–24.

Isakov, E., Mizrahi, J., Graupe, D., Becker, E. and Najenson, T., 1985, Energy cost and physiological reactions to effort during activation of paraplegics by functional electrical stimulation. *Scandinavian Journal of Rehabilitation Medicine*, Supplement **12**, 102–107.

Isakov, E., Mizrahi, J. and Najenson, T., 1986, Biomechanical and physiological evaluation of FES-activated paraplegic patients. *Journal of Rehabilitation Research and Development*, **23(3)**, 9–19.

McBeath, A.A., Bahrke, M.S. and Blake, B., 1980, Walking efficiency before and after total hip replacement as determined by oxygen consumption. *Journal of Bone and Joint Surgery* **62(A)**, 807–810.

Mizrahi, J., Susak, Z., Heller, L., and Najenson, T., 1982a, Objective expression of gait improvement of hemiplegics during rehabilitation by time-distance parameters of the stride. *Medical and Biological Engineering and Computing*, **20**, 628–634.

     Ergonomics in Rehabilitation

Mizrahi, J., Susak, Z., Heller, L. and Najenson, T., 1982b, Variation of Time-Distance parameters of the stride as related to clinical gait improvement in hemiplegics. *Scandinavian Journal of Rehabilitation Medicine*, **14**, 133–140.

Mizrahi, J., Braun, Z., Graupe, D. and Najenson, T., 1985, Quantitative weight-bearing and gait evaluation of paraplegics using functional electrical stimulation. *Medical and Biological Engineering and Computing*, **23**, 101–107.

Mizrahi, J., Susak, Z., Isakov, E., Solzi, P. and Najenson, T., 1986, Optimization of stimulus parameters for reduced fatiguability during standing by functional electrical stimulation of paraplegics. *IV Mediterranean Conference on Medical and Biological Engineering*, September 1986, Sevilla, Spain, pp. 353–356.

Mortimer, T. and Peckham, H., 1973, Intramuscular electrical stimulation. In *Neural Organization and its Relevance to Prosthetics*, edited by W.S. Fields and E.L. Leavitt (New York: Intercontinental Medical Book Corporation), pp. 131–145.

Pagliarulo, M.A., Waters, R., Hislop, H.J., 1979, Energy cost of walking of below-knee amputees having no vascular disease. *Physical Therapy*, **59**, 538–542.

Patterson, R.P. and Fisher, S.V., 1981, Cardiovascular stress of crutch walking. *Archives of Physical Medicine and Rehabilitation*, **62**, 257–260.

Saksena, F.B., 1979, Nomogram to calculate oxygen consumption. *British Heart Journal*, **42**, 282–285.

Saunders, J.B. de C.M., Inman, V.T. and Eberhart, H.D., 1953, The major determinants in normal and pathological gait. *Journal of Bone and Joint Surgery*, **35(A)**, 543.

Smith, P.A., Glaser, R.M., Petrofsky, J.S., Underwood, P.D., Smith, G.B. and Richard J.J., 1983, Arm crank vs. handrim wheelchair propulsion: Metabolic and cardiopulmonary responses. *Archives of Physical Medicine and Rehabilitation*, **64**, 249–254.

Susak, Z., Levy, M., Mizrahi, J. and Isakov, E., 1986, The duration of isometric muscle contraction while using FES with stimuli of different parameters. In *2nd Vienna International Workshop on Functional Electrostimulation*, Vienna, Austria, September 1986, pp. 75–78.

Waters, R.L., Perry, J., Antonelli, D. and Hislop, H., 1976, Energy cost of walking of amputees: The influence of level of amputation. *Journal of Bone and Joint Surgery*, **58(A)**, 42–46.

Waters, R.L., Hislop, H.J., Perry, J. and Antonelli, D., 1978, Energetics: Application to the study and management of locomotor disabilities. *Orthopedic Clinics North America*, **9**, 351–377.

Waters, R.L., Campbell, J., Thomas, L., Hugos, L. and Davis, P., 1982, Energy costs of walking in lower-extremity plaster casts. *Journal of Bone and Joint Surgery*, **64(A)**, 896–899.

PART VI
ASSISTIVE DEVICES

16.
Aided communication for handicapped children

Stephen von Tetzchner
Central Institute of Cerebral Palsy
Oslo, Norway

Abstract

Many disabled children have to use communication aids because they lack the ability to speak or use sign language. The developmental course of aided language is determined by the children's handicap, characteristics of available aids, and environmental effects of the handicap and mode of communication. Children who develop aided language typically have much slower rate of production than children who develop speech. They have a late onset of expressive language and severely limited vocabularies. They are dependent on the listeners' attention and patience to interpret their utterances, and are at risk in developing a passive communicative style. It is possible to enhance an earlier onset of expressive language and the development of more proficient language use by giving the children early opportunities for aided play, providing them with more independent aids based on computer technology, influencing the interaction patterns of disabled children and their caregivers to give the children better opportunities for active and meaningful interaction, and teaching the children effective language strategies. To obtain this, more knowledge about aided communication is needed, as well as new perspectives on aid construction.

Introduction

The causes of communication handicap include CVA aphasia, cerebral palsy, multiple sclerosis, Parkinson's disease and muscular dystrophy. In the USA, there are more than one million children and adults who need communication aids (Bureau for the Education of the Handicapped, 1976). In Sweden, one in every thousand children, between 4 and 16 years of age, has both motor and speech handicaps, mainly due to cerebral palsy. About half of the children have no functional speech (Lagergren, 1981). The figures clearly demonstrate the magnitude of the need, and the importance of developing adequate strategies for language intervention with adults and children who need aided communication.

There is a dearth of knowledge about aided language interaction and conditions that influence language acquisition of children suffering from communication handicap. Such knowledge is necessary for communication aids to be implemented in an optimal way. The developmental course seems, to some degree, to be caused by children's disabilities but also by the characteristics of existing communication aids, as well as by environmental effects of the children's disabilities and mode of communication. This paper discusses some of the factors which may influence the acquisition and use of aided communication among physically handicapped children.

Communication aids

Traditional aids

Traditional aids are based on direct selection or scanning. In direct selection, the user may type on a keyboard or point to a board which displays letters, words, graphic representations of morphemes (e.g. Blissymbols, Pictograms), photographs, or drawings. Typing or pointing may be performed with the hands, feet, a mouthpiece, a headstick, a lightpointer mounted on the head or other parts of the body, or by use of gaze direction.

Scanning may be directed or automatic. In directed scanning, an indicator (e.g. an arrow or a light) is moved to the desired place on a communication board with the help of one or more remote switches. The switches may be operated with, for instance, the hands, feet, eyes, head, or by sucking and blowing. In automatic scanning, the indicator moves at a pre-determined speed, and the user activates the switch when it reaches the desired place. Scanning may also be performed by a helper. The user indicates when the helper points at the right place or says the intended letter name or word. In both direct selection and scanning, the board may consist of numbers of other kinds of code units, which are indicated by the user, and deciphered by the listener.

Use of direct selection presupposes fairly good motor coordination or range of movements, and directed scanning the ability to perform repeated movements. Automatic scanning requires fewer movements, but the user must be able to coordinate his or her own movements with the movement of the indicator to make it stop at the selected place. The number of items on the board is often limited. In direct selection, motor performance may require items of a certain size. In scanning, the indicator may take a long time to traverse the board. Most traditional aids are dependent, i.e. the listener[1] must put the language units together (e.g. letters selected on a letter board) and/or decipher the combinations of morphemes. For example, a typewriter may be an independent traditional aid (for reviews of different types of traditional communication aids, see Charlebois-Marois, 1985; Harris and Vanderheiden, 1980).

Traditional aids are widely used and serve important functions in many people's lives, but they have serious shortcomings: letters are slow if used independently, i.e. without a person guessing words and sentences before the aided communicator is finished spelling; it may take several minutes to spell a single word by pointing to

---

[1] In this paper, *listener* refers to a person who is the receiver of an utterance expressed by an augmentative communication system user.

letters on a board, or wait for an automatic scanning indicator to move to the different letters. For long utterances, communication may break down because the listener has difficulty in keeping track of the letters and words. Sustained attention to the pointing may be difficult, partly because one typically watches the face of a communicating person. When attention fluctuates, one or several letters may be lost, leading to poor guessing, misunderstandings, and frustration. For some users, it is a physical effort to get a message across, and exhaustion may make conversations short. To different degrees, such problems are attached to most of the traditional communication aids.

New generation communication aids

New generation communication aids are based on visual display and synthetic speech. Some aids have both, which makes adaptation to difficult light or sound conditions possible.

Visual display communication utilizes the traditional principles of direct selection and scanning, but with increased efficiency. In scanning, letters, words, or sentences are displayed on a screen and the communicator moves the cursor for selections. For example, for a user to write *camera*, he may write *c*, and then select an alphabetic word list. All stored words beginning with *c* are shown on the screen. If there are too many words beginning with *c*, the user may write *ca* before selecting the word list, and only words beginning with *ca* are displayed. In this way, the user only has to select one or two letters and then a word instead of spelling each letter in the word.

Some communication aids are predictive and supply one or more words, according to the user's own frequency of use, every time a letter is selected. If the user does not want the displayed word(s), another letter is selected. Every time a letter is added, new words are displayed. Preliminary results indicate that this procedure may decrease the number of selections to be made by 40 per cent (Beukelman *et al.*, 1985).

In visual display aids, the words remain on the screen until the users clear it. This makes it less crucial for the listener to attend constantly to what the user writes and communication breakdowns due to lack of attention are less likely to occur than with traditional aids. Both listener and user may be able to pay more attention to non-verbal aspects of communication. The user may, for example, be able to check the listener's gaze direction, facial expression, body movement, or other indications of how the message is received. It may be easier to disregard wrong or unwanted guesses. Utilizing the computer's memory, the user may display premade sentences one at a time, giving, say a joke, the proper timing.

There are two types of synthetic speech. In text-to-speech, words are produced according to complex letter-to-sound rules, including rules for co-articulation, i.e. how sounds near each other influence each other's articulation. The quality of text-to-speech is rapidly improving, but it is still rather poor. It may be difficult to understand when the listener doesn't know what is being talked about. Not being able to have a personalized voice make some users reluctant to use text-to-speech in ordinary communication. The rules used in text-to-speech are also different for each language, which makes it necessary for each language community to have their own text-to-speech aids.

Digitized speech is based on an input of real speech, which is digitized and stored in a memory chip. The advantage of digitized speech over traditionally recorded speech is fast retrieval. There is no need for spooling, and hardly any time is lost in search. A limited vocabulary of digitized speech may be stored in a "talking aid" especially designed for that use, or be a complementary feature of microcomputers used as communication aids.

Synthetic speech has major social advantages over visual display communication. The user can initiate a conversation or interrupt in an ordinary manner. He gets immediate feed-back, both from his own speech and from the listeners' reactions. In vocabularies of digitized speech, the user's local dialect may be apparent, which may be an important factor for group membership and identification of cultural background. In communication aids for children, the immediate feed-back of synthetic speech may be the most important feature. It may facilitate the child's understanding of what it is doing, and comparisons of its own sound productions to those of others may facilitate acquisition of new words.

A major drawback of computer based communication aids has been lack of mobility. The need to communicate exists in a variety of situations, and a stationary communication aid is not of much use when the communicative need arises far away from the work station at home or at school. Traditional aids are more mobile and can be used in most places. Today, many advanced communication aids are portable. Although users of communication aids ought to have back-up strategies, for example in case of equipment failure, new generation aids should be functional in nearly all types of communication situations. Further, in emergency, most new generation aids can be used as traditional aid, for example with a helper doing the scanning if the batteries are low on power.

In the majority of cases, the most functional and easy to use aid is based on high technology. However, the function of an aid in everyday life is dependent on proper implementation, including the teaching of appropriate linguistic and communicative strategies. Lack of resources for the implementation and follow-up may be a limiting factor for distribution of technically advanced aids (Parnes, 1985). In many cases, to provide technology seems to be an end in itself, without enough attention being given to the acquisition of pertinent skills (Goossens and Kraat, 1985). In such cases, new generation aids may have a negative rather than a positive effect on the life situation of aided communicators, because they fail to live up to their own and others' expectations.

Characteristics of aided language

Rate of communication

Aided communication is extremely slow compared to speech. A normally speaking person has a production rate of 120–180 words per minute. In scanning, the output may be 2 words per minute while a person with good motor skills using a device with direct selection or morse may produce 25 words per minute (Beukelman *et al.*, 1985; Yoder and Kraat, 1983). Production rate determines the form and structure of aided language. It is highest when larger units of language than letters are applied, but the size of communication boards severely restricts the number of words. If there are too few words available, the time it takes to get a message across to a listener may increase because analogies of spoken utterances become long, and may have to be repeated or rephrased several times, or conversational topics become limited.

The importance of production rate has implications for adaptation of communication aids to individual users. There is a tendency to favour the activation mode that is most similar to usual writing. For example, the hands may be preferred because this looks more 'natural' to the listeners, even when the user can perform faster with head or feet.

The aided communicators' reduced rate of production makes interaction with natural speakers asymmetrical, with the latter group dominating the communication (Kraat, 1985; Light, 1985). Aids which can increase the users' speed and efficiency will make the interaction more symmetrical, and ease the communicative situation. Thus, several factors indicate that efforts to increase the rate of language production should have a high priority in the construction of communication aids.

Aid input and output

The input and the output of communication aids are parts of the users' language system. Input refers to how the aid is controlled and the language units selected. When communication aids are adapted to individual users, a major part of the process is related to aid input. Aid input is functionally parallel to articulation of speech. For most speaking persons, articulation is an automatized process which is given no attention, unless a rare or difficult word or a foreign language is used. For the aided communicator, control of input may be a cumbersome process even in everyday communications. The physical effort of producing an utterance may take attention from other tasks. If a limited vocabulary is used to formulate a variety of meanings, considerable demands are put on ther user's imagination and ability of divergent thinking.

Aid output is the end product, i.e. the language units presented to the listener. If an aided communicator is using some sort of code (e.g. numbers or Minspeak), the structure of the input may have little relation to the output the receiver sees or hears. The user may, for instance, type *65* and the listener sees 'I am hungry'.

Most communication aids are based on script. The production units are letters, words, and phrases, or combinations of these. In some countries, like Japan or China, morphemic script is in common use (Downing, 1973). In Western countries, graphic representations of morphemes are utilized to make aids accessible to non-reading persons. The most commonly used system is Blissymbolics (Bliss, 1965), but several other systems are also in use (for a review, see Bloomberg and Lloyd, 1986). Investigations of aided communication have mainly been limited to speakers of English or languages that are structurally similar to English (e.g. Norwegian and Danish). It seems important to encourage cross-cultural studies, since they may provide information about linguistic processes in aided communication. For example, the morphemic structures typical of many Asian languages, like classifiers (Allen, 1977), may make it easier to apply written systems like Blissymbolics or Japanese Kanji in these language communities than in many Western countries.

The role of listeners

In aided communication, the listener's role is different from usual communication. The listener actively shapes the aided communicator's utterances. If an aided communicator is spelling, the listener typically guesses words and sentences before they are completed. This seems to speed up communication when the listener's guesses are right. False guesses, however, may slow down communication and lead to frustration. It may also lead to the listener directing the course of the conversation by asking sub-questions, sometimes even before the aided communicator has finished the first utterance (Harris, 1982; Kraat, 1985).

If some kind of pictorial or morphemic system is used, the listener will have to interpret the use of pictures or words, or combinations of pictures, letters, and words. Typically, feedback is given in a step-by-step manner. Many characteristics of aided communication have probably been caused by the dependency on helpers who can remember language elements and interpret the utterance when it is completed. New generation communication aids with screen displays make the users less dependent on the listeners' attention and memory. Synthetic speech make at least single word feedback unnecessary because the user can hear whether the intended word is spoken. Recent researchers have, to some degree, focused on how to make the communicative situation less demanding for the listeners (e.g. Larson and Woodfin, 1986); in choosing communication aids, characteristics of the user's most frequent communication partners should be taken into consideration.

Aided language development and intervention

The environment of physically handicapped children

Children with physical handicaps are at great risk of having a non-optimal language environment. Cerebral palsied children, for example, make much greater demands on their mothers than non-handicapped children. Feeding may be difficult and take up a large part of the time the mother and child spend together. What may be considered mainly as liabilities in a non-handicapped child, seem to be regarded as assets in a cerebral palsied child. These assets include passivity, lack of desires and demands, and full obedience. "She'll sit there and be an angel", "She fell asleep when I wanted her to", "She goes into bed like a good little girl", and "She is not getting into things" are typical descriptions given by the mothers of 2–4 year old children with cerebral palsy (Shere and Kastenbaum, 1966). Although there have been major changes in the way children are regarded, the 20 year old descriptions resemble characterizations made by parents of physically disabled children today.

Children with severe motor handicaps have few activities their parents can react to in a consistent way. They may be unable to smile or vocalize, and crying may be absent or distorted. Signals may be unclear, inconsistent, and in danger of being misinterpreted. For example, the lateral head movement associated with a tonic neck reflex would be easy to misinterpret as rejection of an individual or disinterest in a topic (Bottorf and DePape, 1982).

When present, crying may be less functional than among normally developing children. Parents tend to react when their children cry, and the crying may be interpreted in several ways, depending on situational factors, like time since last feeding or change of diaper (Newson and Newson, 1963). The crying of disabled children, however, may often be caused by conditions the parents can do nothing about, for example muscle aches due to high muscle tonus, or pain in the intestines created by the twisted body. In such cases, soothing will be difficult, the parents may be frustrated by their inability to calm the child, and crying may fail to function as a learning situation for the child. High-pitched crying may provide constant auditory irritation to the caregivers. Such conditions may explain why mothers of children with cerebral palsy interprete passivity as an expression of happiness (Shere and Kastenbaum, 1966).

In addition, the children may lack other kinds of vocalizations, made by the normally developing children, to stimulate interaction.

The transition from pre-linguistic communication may be understood within the framework of overinterpretation theories (Bjerkan *et al.*, 1983). These theories claim that parents and other adults interpret the child as if it was communicating something before it is reasonable to assume that the child has learned this (Ryan, 1974; Lock, 1980). By consistent interpretations of the child's activities, adults create an important learning condition for the child who is learning to use language.

Thus, it is important that parents interpret and react to the child's activities in a consistent way, and this is possible if the child has a repertoire of activities that the parents can interpret as communicative. But children differ with regard to how easy it is to attach meaning to their activities, which corresponds to differences in what Schaffer (1971) calls ''readability''. Children with a low level of activity and/or a limited repertoire of ''legible'' activities (Martinsen, 1980), for example autistic, physically disabled, or blind children, are at risk for having a less optimal environment than normally developing children (Ryan, 1977; Bjerkan *et al.*, 1983). Young disabled children have few activities that their parents conceive as communication.

Children show their interests for objects by moving towards them, pointing or looking at them, or showing them to parents. In contrast, many physically disabled children may not even be able to turn around and scan the environment to see what is talked about. They lack the mobility to explore their surroundings, move to or away from their caregiver, and indicate interest in their surroundings. Reflex patterns may interfere with the child's attempt to react to events and persons in the environment, or to initiate interaction (Morris, 1981; Yoder and Kraat, 1983).

The experiences of physically handicapped children will often be extremely limited. They are dependent on caregivers to introduce objects that can be experienced. Mothers of young cerebral palsied children, however, do not provide their cerebral palsied children with toys, or place the toys out of range of accessibility in order to promote motor performance. Further, they tend not to speak to and do something with the child at the same time (Shere and Kastenbaum, 1966). Thus, the children may have few cues to the relationship between objects or activities and words, and characteristics of the objects or activities which are talked about. Even when the parents are talking about something present in the room, the child may not have sufficient situational cues to understand what they are talking about. It is called ''verbalisms'' when a person is able to define or name correct uses of objects he cannot identify when the objects are put into his hands. Verbalisms is a typical characteristic of the language of blind children, and reflects their experiential world (von Tetzchner and Martinsen, 1980). For instance, a blind child may say that a wrench is a tool and correctly explain how it is used, but call it a hammer when asked to identify it. Similar phenomena can be observed among children with cerebral palsy. For example, a 4-year old boy was asked what the cow says, and replied ''Moo''. Later, when asked to point to a picture of a cow, he was unable to do so, although he could point to other pictures when asked to do so.

The effects of young disabled children's limited repertoire of legible activities on the language environment has implications for development of communication aids. Communication aids to be used by infants and young children will have to be designed in such a way that they contribute to make the language environment more functional. The aids must replace the legible activities the children lack, and provide ways of expressing interest and preferences, which adults and older children in the environment can interpret as communicative and react to. Ideally, infant aids should be designed so that the young aided communicator could, with or without intention, trigger activities that the parents would interprete as signs of interest and communication, thereby creating a basis for the disabled child to learn in a way similar to how children learn non-aided language.

Lack of models who use aided language seems to be another typical feature of the language environment of aided communicators. Even adult-aided communicators may have difficulties communicating with other aid users without an interpreter. Although the listener plays an active role in aided communication, young aided communicators create their own language. There is a wide discrepancy between the language spoken in the environment and the aided language used by infants and young children. Disabled children do not communicate with adult-aided communicators. Neither parents nor professionals use the child's aid to demonstrate how to say something or to expand on the child's utterance. The lack of models seems to make language acquisition more difficult for disabled children than for normally speaking children.

Bottorf and DePape (1982) suggest putting name tags corresponding to present or new words in the aid, on or near important objects in the child's environment to make the child used to non-vocal representations. The labels should be used by family members to supplement speech and to some extent serve as a kind of language model. The strategy may also function to attract the child's and the parents' attention towards object names, and to increase the probability that the child is directed towards the object or activity that the parents are talking about.

Parents and other adults may engage in ''conversations'' with disabled children where they both ask questions and provide the answers. This may have some model value if the questions are yes-no questions and the child is able to express some kind of *yes* and *no*, but most of the time parents use words outside their child's expressive vocabulary. Further, because of the limited vocabulary, parents of disabled children may depend mainly on situational cues for interpreting their children's utterances at an age when parents of non-handicapped children rely more on the words used. This delayed reliance on situational cues may contribute to the development of a passive communicative style and hinder development of less context-dependent language use. The children may have limited possibilities of influencing this situation since their communication skills are assessed according to interpretations made by the listeners.

Implementation of early aided communication is more than a matter of adapting aids to young users. The professionals must try to change the language environment in such a way that it contributes positively to the child's development of communicative skill and independence. It is important that this be done when the child is at an early age. The consequences of early dependency and limited communication skills may, otherwise, not be easily reversed.

Acquisition of communicative style

The passive communicative style is a pervasive characteristic of aided communicators (Hardy, 1983; Harris 1982; Lewis and Ripich, 1983; Kraat, 1985). Both limited response to children's initiations and a tendency not to give the children enough time to initiate interaction are likely to increase the probability that physically disabled children will develop passive styles. Deprivation of contingent experiences may cause lack of initiative and development of social skills (Brinker and Lewis, 1982).

Many aid users do not utilize the potential of their communication aids. They may use it only part of the day, and fail to take advantage of opportunities for expressive communication (Light, 1985; Culp *et al.*, 1986). One reason for this may be the aid users' lack of experience with true speaker–listener conversational exchanges in which the roles of initiator/respondent are switched forth and back. Instead, the aided communicator occupy the listener-respondent role in the majority of interactions (Glennen and Calculator, 1985).

Mothers of 4 to 6 year old children with cerebral palsy initiate 80 per cent of conversations with their children. They treat silence as an indication of a possible breakdown in the flow of conversation, and are likely to start to speak within 1–2 seconds of silence. Silences greater than 1 second are followed by caregiver talk 92·5 per cent of the time. Most of the children's turns are responses, and the average time to respond is significantly shorter than the average time to initiate (0·69 vs. 2·61 seconds), indicating that the passive communicative role, which is typical of aided communicators, is apparent at this early age. Further, the mothers tend to respond more often to the children's responses than to their initiations (Light *et al.*, 1985). Similarly, teachers initiate interaction by asking questions, and the children respond. Children 6–7 years of age, have approximately the same number of turns as their teachers, but the teachers' turns are much longer. The children never express more than one unit per turn. Furthermore, the teachers often ask new questions without giving enough time to reply to the original question, as in this example (Harris, 1982):

T(eacher):   What did you do last night?
C(hild):   (Begins to formulate a response using a communication board).
T: Did you go home?
T: Did you watch tv?
T: Did you see Walt Disney?
T: Did your brother come?

Also in other situations, children who use aids are often deprived of opportunities for communication. When somebody tries to speak with the child, parents or professionals typically answer on the child's behalf even when it is evident that the child understands what is said and has the means to respond.

The communicative styles developed by cerebral palsied children and their caregivers may be linked to the aided communicators' dependency on the listener to give feedback about words and to interpret the utterance. Independent feedback may be of special importance for language acquisition. Early and extended dependency on the environment may create a communicative style which is difficult or impossible to change. For example, about three years ago, a ten year old boy got a computer with a programme which enabled him to write by directed scanning of letters and words with two switches. Until that day, his expressive language had consisted of looking up to indicate affirmative and down as negative. Today, the boy has good technical mastery of his aid, yet he does not initiate communication. He is entirely dependent on someone to prompt him, even in situations where it is evident that he has something to say.

Glennen and Calculator (1985) found that they could teach users to take conversational initiative, but the skill remained dependent on the presence of situational cues and did not transfer to other board uses. This indicates that aided language, the way it is taught, does not provide children with an optimal tool for developing interactive skills. It is possible, however, to foster independence and a less passive style if strategy development begins early and is incorporated throughout the process of developing an augmentative communication system (Bottorf and DePape, 1982).

The physical characteristics of most aids make communication with other aid users difficult; peer interaction is extremely rare (Harris, 1982). Even young adult aid users in daily contact with other aid users are not used to discuss how they can ease communication between themselves and thus create a richer language environment. However, when asked to, they may invent new strategies, like, pointing at the other person's board, or turning the board upside down so the listener can see better (von Tetzchner, 1986). Another way of enhancing peer interaction would be to connect two or more children (not necessarily all handicapped), for example by using a split computer screen or connecting two Cannon Communicators so the users can write on each others paper strip. Besides giving new opportunities for communicating, it may increase communicative independence since the users would have to solve communicative breakdowns or misunderstanding sharing the same tools. It may also be an early training in use of telecommunications for children who live in residential schools to communicate with their parents without help.

In addition to giving the child an independent aid and enough time to initiate and sustain interaction, it is important to provide subjects to talk about. Because of the physical limitations on independent experiences, most conversations may centre around events that are known to the listener. Conversations become social rituals rather than an exchange of information and views. This, in addition to expressing basic needs and answering simple questions, may become tedious, even among children.

Availability of communication is an obvious, but often neglected point with regard to acquisition of communicative style. Communication aids are typically put at the back of the wheel chair, and only made available to the child when the adult wants to ask something or believe the child has something to say. Boards are often made of cardboard which are difficult to lean on for the user who needs physical support for pointing. For example, a woman failed to learn Blissymbols until she got a wooden board strong enough to support her.

Onset of aided language

Disabled children often do not get communication aids until they start school (Morris, 1981). Shere and Kastenbaum (1966) mention no use of communication boards or other kinds of communication aids among 2 to 4 year old children with cerebral palsy. Udwin and Yule (1976) report an average age of onset of just over 5 years for 20 children using Blissymbols. The late onset may be linked with the fact that the children on the average scored below their age level on the Colombia Mental Maturity Scale and the Reynell Comprehension Language Scale. For 8 children with cerebral palsy and no known cognitive handicap, onset of Blissymbols use ranged from 2:11 to 6:1 years, with 5 of the children between 3:5 and 3:9 years. No use of aids before the onset of Blissymbols is mentioned (Light, 1985; Light *et al.*, 1985). A cerebral palsied girl got a picture board when she was 2 years old. The board was first replaced with line drawings and then written words. At 5 years of age, the girl could respond with a 200–word communication board (Beukelman *et al.*, 1985). Many children start with picture boards, but there is very little data available on the onset of these early aids. Seemingly, children tend to be three years or older before communication aids are put to functional use.

The implementation of communication aids contrasts with the willingness of parents to believe that children have begun to use language. Physically handicapped children have to demonstrate that they are able to use language, while among other children the use is taken for granted. Aids are implemented to disclose communicative skills rather than to develop them.

The late onset of aided communication puts unnecessary restrictions on non-vocal children's development. It has been suggested that aid implementation should start between 18 and 24 months (Morris, 1981). The situation would improve if children could get their first aid at 18–24 months of age. On the average, normally developing children have an expressive vocabulary of 10 words when they are 15·1 months old, and of 50 words when they are 19·6 months old (Nelson, 1973). It is not evident that physically disabled children have a more limited vocabulary at this early age. It has yet to be demonstrated that acquisition of aided language is more difficult than the acquisition of non-aided language. Some language delayed children learn to use aided language (Premack's symbol system), although they had failed to learn non-aided forms (signs or speech) (Deich and Hodges, 1977). Even the general claim that augmentative communication for mentally retarded children should wait until they have reached Piaget's sensorimotor stage 6 (Kahn, 1975) is not supported (Smith and von Tetzchner, 1986).

Naturally developing speech and signs are based on spontaneous vocalization and actions, while today's aids cannot be applied spontaneously before the child understands their use. It is not an easy task to teach infants to follow instructions, especially when they are physically handicapped, and therefore it is not surprising that aided communication skills appear to be more difficult and must be learned at a later age than signs and speech. It is necessary to make aids which are not dependent on extensive instructions to create opportunities for development of social and communicative skills at the earliest possible age. This is a new demand on communciation aids. Ideally, and in line with over-interpretation theories, aids should be designed so that the young disabled child could, with or without intention, trigger activities that the parents could interpret as signs of interest and communication. Admittedly, such aids are not easy to construct. They must have an input mechanism, for example a switch, and some kind of cue that signal which word will be expressed. The words may be triggered by movements of the head, leg, arms, etc. Visual or tactile cues may be combined with switch activation to produce synthetic speech. For example, vibrations of different frequencies from a device strapped to the arm, or a light moving behind simple symbols or drawings on the wall, a table, or a wheelchair may provide the child with information about what word will be spoken on the speech aid on his chest when he activates the switch. The aid should not be triggered too often by random movements, but often enough to give the child frequent practice and the chance to learn the distinguishing cues for the different words that are spoken, and the consequences that each of them has.

Disabled children may have some ways of signalling their needs and interests before they get a communication aid. It is important to maintain these already acquired communication skills. The first expressions of like and dislike (typically interpreted as *yes* or *no*) may develop through a process of overinterpretation similar to usual language development. Many disabled children vocalize, gesture, and point or look at objects in the environment (Culp *et al.*, 1986). Such activities may be regarded as augmentative communication since they in many situations replace part of an ordinary vocabulary. The aim of implementing communication aids is to extend communication and provide new skills, and negligence of incorporating existing non-aided skills may partly explain the reluctance many children have to using their aids. It may have taken years to acquire the skills, and simply replacing old functions with new forms does not necessarily make the child a better communicator. A child may be more motivated to use his aid if he is taught how the aid provides him with means of expressing things he could not say before, and to communicate in new situations.

Aided play

Early aids replace the functions of pointing, gestures, and/or speech. Similarly, aids for play and exploration may be developed and used both before and parallel with aided communication. Aided play, for example manipulating toys or playing simple games with another person by using switches, may give the child and his parents new opportunities for meaningful activities and enhance the child's communicative development in several ways.

In child–parent interaction, a play aid may enable the disabled child to signal its interests and toy preferences. The parent can focus on the object of the child's attention, and thereby create shared attention based on the child's initiative and signals. A computer may be used to simulate play objects typically used by 10 to 15 month-old children. Cars may be driven across the screen, or made to disappear. Blob, a computer program for the BBC (Douglas and Detheridge, 1986), has a kind of peek-a-boo game where a partly hidden animal appears from behind a brick wall when the child activates a switch. In the scanning mode, a star moves at a pre-set speed over a wall with a partly hidden cat, teddy bear, and Blob. The animal under the star appears when the switch is activated. The game may be extended to use in communicative interaction by giving the animals different functions. For example, there may be complementary pictures where the cat drinks milk, the teddy bear drinks juice, and Blob drinks water. The child gets milk, juice, or water, depending on which animal is under the star when he activates the switch. In other situations, the animals may be doing different activities, like playing music, reading, or playing a computer game.

Play aids may help the child to learn that things have names, and new words may be introduced by the parents while they and the child are playing with objects and animals on the screen. The initial goal, however, is to establish interaction rather than building up language.

The first words

Early communication aids for children tend to be based on pictures, and conventional writing systems are introduced late. There is no available data on the content of early picture boards or how they are used. Research has mainly focused on communication aids with orthographic script or Blissymbols. Pictures are often not taken seriously although they form the basis of expressive language development. The pictures used by young aided communicators are early language forms, comparable to the first words uttered by speaking children, and may be regarded as verbal, but not vocal.

The function of the first aided words is to make children understand that they can use language to obtain objects, get attention or help, and control the environment in different ways. All children have a wish to communicate and if they do not communicate, it is because they do not know how to do it or that thay can do it (Gunning, 1983). But children differ with regard to interests; there is no set of words that would suit all children. The first words must be chosen among objects and activities that the child has already showed an interest in or a preference for.

Early vocabularies are often of limited functional value to the child and tend to be used to make the child respond. It is usual to initiate aided language by teaching the child to say *yes* and *no* by pointing or using a scanning device even when the child's expression of likes and dislikes are quite well understood. This is only a formal replacement which may even make communication less efficient. There is no reason why a child should point on a board or use a scanning device instead of nodding his head. Of course, when his language proficiency improves, it may be useful to know how to indicate *yes* or *no* in a formal manner. For the first words, it is a better strategy to introduce two words (letters, symbols, pictures) at a time, and to teach the child to indicate the preferred one. This would expand the child's vocabulary and may support development of active strategies.

For many children, Blissymbols represents the first introduction to a systematic vocabulary. Blissymbols mostly denote generic categories like *building*, *animal*, or *man*, and use is based on analogy. To name something specific, two or more Blissymbols have to be combined. For example, *building* and *letter* may be combined to indicate "post office". The system makes words (which tend to be acquired early) more complex to express than words which tend to be acquired at a later age. For example, children typically use *cat* and *dog* before they learn to use *animal*, and only gradually do they learn that cats, horses, dogs, and cows are animals (Macnamara, 1982). It is possible to express "cat" by combining several Blissymbols, or *animal* + C, but this is more complicated than expressing "animal". The implication may be that Blissymbols can not be fully used until the child is a fairly competent language user.

The possibility of multiple layers of new generation communication aids (Baker, 1986) may imply natural strategies for developing aided language vocabularies. One approach would be to make aids based on semantic tree structures where the child starts with the medium sized branches (dog, cat) rather than larger (animals, birds) or smaller (Persian, Siamese cats) branches. Multiple layering is also a way of providing easy access to large vocabularies. Sixteen items and two layers give 156 (16 × 16) words. If another layer is added, vocabulary size increases to 4096 (16 × 16 × 16) words which can be accessed with three selections.

Vocabulary growth

In aided communication, vocabulary growth refers to both qualitative and quantitative aspects. The user may learn to spell and will then, in principle, have an unlimited vocabulary. Most young aid users, however, use morphemic writing or pictures which may make vocabulary size a complicated issue. Board items to be used with children must be fairly large, and even then it may be difficult to see which picture or word is indicated. If the board is too big, some words may be difficult to reach. If it takes too long before the scanning marker reaches the target, the child may become impatient. Furthermore if the vocabulary is small the child may lack ways of expressing himself. If it is large, retrieval may be too slow for efficient use.

Limited vocabularies make it necessary to develop preferences. It is important that the vocabulary mirrors developmental changes. As the child grows older and gets new interests, he will need new words to be able to talk about new topics. It may be a severe mistake to follow an additive principle. Some words may be removed to make room for words which are more functional for the child's age and current interests.

The words on a communication aid influences the topics communication partners initiate. If the parent or teacher who selects new words fails to remove old ones, the vocabulary may contribute to infantilization of the user. New topics will seldom be introduced by communication partners, and the user may be deprived of language experiences which are necessary for the development of more advanced conversational skills. Word selection is not an issue in the development of normal speech because there seems to be nearly unlimited storage, but it may be of extreme importance in the development of aided language and conversational skills.

Physical growth may make the child able to reach more words and improved motor control may make it possible to use smaller items or a faster scanning device. New retrieval modes may be introduced as the child gets older making access faster and productive vocabulary larger (Beukelman and Yorkston, 1984). It may be necessary to change word positions according to the child's change in use of vocabulary. Traditionally boards are organized according to syntactic and semantic categories that may slow down production rate in comparison to arrangements where the most frequent words are the most easy to access (Goossens and Kraat, 1985). For example, a fifteen year old Blissymbols user who relied heavily on spelling was able to move his hand quite quickly and with good motor control at the mid-line in front of him, but the letters were placed in alphabetic order in two rows under the Blissymbols. Frequently used letters like *a* and *b* were difficult for him to reach and likely to be confused by the listener due to imprecise pointing. The board arrangement made communication unnecessarily hard and exhausting.

Many young aided communicators have a well developed comprehension of spoken language, but their expressive vocabularies are severely limited. Vocabulary growth is very slow and typically vocabulary size ranges between 100 to 350 words/symbols in 5 to 12 year old aid users (Beukelman, *et al.*, 1985; Glennen and Calculator, 1985; Light, 1985; Magnusson, 1983). The figures may be somewhat misleading since some of the older children may use letters as well. It should be noted, however, that many physically disabled children develop poor reading and writing skills and some fail to acquire even basic reading skills. Children who use Blissymbols may change nouns to verbs, or indicate ''opposite'', which may make their functional vocabulary correspond to a somewhat larger spoken vocabulary. Still, many topics are not easily communicated due to the general content of the vocabulary.

A vocabulary must be structured in an efficient way which the user can remember and utilize. The user must have possibilities of in-depth discussions. Common words may be of little use without the less common words which often are critical for the message (Michaelis *et al.*, 1977). It would be useful to teach children specialized vocabularies in different situations in addition to a more general vocabulary for use in situations where a topic is not yet agreed upon. For example, a child may have one board (manual or computer display) for use in shopping, another for eating out, playing, walking in the woods, swimming, watching a soccer game, etc. (Mills and Higgins, 1984). He may also have a ''work station'' with a complete vocabulary for writing letters, documents, poetry, etc. The child may not have his full vocabulary available at all times, but one doesn't need all the words one knows in all situations and the words that are available have been selected to be efficient in that particular situation. Also, their knowledge about a wider range of vocabularies may increase the children's participation in communication situations where these words are needed.

For children who have learned to spell, Blissymbols may be replaced with syllables, frequent consonant clusters, or words selected to function optimally with spelling. For example, words shorter than three letters may have to be avoided because they may make little difference with regard to production rate. Function words are often redundant and may be left out in colloquial communication with familiar listeners (Morningstar, 1981; Jennische and Thunnell, 1982) and, therefore, they should not take up part of a supplement vocabulary.

The transition from Blissymbols to orthographic script is a crucial point in the development of aided language. It normalizes communication and makes it possible to have an alphabetically structured memory which is the most efficient way of storing larger language units. In conversations with Blissymbols users one may find wide variation in how Blissymbols are used together with letters. Some use Blissymbols to speed up spelling while others rarely use the letters on the boards. However, even the most proficient spellers seem to lack adequate strategies for combining Blissymbols with spelling. For example, one 20 year old boy spelled *han* (he) although it would have been faster to point to the corresponding Blissymbol. Another user pointed to *animal* before he spelled *cow*. Blissymbols are not designed to be used in conjunction with spelling and may not function optimally in this way. Still, as soon as letters are introduced, children should be taught how to use them in the most efficient way in conjunction with Blissymbols. It may also be a good strategy to introduce the initial letter of words together with Blissymbols before the child is taught to read. This may expand the vocabulary because the letters can be used in combination with Blissymbols. *Animal* + *h* may mean ''hare'' while *animal* + *m* may mean ''moose''. The strategy may make parents point to the correct letter creating natural modelling situations.

When decisions on aids are made, the characteristics of likely communication partners must be taken into account. For aided communicators in daily contact with Blissymbols users, a re-structuring of the Bliss board may be the best solution since this will make interaction with the other Blissymbols users possible. In addition, the better communicator may serve as a language model for the others. The user may, however, have a communication aid which is not based on Blissymbols for use in other settings.

Descriptions of aided language seem invariably to refer to language production. It is readily assumed that the comprehension of aided speakers is similar to children developing non-aided language. If a disabled child shows poor understanding of language, this is attributed to brain damage or lack of relevant life experience. However, the total language learning situation of aided communicators is radically different from normal development. Naturally speaking children learn the conventional use of words both by hearing the words used by others and by using old and new words themselves in a variety of contexts; what have been termed ''Talking to learn'' (Clark, 1982). Young aided communicators are more or less cut off from exploring words through use because their vocabularies are rarely changed. On the contrary, they are constantly trying to use their limited vocabulary in unconventional ways in order to expand the range of meanings they can express. The impact on their understanding of how words are used is not known, nor whether their language comprehension differs from that of other children of the same age.

Combination of words

Young aided communicators use word combinations to express syntactic relations and to make analogies to words which are not in their aid's vocabulary.

Children with limited vocabularies may have to create combinations to say even quite common words. It is not known to what extent each child develops a consistent system and whether combinations are understood consistently by the listeners. The analogies made by a child may be regarded as neologisms. In general, children using aids do not interact and do not create common use of combinations among themselves. For children using Blissymbols, establishing conventions, i.e. agreeing how the symbols are to be used, would be extremely difficult. The users do not have the means to provide feedback to each other about how to gloss a particular combination of Blissymbols because this would imply that they have another common language system. Thus, young aid users are dependent on help in establishing common use. Idiosyncratic Blissymbols combinations may become conventional only if they are identified by a person who can teach them to other persons using the same system. From a different point of view, by identifying and distributing neologisms made by the children, teachers may utilize the children's creativity and signal an accepting attitude towards their use of language.

There is almost no data on syntactic features of aided communication. There seems to be a tendency to use short utterances. Lewis and Ripich (1983) found a mean length of utterance of 1·45 for two adults using Blissymbols. Only 30 per cent of the utterances produced by 3 to 9 year old children using Blissymbols were combinations (Udwin and Yule, 1986). All the combinations showed the word order of written English, which may not seem surprising since this is the word order that children are taught. Still, aided communicators have been reported to utilize special language strategies related to characteristics of communication aids. Patterson and McFreter (1980, reported by Morningstar, 1981) found that children omit symbols lacking in semantic content and point to the most informative symbol first. The study indicates that children with comnmunication aids may use a word order that is different from the one used in speech and presumably fitted to their mode of communication. Such adapted strategies should be encouraged and refined by teaching and applying strategies found in pidginization (von Tetzchner, 1985).

Early communication is typically taught with great stress on syntax and grammar (Harris, 1982; Beukelman, *et al.*, 1985). When the output is script, the bias towards formal language is emphasized. Many researchers are concerned with how aided communicators' ''ungrammatical'' utterances can be translated by high technology aids into correct written language and treat deviances from correct written language as syntactic errors. It may be more constructive to regard the ''errors'' as expressions of the communicator's language. A closer investigation may disclose linguistic features related to characteristics of the aid. For example, pointing twice at a word for affirmation. Further, the most effective strategy should be encouraged even when it is at odds with spoken language use (Goossens and Kraat, 1985; von Tetzchner, 1985).

Concluding remarks

The ultimate goal in aided communication is to teach handicapped children proficient and independent language use. To approach this goal, it is necessary to provide them with aids at the same early age as other children begin to develop language. The communication aids must be functional, and enhance initiative and peer interaction. Implementation must be based on current knowledge about natural and aided language, and developmental processes in general. It will be necessary to influence the interaction patterns of young aided communicators and their parents, as well as the language intervention strategies of professionals. Both traditional and new generation aids have shortcomings that influence the users' possibilities of communication. As quality improves, synthetic speech is likely to become the standard output, and the major obstacles to efficient communication are related to user input. Production rate and vocabulary access are the major concerns, and even the technologically most sophisticated aids are probably far from a final solution. The new scanning devices that are constructed have little impact on the users' communication skills. At present, there seems to be most to be gained from using better instructional strategies and teaching language strategies that are adapted to aid and user characteristics. The characteristics of communication aids influence aided language, and the language strategies determine how the aids are used. When more knowledge about aided communication has been gathered, it may be possible to look at aid construction from new perspectives, and to improve disabled children's language and interaction strategies.

References

Allen, K., 1977, Classifiers. *Language*, **53**, 285–311.

Baker, B., 1986, Using images to generate speech. *Byte*, **11**, 160–168.

Beukelman, D.R. and Yorkston, K.M., 1984, Computer enhancement of message formulation and presentation for communication augmentation system users. *Seminars in Speech and Language*, **5**, 1–10.

Beukelman, D.R., Yorkston, K.M. and Dowden, P.A., 1985, *Communication augmentation: A casebook of clinical management* (London: Taylor and Francis).

Bjerkan, B., Martinsen, H., Schjolberg, S. and von Tetzchner, S., 1983, Communicative development and adult reactions. Paper presented at 2nd International Conference on Social Psychology and Language, Bristol, U.K., July, 1983.

Bliss, C.K., 1965, *Semantography-Blissymbolics* (Sidney: Semantography Publications).

Bloomberg, K.P. and Lloyd, L.L., 1986, Graphic/aided symbols and systems: resource information. *Communication Outlook*, **7**, 24–30.

Bottorf, L. and DePape, D., 1982, Initiating communication systems for severely speech-impaired persons. *Topics in Language Disorders*, **2**, 55–71.

Brinker, R.P. and Lewis, M., 1982, Making the world work with microcomputers: a learning prosthesis for handicapped infants. *Exceptional Children*, **49**, 163–170.

Bureau for the Education of the Handicapped, 1976, *Conference on Communication Aids for the Non-vocal Severely Handicapped*, Alexandra, Virginia, December, 1976.

Charlebois-Marois, C., 1985, *Everybody's technology: a sharing of ideas in augmentative communication* (Montreal: Charlecoms).

Clark, R., 1982, Theory and method in child-language research: Are we assuming too much? In *Language development*, edited by S. Kuczaj, II (Hillsdale, N.J.: Lawrence Erlbaum), **1**, pp. 1–36.

Culp, D.M., Ambrosi, D.M., Berniger, T.M. and Mitchell, J.O., 1986, Augmentative communication aid use — a follow-up study. *Augmentative and Alternative Communication* **2**, 19–24.

Deich, R.F. and Hodges, P.M., 1977, *Language without speech*. (London: Souvenir Press).

Douglas, J. and Detheridge, T., 1986. *Blob. Switch operated programs for children* (London: Widgit Software).

Downing, J., 1973, *Comparative reading* (New York: MacMillan).

Glennen, S.L. and Calculator, S.N., 1985, Training functional communication board use: a pragmatic approach. *Augmentative and Alternative Communication*, **1**, 134–142.

Goossens C. and Kraat, A., 1985, Technology as a tool for conversation and language learning for the physically disabled. *Topics in Language Disorders*, **5**, 56–70.

Gunning, M., 1983, Facilitating communication in young children with Cerebral Palsy at the Cheyne Centre for Spastic Children. Paper presented at the International Cerebral Palsy Society Workshop on Communication, Stockholm, Sweden, October, 1983.

Hardy, J.C., 1983, *Cerebral Palsy* (Englewood Cliffs, New Jersey: Prentice-Hall).

Harris, D., 1982, Communicative interaction processes involving nonvocal physically handicapped children. *Topics in Language Disorders*, **2**, 21–37.

Harris, D. and Vanderheiden, G., 1980, Augmentative communication techniques for non-vocal severely handicapped children. In *Non-speech language and communication*, edited by R.Schiefelbusch (Baltimore: University Park Press).

Jennische, M. and Thunnel, G., 1982, In *Mina ord i bild-fallstudie av Bliss*, edited by M. Magnusson (Stockholm: Handikappinstitutet), pp. 51–53.

Kahn, J.V. 1975, Relationship of Piaget's sensiromotor period to language acquisition of profoundly retarded children. *American Journal of Mental Deficiency*, **79**, 640–643.

Kraat, A., 1985, *Communication interaction between aided and natural speakers* (Toronto: Canadian Rehabilitation Council for the Disabled).

Lagergren, J., 1981, Children with motor handicaps. *Acta Paediatrica Scandinavia*, suppl. 289.

Larson, J. and Woodfin, S.T., 1986, Training communication partners of augmentative communication users. Paper presented at '*Communicating across Boundaries*', Cardiff, September, 1986.

Lewis B.A. and Ripich, D.N., 1983, Pragmatic language of cerebral palsied adult speakers and augmentative communication device users in a group interaction. Paper presented at the 37th Annual Meeting of the American Academy for Cerebral Palsy and Developmental Medicine, Chicago, October, 1983.

Light, J., 1985, *The communicative interaction patterns of young nonspeaking physically disabled children and their primary caregivers* (Toronto: Blissymbolics Communication Institute).

Light, J., Collier, B. and Parnes, P., 1985, Communicative interaction between young nonspeaking children and their primary caregivers: Part I — Discourse patterns. *Augmentative and Alternative Communication*, **1**, 74–83.

Lock, A., 1980, *The guided reinvention of language* (London: Academic Press).

Macnamara, J., 1982, *Names for things* (London: MIT Press).

Magnusson, M., 1983, *Mina ord i bild — fallstudie av BLISS* (Stockholm: Handikappinstitutet).

Martinsen, H., 1980, Biologiske forutsetninger for kulturalisering. *Tidsskrift for Norsk Psykologforening, Monografiserien*, **6**, 122–129.

Michaelis, P.R., Chapanis, A., Weeks, G.D. and Kelly, M.J., 1977, Word usage in interactive dialog with restricted and unrestricted vocabularies. *IEEE Transactions on Professional Communication*, PC–20, 214–221.

Mills, J. and Higging, J., 1984, An environmental approach to delivery of microprocessor-based and other communication systems. *Seminars in Speech and Language*, **5**, 35–45.

Morningstar, D., 1981, Blissymbol communication: Comparison of interaciton with naive vs. experienced listeners. Unpublished M.Sc. thesis (University of Toronto, Canada).

Morris, S.E., 1981, Communication/interaction development at mealtimes for the multiple handicapped child: implications for the use of augmentative communication systems. *Language, Speech, and Hearing Services in Schools*, **12**, 216–232.

Nelson, K., 1973, Structure and strategy in learning to talk. *Monographs of the Society for Research in Child Development*, **38**.

Newson, J. and Newson, E., 1963, *Infant care in an urban community* (London: Allen & Unwin).

Parnes, P., 1985, Augmentative communication: a model for service delivery. Paper presented at "Communication through Technology for the Physically Disabled", Dublin, September, 1985.

Ryan, J., 1974, Early language development: Towards a communication analysis. In *The integration of a child into a social world*, edited by M.P.M. Richards (London: Cambridge University Press), pp. 185–213.

Ryan, J., 1977, The silence of stupidity. In *Psycholinguistic series*. Vol. 1: *Development and pathological*, edited by J. Morton and C. Marshall (London: Elek Science), pp. 99–124.

Schaffer, H.R., 1971, *The growth of sociability* (Harmondsworth: Penguin).

Shere B. and Kastenbaum, R., 1966, Mother–child interaction in cerebral palsy: Environmental and psychosocial obstacles to cognitive development. *Genetic Psychology Monographs*, 1966, **73**, 255–335.

Smith, L. and von Tetzchner, S., 1986, Communicative, sensiromotor, and language skills of young children with Down syndrom. *American Journal of Mental Deficiency*, **91**, 57–66.

Udwin, O. and Yule, W., 1986, An evaluation of nonspeech modes taught to cerebral palsied children: Final report to the education Advisory Committee of the Spastics Society (London, U.K.).

von Tetzchner, S., 1985, Words and chips — pragmatics and pidginization of computer-aided communication. *Child Language Teaching and Therapy*, **1**, 295–305.

von Tetzchner, S., 1986, Samtaler om kommunikasjon. *CP-Bladet*, **32**, 17–20.

von Tetzchner S. and Martinsen, H., 1980, A psycholinguistic study of the language of the blind: I. Verbalism. *International Journal of Psycholinguistics*, **19**, 49–61.

Yoder D.E. and Kraat, A., 1983, Intervention issues in nonspeech communication. In *Contemporary issues in language intervention*, edited by J. Miller, D.E. Yoder, and R.L. Schiefelbusch (Rockville, Maryland: American Speech and Hearing Association), ASHA report 12.

17.

Sensory augmentation for enhanced control of FNS systems

Ronald R. Riso

Departments of Biomedical Engineering and Orthopedics
Case Western Reserve University, USA

Address of author: Rehabilitation Engineering Program
c/o Department of Orthopedics
Cleveland Metropolitan General Hospital
3395 Scranton Road
Cleveland, Ohio, 44109, USA

Abstract

Sensory systems that provide cognitive feedback of limb function and stimulator status are being developed for use with functional neuromuscular stimulation (FNS) systems that restore purposeful movements in otherwise paralysed limbs. Two systems based on electrical stimulation of the skin sense are being incorporated into FNS grasp restoration systems. One sensory system, designed for spinal injured patients with C6 level function, uses a single, subdermally placed electrode to stimulate the skin. This simple sensory display provides the C6 patient with machine status information and a frequency encoded signal related to the grasp system control signal which the patient generates through voluntary shoulder movements. The other sensory feedback system is for patients with C5 level function and utilizes an array of five cutaneous electrodes to display machine status information, a spatially encoded signal that tracks the output of the patient's shoulder command controller, and a six level frequency encoded signal that is proportional to the grip strength produced during active grasping. Investigators working with FNS based gait assist systems for paraplegics and prosthetic limbs for leg amputees are also studying ways to substitute for lost sensation. This includes kinesthetic and tactile information such as knee angle, foot-to-floor contact and weight distribution under the feet. This paper reviews the current status of these efforts.

Introduction

Functional Neuromuscular Stimulation (FNS) systems are being developed to augment or restore purposeful movements in otherwise paralysed limbs. FNS systems enable stroke patients and quadraplegic individuals with high cervical level spinal injuries to grasp and release objects (Rebersek and Vodovnik, 1973; Vodovnik *et al.*, 1978; Peckham *et al.*, 1980a, 1980b); hemiplegic patients to have improved gait (Liberson *et al.*, 1961; Waters, 1977; Stanic *et al.*, 1978) and paraplegic patients to stand or walk (Brindley *et al.*, 1978; Kralj *et al.*, 1980; Cybulski *et al.*, 1984; Holle *et al.*, 1984; Petrofsky and Phillips, 1983; Petrofsky *et al.*, 1986; Marsolais and Kobetic, 1987). A prominent difficulty common to all of these efforts is the necessity to deal with the losses of tactile and kinesthetic sensibility that are present in the majority of these patients. The absence of sensory information as basic as contact of the fingers with a grasped object, or that one's foot is in contact with the floor so that weight transfer can be executed, can severely limit the effectiveness of FNS systems. Without the benefit of substitute or augmentative sensory feedback devices, patients mainly have to rely on direct visualization of their FNS-assisted limb for feedback control. This can be tedious at best, and in certain circumstances, such as in dim illumination, can be totally ineffective.

Investigators from several laboratories have sought to alleviate these problems to improve the function of artificial limbs for amputee patients, and have pioneered a variety of techniques to provide sensory feedback information including stimulation of the skin sense (Mann and Reimers, 1970; Kato *et al.*, 1970; Prior *et al.*, 1976; Shannon, 1979; Scott *et al.*, 1980; Kawamura *et al.*, 1981; Lovely *et al.*, 1985) and direct electrical stimulation of peripheral trunk nerves (Clippinger *et al.*, 1974; Mooney and Reswick, 1976; Anani *et al.*, 1977). Some of these techniques can be used to enhance the control of neuromuscular prostheses that are based on FNS.

The skin sense is an attractive channel to input information to the individual, because, unlike vision and audition, it can be accessed in a manner that is totally private to the individual, and because it is usually under-utilized. Although electrical or mechanical means can be used to stimulate the skin, it is more convenient to use electrical stimulation for applications to FNS systems since they already involve the use of electrical stimulators and electro-neural interfaces. Also, it is possible to implant both the stimulator and the stimulating electrodes to provide a highly cosmetic sensory system (Riso, *et al.*, 1983, 1987; Smith *et al.*, 1987).

The issues of privacy and cosmetic acceptability are similarly well met if electrical stimuli are delivered to peripheral trunk nerves, but this approach is not practical for spinal injured patients who have high level lesions. For such patients, stimulation of the skin sense above the level of anaesthesia using some form of electocutaneous display is probably most appropriate.

This paper reviews the current status of efforts to incorporate cognitive feedback into upper and lower extremity FNS systems. Hopefully, this information will be useful to those who would like to expand the applications of the current sensory feedback techniques and to those who are working to advance the technology in this field.

Sensory augmentation for use with FNS grasp restoration systems

Sensory systems that provide cognitive feedback for use with neuroprostheses that restore functional grasp in quadriplegic patients have been developed by investigators at Case Western Reserve University (Riso *et al.*, 1987). Two systems based on electrical stimulation of the skin sense, or electrocutaneous communication as this technique is called, are being evaluated. One system utilizes only a single sensory electrode site and is for use with patients whose spinal injury has left them with cervical level 6 (C6) voluntary function, and who consequently usually retain some touch and pressure sensibility in their thumb. The other sensory system utilizes an array of five sensory electrodes and is intended for use by quadriplegic individuals whose residual functional ability is limited to cervical level 5 (C5). The C5 patients, in addition to their more severe motor deficits during attempted voluntary grasping (i.e. they lack voluntary wrist extention which C6 patients retain, Malick and Meyer, 1978), have the additional liability of not having any appreciation for pressure or even skin contact over their thumb or anywhere else on their hand. To help compensate for this, the multi-electrode sensory feedback system incorporates means to provide C5 patients with coded information about grasp force derived from an artificial thumb-mounted sensor.

Although the single electrode sensory system provides a sub-set of the information furnished with the multi-electrode system, details of how similar information is coded for each of the two systems differ in some regards. In either case, to appreciate the rationale behind the choice of the information displayed and the manner in which it is displayed, it will be helpful to first describe the algorithms that the patients utilize in controlling the Case Western grasp restoration systems. This is presented below, and is followed by descriptions of the two sensory systems with the order of presentation arranged so that the multi-electrode system is considered first.

Control algorithms for the CWRU grasp restoration neuroprosthesis

As illustrated in Figure 1, a patient generally obtains voluntary control of the FNS hand prosthesis system by means of a chest-mounted push-button switch which serves mainly to turn the system *ON* and *OFF*, and by a two-axis proportional joystick transducer (Neuman and Buckett, 1982) that monitors voluntary shoulder movements along two orthogonal axes. Through protraction-retraction of the shoulder opposite to the FNS-assisted hand, the patient can proportionally grade the contractions of the forearm muscles to open or close the hand and to regulate grip strength. Rapid vertical shoulder movements are also transduced and are converted to logical signals to serve other control functions. The FNS system provides several user-selectable functions which include a choice of either *PALMAR* or *LATERAL* hand grasp modes; the capability to specify any arbitrary position of shoulder rotation to the the *ZERO* or start position of the shoulder controller command movements; and the ability to latch or *LOCK* the grasp at any desired level of force. (Peckham *et al.*, 1980a,b).

Multi-electrode sensory system for use with C5 quadriplegic patients

A multi-electrode cutaneous display is used to furnish C5 patients with tactile feedback and with other information pertinent to the operation of the FNS hand system. A possible arrangement of the multi-element electrocutaneous sensory display

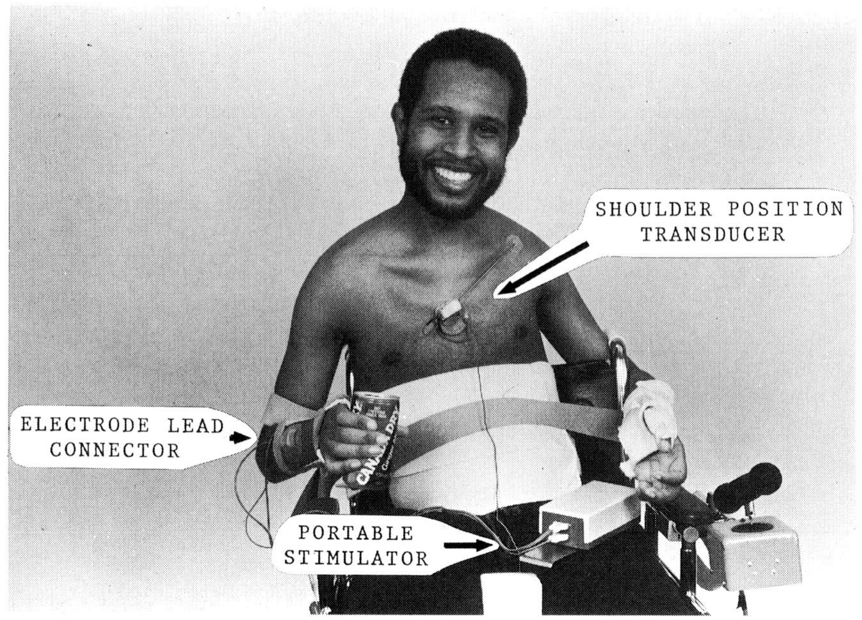

Figure 1

Quadriplegic patient with C5 level function using a functional neuromuscular stimulation-based grasp restoration neuroprosthesis developed at Case Western Reserve University. The patient controls the opening and closing of his FNS-assisted hand and regulates the grip strength by movements of his opposite shoulder. Increased shoulder protraction causes proportional increases in the contraction strength of the nger flexor muscles. Shoulder retraction relaxes the grasp and opens the hand through activation of nger extensor muscles. Logical control signals are derived from rapid vertical shoulder movements.

is to place the electrodes in a single line across one side of the patient's upper chest (Figure 2). The electrodes should be placed at least high enough on the chest to be above the level of skin anaesthesia that exists because of the spinal injury. For the C5 quadriplegic this requirement can usually be met by placing the sensory electrodes within or above the C4 sensory dermatome which should possess normal sensibility. The electrodes should be separated from each other by about 35 to 40 mm to ensure that the patient can easily discriminate their absolute positions. If the patient's chest is too narrow for the five electrodes to fit in a single line, an arch pattern could be used.

The information provided to the patient via the multi-element electrocutaneous display includes: coded messages to enable the patient to switch his FNS hand function between *PALMAR* and *LATERAL* grasp modes; a spatial position analogue of the prosthesis proportional command signal; and a frequency coded signal that is proportional to the amount of prehensile force developed during active grasping.

The sensory feedback algorithms may best be explained by discussing each of them in relation to the steps described below (see Figure 3) that an FNS system user normally performs to grasp or release an object.

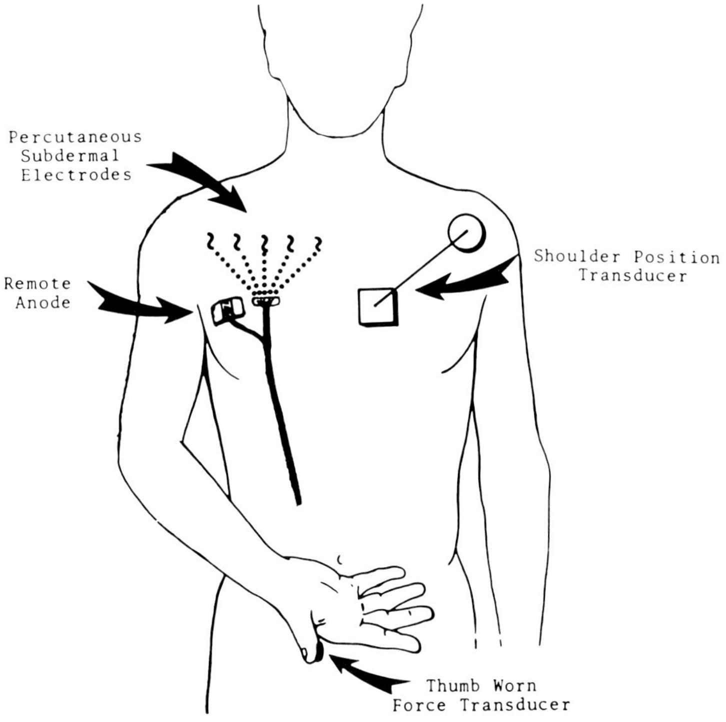

Figure 2
An electrocutaneous display is used to provide the patient with grasp force and other sensory information about the function of the neuroprosthesis during grasp activities. The sensory feedback display consists of five percutaneous coiled wire electrodes placed subdermally in the upper region of the patient's chest. Electrodes are spaced about 35 mm apart so that the patient can readily identify their individual positions when they are activated.

Selection of the mode of grasp

The FNS system is in an *IDLE* state at the beginning of the system "start-up" procedure and there is no activity in the sensory display. By depressing the chest-mounted switch, the user causes the system to enter the *SELECT GRASP* state. During this state the user maintains the chest switch in the depressed condition, causing the FNS system to toggle every two seconds between the *PALMER GRASP* and *LATERAL GRASP* modes. The user then releases the chest switch to select the grasp mode that was most recently displayed. Alternation between the two modes of grasp during the *SELECT GRASP* state is signalled to the user by activating the most *medial* sensory electrode from the array at a *low* frequency (7 Hz) for palmar grasp and by activating the *lateral* sensory electrode at a *high* frequency (30 Hz) for lateral grasp.

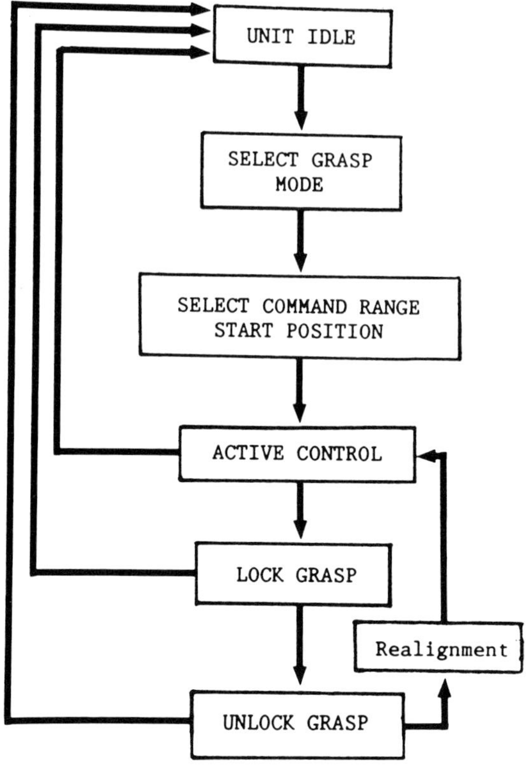

Figure 3

Flow chart describing patient control algorithms for the FNS grasp restoration system developed at Case Western Reserve University. With the system in the IDLE state, the user depresses a chest-mounted switch to begin the system start-up procedure which consists of first selecting a mode of grasp and then specifying a particular position for the controller range. With this accomplished, the user is given active proportional control of his hand grasp function. At any time, the user may enter a state of LOCK GRASP and disengage the proportional controller. Voluntary control may be regained after performing a realignment procedure. The system can be returned to the idle state from either the ACTIVE, LOCK GRASP or UNLOCK GRASP states.

Selection of the zero command position

As soon as the chest switch is released, a 3 second interval commences during which the user must move his shoulder to the position that will correspond to the *ZERO* or start position of the command range. The position of the shoulder at the end of the 3 second period is automatically registered by the FNS system to be the zero position for the command range. The shoulder control scheme is designed so that the user's full command range generally corresponds to 50 per cent of the total possible shoulder protraction-retraction excursion. This arrangement eliminates the need for the user to produce large changes in shoulder position which can be fatigueing and can interfere with postural stability while seated in a wheelchair. Occasionally, if the user must change his posture greatly to grasp an object that is unusually far away or overhead, he may need to return to the system start-up procedure and specify a different shoulder position to be the zero of the command range.

During the selection of the zero point, there is some risk that the user might select a shoulder position that is too protracted. Since the gain of the shoulder movement transducer is constant, he would then not be able to produce the upper limits of the command range even when he fully protracts his shoulder. This problem is avoided by providing the user with feedback about the absolute position of his shoulder during the timing interval for the specification of the zero point as follows: each of the five positions in the electrocutaneous display is used to code for one-fifth of the range of shoulder motion, and at any given moment, one of the electrodes is active in accordance with the instantaneous position of the user's shoulder. During specification of the zero point, the user can be assured that the full command range will be available by ascertaining that the zero point selection is made only when either the 1st (most lateral), 2nd or 3rd electrode sites are active.

The sensory system also provides the user with a confirmation of the shoulder position that the FNS system registers as the zero point. This is accomplished by briefly increasing the frequency (to 30 Hz) of the electrode that was active within the display at the time that the zero point selection period expired. The high frequency signal then ceases, and activity shifts to the 1st electrode of the array (at a base frequency of 4 Hz), which from then on represents the shoulder position corresponding to the start of the command range. At this time, the remaining four electrode sites are automatically remapped relative to the absolute position of the shoulder, so that the five electrode sites together represent the full command range, which, as previously stated, is usually established to be 50 per cent of the user's possible range of shoulder movement. At this point, the user proceeds to the state of *ACTIVE CONTROL*.

Active Control

During the state of *ACTIVE CONTROL*, voluntary shoulder movement in the direction of increased protraction increases the command signal and causes the hand to begin closing. The proportional command signal as determined by the position of the shoulder is continuously displayed to the user by activating successive electrodes in the feedback display in a manner proportional to the output of the shoulder position transducer.

Locking and Unlocking the Grasp

The final aspect of what may be referred to as *machine state feedback* concerns the initiation of the *LOCK GRASP* state and the user's subsequent task of *UNLOCKING* the grasp and regaining voluntary proportional control. *LOCK GRASP* refers to a condition during which the user can maintain the muscle stimulation parameters according to an arbitrary level of command and disengage the shoulder position controller. This is useful when an object is to be held for an extended period since the user can maintain the grasp at a fixed strength while not having to devote attention to its control.

To enter the lock grasp state, the user raises his shoulder rapidly at the extent of protraction and, hence, level of grasp force that is desired. After the lock grasp state is entered the feedback signal automatically turns OFF to keep the skin from becoming excessively accommodated and to serve as a message confirming the entry into the lock grasp state.

Unlocking the grasp is not as easily performed as locking the grasp and can present some difficulty. In order to avoid an abrupt change in the level of grasp force upon the resumption of voluntary control, the user's shoulder must be returned to the same position that it was in when the lock grasp state was entered. However, the user may not be able to find the "realignment position" readily. To assist in this matter, the FNS sensory system provides a realignment error signal via the electrocutaneous display, as follows: to initiate the unlock procedure, the user makes a rapid vertical shoulder movement, and a sensory electrode at either the medial or lateral end of the display is activated as an indication to the user to further protract or retract his shoulder, respectively, to achieve realignment. Successful realignment is signalled by brief activation of the electrodes at both ends of the display simultaneously at a standard frequency of 7 Hz. If no adjustment is necessary because the shoulder position at the time of the rapid vertical movement to initiate the *UNLOCK* was already at the realignment position, then no error signal is given and the sensory system immediately displays the signal for successful realignment. In either event, the user must maintain the realignment position for at least 0·5 second after which the unlock task is automatically completed; a confirmation of this event is given by momentarily increasing the frequency of the endmost electrodes of the display to 30 Hz; and the user is returned to the state of *ACTIVE CONTROL*, in which only one electrode from the display is active in accordance with the instantaneous position of the shoulder. Failure to maintian the realignment position for the 0·5 second interval results in a return to the realignment error signal state.

Grasp force feedback

The multi-electrode sensory system provides feedback about the strength of the grasp whenever the user is in the state of voluntary active control and is engaged in grasping activities. Grasp force information is encoded by a frequency modulation code which consists of six discrete levels within the range of 4 to 55 Hz. This information is super-imposed on the shoulder position–command feedback signal by varying the frequency of the electrical stimuli to whichever one of the sensory display electrodes is activated at each moment.

Alternative spatial position feedback for FNS grasp restoration systems

An alternative strategy for sensory feedback that might be useful is illustrated in Figure 4. The sensory display would consist of the same five element electrode array

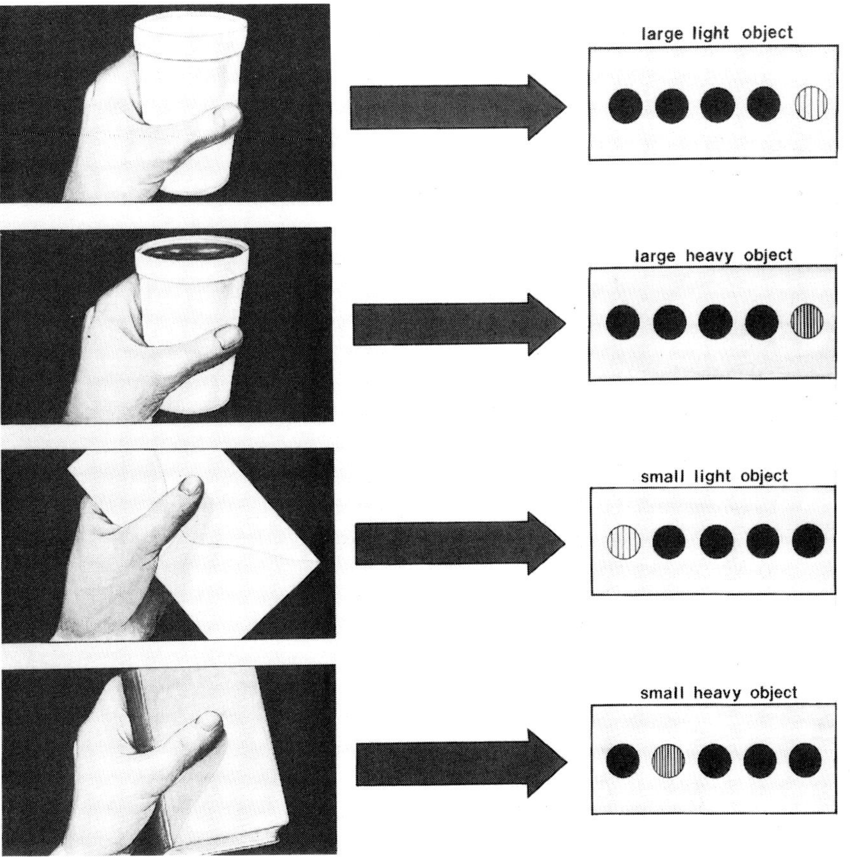

Figure 4

An alternative multi-electrode sensory feedback system that could be used with future FNS grasp restoration neuro-prostheses. This sensory system would provide information about the extent of opening of the hand by keeping one electrode from the array active at any given moment, and using the position of the active electrode to encode the absolute position of the hand. Grasp force information would be displayed simultaneously, by varying the stimulation frequency of whichever electrode of the display is active at each moment. Highly cosmetic and unencumbering sensors that can be worn on the hand will be used to obtain the grasp force and hand position information.

previously described, but the spatial position channel will furnish information about the extent of hand opening (i.e. the thumb to forefinger separation) instead of the output of the shoulder controller.

Proprioceptive information about finger opening could be a useful adjunct to grasp force information, however, such feedback will only become practical when highly cosmetic and reliable transducers become available to monitor finger position. Such transducers are presently under development (Neuman, 1987) and hold promise for meeting these needs.

Single electrode sensory system for use with C6 quadriplegic patient

A sensory system compatible with FNS systems that are totally implanted has been developed for use with C6 quadriplegic patients. This sensory system was designed to utilize only a single electrode sensory display, because the implanted stimulator (Smith *et al.*, 1987) that is used with it has a total of eight independent outputs, and seven of them are required for muscle stimulation to produce the grasp patterns. The single electrode sensory display consists of a monopolar, platium-iridium, disk-shaped electrode placed subcutaneously near the implanted stimulator in the patient's upper chest. The stimulator is controlled and powered by means of a transcutaneous radio frequency coupling.

To facilitate the user of the implanted FNS system with selection of the grasp mode, the single electrode sensory system provides ''machine state'' information as follows: as the FNS system toggles between lateral grasp and palmar grasp modes, the activity of the sensory electrode alternates between a low frequency (4 Hz) and a high frequency (20 Hz) signal respectively.

During the state of *ACTIVE CONTROL*, the single sensory electrode provides a feedback signal consisting of stimulation at one of 5 fixed frequencies (4, 10, 20, 35, or 55 Hz) in conjunction with the output of the user's shoulder position command transducer. Whenever the user enters the *LOCK GRASP* state, the sensory electrode is inactive. When the user desires to return to the state of active control a ''review'' of the previous command level is provided by displaying a 2 second signal at the frequency corresponding to the command level that was present when the user entered the lock grasp state.

Sensory feedback for lower extremity FNS systems

Sensory feedback can be incorporated into lower extremity FNS system by applying the same technology used for upper extremity FNS systems. However, there are presently only a few reports of investigators' experiences in this area. These are described below, after which the scope of the discussion is expanded to include some of the developments of researchers who have sought to provide sensory feedback to amputee patients who use artificial legs. It is useful to consider these efforts as well, since the FNS and artificial limb prostheses situations share considerable common ground.

An array of electrodes applied to the left and right sides of the patient's trunk (Figure 5) could be used to signal foot-to-floor contact, relative limb loading, and the distribution of forces at strategic points on the bottom of the feet. This approach is under development at the Veterans Administration Medical Center in Cleveland (Riso, Kobetic and Marsolais, work in progress) as part of their efforts to restore standing and walking in paraplegic patients.

For each of the patient's feet, the heel, and the medial and lateral aspects of the fore-foot, are each represented by one of the three spatially-distinct electrodes applied to a region of the patient's body where cutaneous sensation is intact. Individual load cells incorporated into the shoe soles measure the forces at the respective positions under the feet, and this information is presented to the patient via the electrocutaneous display.

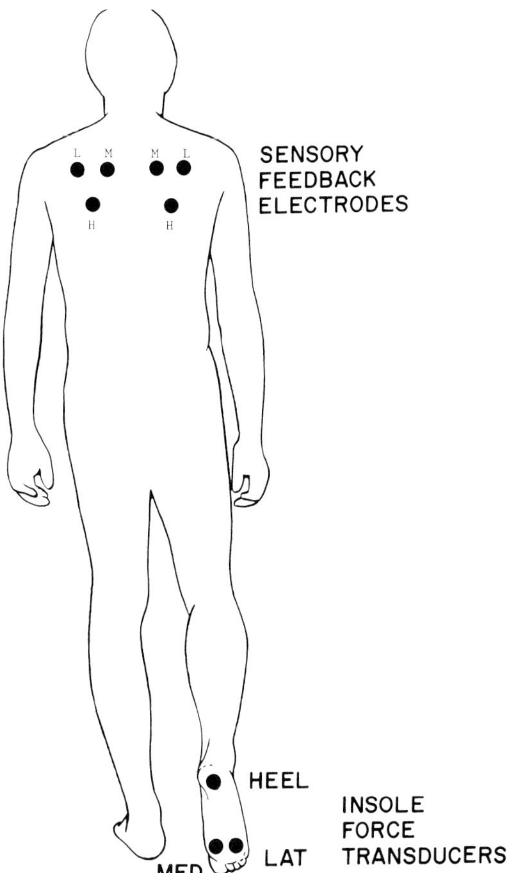

Figure 5

Spatial coding scheme used to display foot–floor contact and the distribution of forces on the bottoms of the feet for cognitive feedback during paraplegic walking and standing. This sensory system, which is based on electrocutaneous communications techniques, is being developed for use by spinal injury patients at the Cleveland Veterans Administration Medical Center who use FNS neuroprostheses for standing and walking. Each of the three electrodes on either side of the patient's back will provide contact and force information from the respective sensors mounted on the soles of the patient's shoes. (L, Lat = Lateral; M, Med = Medial; and H = Heel).

The informational content of this feedback algorithm is considerably greater than that of the upper extremity sensory feedback system described previously, even though the information that is input to the individual is force in both instances. The difference is that for the lower extremity application, the patient must attend to several different force feedback signals simultaneously. Success in this endeavor will depend on the details of the coding strategy employed. The use of frequency modulation coding could be expected to be particularly difficult to interpret when more than one FM signal is changed simultaneously as was found by Prior and his colleagues (1976), who applied a frequency modulated electrocutaneous signal to the dorsal surface of an amputee's forearm to code for grasp force, and at the same time, displayed the extent of prosthesis hook opening via a similar frequency modulation electrocutaneous signal applied to the ventral surface of the forearm. In attempting to interpret the sensory signals, the prosthesis wearer alternated his attention between the two electrode sites and experienced considerable confusion in so doing.

For the case of lower extremity FNS sensory systems, it is likely that the integration of the three force encoding signals from each foot would be accomplished most successfully if they could be perceived as a Gestalt rather than individually. This will probably be facilitated by using an intensity modulation scheme to encode each of the foot position–force feedback signals, although there are presently no reports of this hypothesis having been tested.

Investigators at Wright State University have incorporated a sensory feedback component into their FNS system for paraplegic walking (Phillips and Petrofsky, 1985a,b; Petrofsky *et al.*, 1986). Their sensory system is conceptually similar to that which was described above, in that force information from transducers mounted in the shoe soles is displayed to the patient via the skin sense; however, some substantial differences from the Cleveland system include the use of vibrotactile stimulation of the skin instead of electrocutaneous stimulation, and the provision of force information from four locations on the bottom of each shoe instead of from only three.

The coding paradigm used with the Wright State vibrocutaneous feedback system involved assigning a separate carrier frequency to each of the four skin stimulation channels that represent the floor contact forces of each of the patient's feet. The amplitude of the vibratory sensation induced at each skin locus was modulated to reflect the magnitude of the corresponding foot-to-floor contact forces. The investigators reported that the use of the sensory feedback system by a blindfolded paraplegic subject (T-4 level, complete) during FNS assisted standing enabled him to maintain balance, and that he could not do so when the system was deactivated. Problems reported included variability in the patient's sensory perceptions which were caused by difficulties in maintaining the skin contactors in their correct locations during each day's use and from one day to the next. Also, the investigators suggested that a simplified scheme may have to be studied in which each skin stimulator would provide only a binary signal instead of a proportional signal, so that each stimulator would either be active at a fixed amplitude or not active at all.

Sensory feedback for use with artificial legs

A sensory feedback system based on surface application of electrocutaneous signals was tried by Kawamura and his colleagues (1981) to provide above-the-knee amputee patients with foot-to-floor contact information from prosthetic legs. The electrocutaneous display consisted of a set of four concentric electrodes applied to the anterior of the patient's thigh (Figure 6A). A system of four independent tape switches, each with a fixed sensitivity and output stimulator channel, was used to drive the four corresponding electrodes. Quantitative testing using experimentally-produced floor tilt indicated that the feedback system enhanced the amputees' abilities to detect foot-to-floor contact and maintain balance.

An alternative feedback system (Figure 6B) was designed by these same investigators to furnish above-the-knee amputees with information related to prosthesis knee angle. This information was signalled as follows: with the knee in full extension the sensory electrodes were all off. As knee flexion exceeded 15°, electrode 1 was switched on. At 45° flexion the sensory activity was switched to electode 2, and at 60° of knee flexion or greater electrode 3 was activated. Studies with the first subject using this paradigm showed that he had no difficulty in using the feedback signal to estimate knee position when his prosthetic knee was moved slowly, but he experienced difficulty in tracking the absolute position of his knee at walking speeds. This result ensued despite the fact that the investigators had already simplified the feedback display to consist of just three electrode positions instead of a more complex four electrode design, which earlier studies had shown was too difficult for subjects to interpret rapidly.

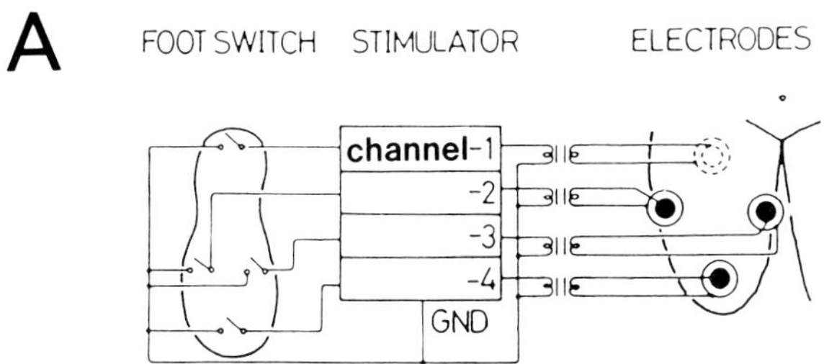

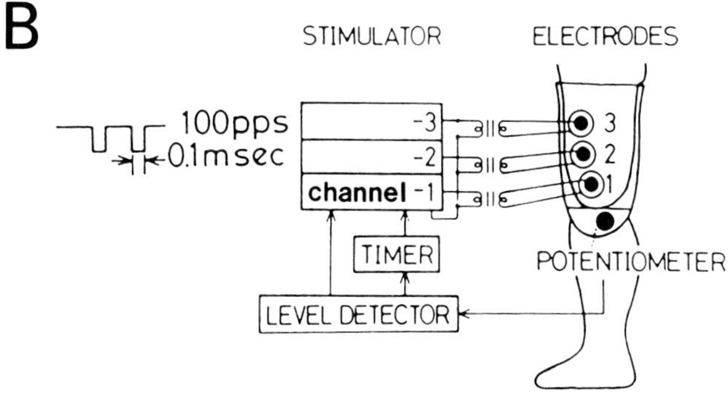

Figure 6
Electrocutaneous sensory feedback strategies used to provide above-the-knee amputees with information about their prosthesis function. The system shown in part A utilizes a set of four concentric electrodes to signal the state of prosthesis foot–to–floor contact from four independent locations under the foot. Part B of the figure shows an alternative sensory display which provides spatial position encoded information related to the state of flexion-extention of the prosthesis knee.

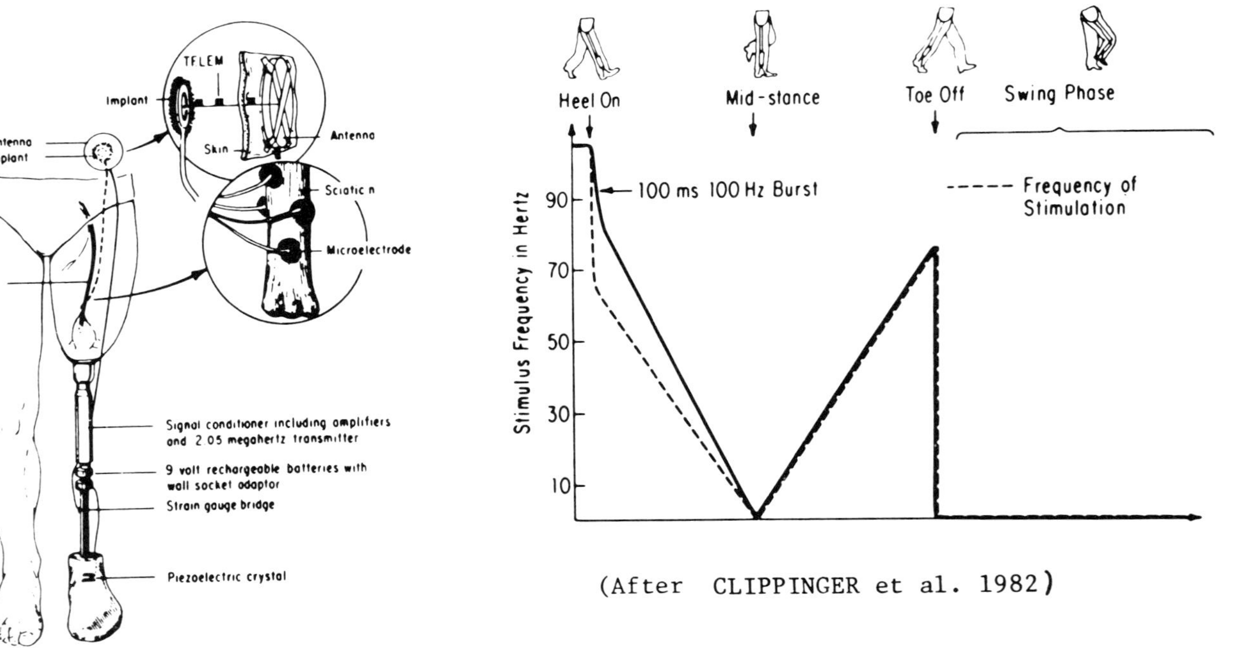

Figure 7

LEFT — *Strain gauges mounted on the prosthesis leg pylon monitor limb load conditions, and after signal conditioning, this information is used to drive the implanted nerve stimulator via an antenna and RF receiver.*

RIGHT — *Schematic showing the relation between the feedback signal and the patient's gait cycle. The patient receives a burst of activity to accent heel strike. Following this, the frequency of the feedback stimulation decreases until mid-stance, then increases again until toe-off. No stimulus is given at mid-stance or during swing phase.*

The potential usefulness of sensory feedback for leg prostheses was also studied by a group of researchers at Duke University (Clippinger *et al.*, 1982), who employed an approach to input prosthesis information that is based on the direct stimulation of peripheral trunk nerves. The sensory system developed by these investigators used an inductively-controlled implanted stimulator and a set of four platinum-iridium "button" shaped electrodes that were sutured around the circumference of the sciatic nerve. Strain gauges were attached to the pylon of the artificial leg to register the magnitude and direction of the bending moment. Subtle differences in prosthesis motion and balance during walking were provided to the prosthesis user by using the strain gauge signal to modulate the frequency of electrical stimulation applied to the nerve. As shown in Figure 7, a contact sensor located in the heel of the prosthesis foot was used to detect the start of the stance phase and to gate the feedback stimulation ON and OFF. The occurrence of *Heel Strike* was signalled to the prosthesis user by a burst of high frequency (100 Hz) stimulation that lasted a maximum of 100 ms. After that initial burst of activity, the frequency of the sensory signal changed in inverse proportion to the magnitude of the pylon bending moment over the range of 0–70 Hz, so that the frequency was maximum at the beginning and end of stance, but at mid-stance, when there was no bending moment, no stimulation was given.

The average period that the sensory feedback system was used by the thirteen patients that were studied was eight months, and one patient used it for six years. The stimulus intensity to evoke adequate sensations remained stable and no problems from infection, or intolerance of the nerve stimulation were reported. Some benefits that derived from applying the stimulation immediately following the amputation surgery were also reported including suppression of pain and faster healing. Amputees' walking speeds increased when using the sensory feedback, and they stated that they felt more confident about walking in darkened areas, on stairs and over uneven or soft terrain.

Summary and conclusions

A few techniques that investigators have devised for using electrical stimulation of the skin or peripheral nerves to provide cognitive feedback for users of FNS systems and prosthetic legs were presented. As an adjunct for the control of upper extremity grasp restoration systems, an electrocutaneous display was described that provides feedback of the user's proportional command signal, the level of grasp force, and information related to the status of the stimulator hardware. With regard to lower extremity FNS systems and prostheses, sensory systems were reviewed that include the distribution of foot-to-floor contact forces, and joint position kinesthesia. While the examples presented represent only a few of many approaches that could be tried, most of the basic tools available with present technology were shown, and supplemental information can be obtained from several reviews of this subject (Korner, 1979; Szeto and Saunders, 1982; Szeto and Riso, 1987).

Developmental efforts of the past twenty-five years have seen the sophistication of FNS systems advance from single channel *ON–OFF* devices such as the foot-drop neuromuscular assist that activates the foot dorsiflexors in hemiplegic stroke patients (Liberson *et al.*, 1961), to highly complex microprocessor-based systems that enable paraplegics to walk and climb stairs (e.g. Marsolais and Kobetic, 1987).

While these accomplishments are laudable, many difficulties must still be resolved before such devices will have widespread clinical utility and patient acceptance. Solutions to some of these problems, such as the need to develop electrodes that can elicit reliable, graded contractions from small muscles independently from neighbouring muscles, and that can withstand years of mechanical stresses without fracturing, may have to await major breakthroughs in materials technology and fabrication.

Other significant problems have to do with system control. Normally, voluntary movement involves control at both conscious and unconscious levels. For example, in normal grasping, grip force is regulated automatically via presumed spinal reflexes that react to signals of pressure and ''incipient slippage'' in the finger tip skin (Westling and Johansson, 1984). A similar strategy that uses artificial wearable tactile sensors is being developed to impart FNS grasp systems with closed-loop control of the grip strength. While such efforts should go a long way towards easing the burdens of operator control, other problems still remain that directly involve the interface between user and his FNS system. These problems seem poorly tractable with available engineering technologies and escalate rapidly as FNS systems are made more versatile. In the case of the CWRU grasp restoration system, the user has the choice of switching between two modes of grasp; but, this seemingly simple function may require several seconds to execute, and such delays may be sufficiently tedious for the patient that the overall effectiveness of the FNS system is compromised. A similar situation applies to the development of FNS-assisted walking. Current efforts aimed at automating the coordination of the muscles used for reciprocal gait seem promising, and researchers are developing ways to incorporate closed-loop feedback, from joint position and foot-ground reaction forces, to enhance the precision of the movements (Crago *et al.*, 1980; Chizeck *et al.*, 1985). However, a major bottleneck still remains whenever the user comes to a curb or stairs, or wishes to step backward.

Improved control techniques are needed that reduce the user's mental attentiveness and response delays whenever it is necessary to ''shift gears''. Control strategies that utilize naturally occurring EMG signals produced by the patient during preparation for walking or in anticipation of changes in the gait (Graupe *et al.*, 1983), and the development of highly cosmetic and reliable hardware, such as the finger worn telemetry based microswitch ring that paraplegic patients can use to produce system commands for walking (Marsolais *et al.*, 1985), are among the efforts being tried to achieve these goals. It is unlikely that the degree of automaticity present in normal control of movements can ever be attained in neuroprostheses; nevertheless, it is hoped that imaginative investigators will be inspired from ideas such as those presented in this review to design sensory feedback algorithms that can improve the ergonomics of present and future FNS systems.

Acknowledgement

The author gratefully acknowledges the investigators and staff of the Rehabilitation Engineering Program at Case Western Reserve University for their contributions in the development and implementation of stimulator hardware and FNS grasp system control algorithms. In particular, appreciation is expressed to James Buckett, Patrick Crago, Anthony Ignagni, Michael Keith, Michael Neuman, Hunter Peckham, Brian Smith and Geoffrey Thrope. The work on sensory systems for use with upper extremity FNS nueroprostheses was supported by NINCDS Contract NO1 –NS–3–2345 and NIHR Grant G 009 300118. The work on sensory systems for use with the Cleveland VAMC FNS system for gait restoration in paraplegia was supported by the Rehabilitation Research and Development Service of the Veterans Administration.

References

Anani, A.B., Ikeda, K., and Korner, E.R., (1977), Human ability to discriminate various parameters in afferent electrical nerve stimulation with particular reference to prosthetic sensory feedback. *Medical and Biological Engineering and Computing*, **15**, 363–373.

Brindley, G.S., Polkey, C.E., and Rushton, D.N., (1978), Electrical splinting of the knee in paraplegia. *Paraplegia*, **16**, 428–435.

Chizeck, H.J., Lalonde, R., Change, J.A., Rosenthal, J.A. and Marsolais, E.G., (1985), Performance of a closed-loop controller for electrically stimulated standing in paralyzed patients. In *Proceedings of the 8th Annual RESNA meeting*, June, (Memphis, Tennessee), pp. 231–233.

Clippinger, F.W., Avery, R. and Titus, B.R., (1974), A sensory feedback system for an upper-limb amputation prosthesis. *Bulletin of Prosthetics Research*, BPR 10–22, 247–258.

Clippinger, F.W., Seaber, A.V., McElhaney, J.H., Harrelson, J.M., and Maxwell, B.M., (1982), Afferent sensory feedback for lower extremity prosthesis. *Clinical Orthopedics and Related Research*, **169**, 202–206.

Crago, P.E., Mortimer, J.T., and Peckham, P.H., (1980), Closed-loop control of force during electrical stimulation of muscle. *IEEE Transaction on Biomedical Engineering*, BME–27, 306–312.

Cybulski, G.R., Penn, R.D., and Jaeger, R.J., (1984), Lower extremity functional neuro-muscular stimulation in cases of spinal cord injury. *Neurosurgery*, **15**, 132–146.

Graupe, D., Kohn, K., Kralj, A., and Basseas, S., (1983), Patient-controlled electrical stimulation via EMG signature discrimination for providing certain paraplegics with primitive walking functions. *Journal of Biomedical Engineering*, **5**, 270–276.

Holle, J., Thoma, H., Frey, M., Gruber, H., Kern, H. and Schwanda, G., (1984), Walking with an implantable stimulation system for paraplegics. In *Proceedings of the 2nd International Conference on Rehabilitation Engineering*, 551–552.

Kato, I., Yamakawa, S., Ichikawa, K., and Sano, M., (1970), Multifunctional myoelectric hand prosthesis with pressure sensory feedback system. In *Proceedings of the the 3rd International Symposium on External Control*, E.T.A.N, Dubrovnik, Yugoslavia, August, pp. 155–170.

Kawamura, J., Sweda, O., Kazutaka, H., Kazuyoshi, N. and Isobe, S., (1981), Sensory feedback systems for the lower-limb prosthesis. *The Journal of Osaka, Rosai Hospital*, **5**, Number 2, 104–112.

Korner, L., (1979), Afferent electrical nerve stimulation for sensory feedback in hand prostheses. *Acta Orthopaedica Scandinavica*, Supplement 178, 1–45.

Kralj, A., Bajd, T., and Turk, R., (1980), Electrical stimulation providing functional use of paraplegic patient muscles. *Medical Progress Through Technology* **7, 3**.

Liberson, W.T., Holmquest, H.J., and Scott, D., (1961), FUnctional electrotherapy: stimulation of peroneal nerve synchronized with swing phase of gait of himiplegic patients. *Archives of Physical Medicine and Rehabilitation*, **42**, 101–105.

Lovely, D.F., Hudgins, B.S., and Scott, R.N., (1985), Implantable myoelectric control system with sensory feedback. *Medical and Biological Engineering and Computing*, **23**, 87–89.

Malick, M.H., and Meyer, C.M.H., (1978), *Manual on Management of the Quadriplegic Upper Extremity*, Harmarville Rehabilitation Center, Pittsburgh, PA.

Mann, R.W., and Reimers, S.D., (1970), Kinesthetic sensing for the EMG controlled "Boston Arm". *IEEE Transactions on Man–Machine Systems*, MMS–11 **(1)**, 110–115.

Marsolais, E.B., Massiello, A., Ko, W.H., and Spear, (1985), Finger switch for a portable microprocessor system to restore walking in paraplegics. In *Proceedings of the 8th Annual RESNA Meeting*, June, (Memphis, Tennessee), pp. 376–378.

Marsolais, E.B., and Kobetic, R., 1987, Walking of paraplegic subjects with computer-controlled electrical stimulation. *Journal of Bone and Joint Surgery*, In Press.

Mooney, V., and Reswick, J.B., (1976), Sensory feedback myoelectric prosthesis, Rancho Los Amigos Hospital, Final Report to VAPC.

Neuman, M.R., and Buckett, J.R., (1982), Thumb force and position sensors. In *Proceedings of the 35th Annual Conference on Engineering in Medicine and Biology*, vol. 24, p.122.

Neuman, M.R., (1987), Force and position transducers for use on the paralyzed hand. Final Report to NIH–NINCDS, Contract # NO1–NS–3–2345.

Peckham, P.H., Mortimer, J.T., and Marsolais, E.B., 1980a, Controlled prehension and release in the C5 quadriplegic elicited by functional electrical stimulation of the forearm musculature. *Annals of Biomedical Engineering*, **8**, 369–388.

Peckham, P.H., Marsolais, E.B., and Mortimer, J.T., 1980b, Restoration of key grip and release in the C6 quadriplegic through functional electrical stimulation. *Journal of Hand Surgery*, **5**, 464–469.

Petrofsky, J.S. and Phillips, C.A., (1983), Computer-controlled walking in the paralyzed individual. *Journal of Neurological and Orthopedic Surgery*, **4**, 153–164.

Petrofsky, J.S., Phillips, C.A., Douglas, R., and Larson, P., (1986), A computer-controlled walking system: the combination of an orthosis with functional electrical stimulation. *Journal of Clinical Engineering*, **11(2)**, 121–133.

Phillips, C.A., and Petrofsky, J.S., (1985a), Preliminary report on balance/stance in paraplegia with and without a cognitive feedback system. Abstract, American Physiological Society Annual Meeting, October, Niagara Falls, New York, 13–18.

Phillips, C.A., and Petrofsky J.S., (1985b), Cognitive feedback as a sensory adjunct to FES neural prosthesis. *Journal of Neurological and Orthopaedic Medicine and Surgery*, **6**, 231–238.

Prior, R.E., Case, P.A., Scott, C.M., and Lyman, J., (1976), Supplemental sensory feedback for the VA/NU myoelectric hand: Background and feasibility. *Bulletin of Prosthetics Research*, BPR–10–26, 170–190.

Rebersek, S., and Vodovnik, L., (1973), Proportionally-controlled functional electrical stimulation of hand. *Archives of Physical Medicine and Rehabilitation*, **54**, 378–382.

Riso, R.R., Ignagni, A.R., and Keith, M.W., (1983), Subdermal stimulation for electrocutaneous communication. In *Proceedings of the 6th Annual Conference on Rehabilitation Engineering*, San Diego, 321–323.

Riso, R.R., Ignani, A.R., and Keith, M.W., (1987), Electrocutaneous communication via subdermal stimulation from chronic implanted electrodes. (In preparation).

Scott, R.N., Brittain, R.H., Caldwell, R.R., Cameron, A.B., and Dunfield, V.A., (1980), Sensory feedback system compatible with myoelectric control, *Medical and Biological Engineering in Computing*, **18**, 65–69.

Shannon, G.F., (1979), A myoelectrically-controlled prosthesis with sensory feedback. *Medical and Biological Engineering in Computing*, **17**, 73–80.

Smith, B., Peckham, P.H., Keith, M.W., and Roscoe, D.D., (1987), An externally-powered, multichannel, implantable, stimulator for versatile control of paralyzed muscle. *IEEE Transactions on Biomedical Engineering*. In press.

Stanic, U., Acimovic-Janezic, R., Gros, N., Trnkoczy, A., Bajd, T., Kljajic, M., (1978), Multi-channel electrical stimulation for correct of hemiplegic gait. *Scandinavian Journal of Rehabilitation Medicine*, **10**, 75–92.

Szeto, A.Y.J., and Saunders, F.A., (1982), Electrocutaneous stimulation for sensory communication in rehabilitation engineering. *IEEE Transactions on Biomedical Engineering*, **29(4)**, 300–308.

Szeto., A.Y.J., and Riso, R.R., (1987), Sensory feedback using electrical stimulation of the tactile sense. In *Rehabilitation Engineering*, edited by J.H. Leslie, Jr., and R.V. Smith, (Boca Raton, FL: CRC Press) (In press).

Vodovnik, L., Kralj, A., Stanic, R., Acimovic, R., and Gros, N., (1978), Recent applications of functional electrical stimulation to stroke patients in Lubljana. *Clinical Orthopedics*, **131**, 64–70.

Waters, R., (1977), Electrical stimulation of the peroneal and femoral nerves in man. In *Functional Electrical Stimulation: Applications in Neural Prostheses*, edited by F.T. Hambrecht and J.B. Reswick (New York: Dekker), pp.55–64.

Westling G., and Johansson, R.S., (1984), Factors influencing the force control during precision grip. *Experimental Brain Research*, **53**, 277–284.

18.

On the dynamic control of myoelectric powered prostheses for the disabled

Koji Ito, Toshio Tsuji and Mitsuo Nagamachi

Human Factors and Bioengineering Laboratory
College of Engineering
Hiroshima University
Japan

Abstract

Dynamic properties of the man–machine system must be considered while designing human–prosthesis systems. In this paper, position control during arm manipulation is discussed. A bilinear mechanism of the musculoskeletal system is proposed to assist the amputee in controlling the position of a myoelectric powered arm. It is shown that if a bilinear structure, such as the one proposed, is added to the human–prosthesis interface the arm position control can be substantially improved.

Introduction

It is generally agreed that an artificial limb should simultaneously satisfy the following conditions (Jabobson, 1982):

(a) easy and natural control,
(b) cosmetically acceptable,
(c) light, quiet, and durable,
(d) efficient enough for compact energy storage,
(e) inexpensive.

Except for the first factor, the above list is primarily hardware oriented. However, ''controllability'' depends largely on how to design the interface between the amputee and the prosthesis. A simple block diagram, Figure 1, illustrates an example of the interface of multiple degree-of-freedom myoelectric prosthesis. The cutaneously measured electromyographic (EMG) signals from residual muscles are used to identify the motions that the amputee intends to begin, and to select a set of prosthesis functions, e.g., wrist flexion/extension, forearm pronation/supination and the like. In parallel, the muscle force is estimated from the same EMG signals driving the prosthesis. On the other hand, the position, velocity, and force informations from the pros-

273

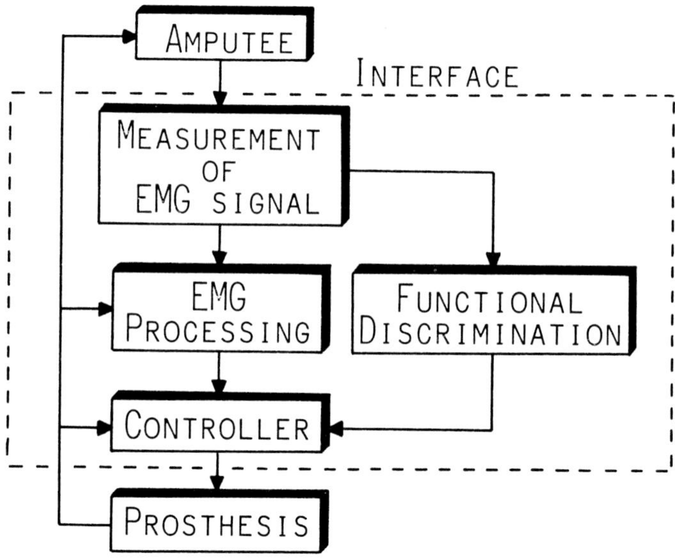

Figure 1. Human–prosthesis interface.

thesis are returned to the amputee through visual or tactile sensation, or electrical stimulation.

The human–limb system includes a highly integrated network of multi-level controllers, actuators and information receptors, connected by efferent and afferent pathways, which allows the musculoskeletal system to modulate a wide variety of dynamic behaviours. Unfortunately, the amputee may have lost not only the abilities of the musculoskeletal system as the effector organs, but also the physiological mechanisms which provide him/her with locomotion, postural control and manipulative skills. Since the information exchange in human–prosthesis system has to be done through residual functions, it is difficult for the amputee to control the prosthesis unless some kind of compensatory mechanism is introduced into the interface design.

In general, the myoelectric prostheses available today adopt on-off control or proportional control. The former uses the myoelectric signal only to turn on or off the actuator of the prosthesis. The latter uses the processed myoelectric signal directly as command signals of the actuator. The lock-unlock control is accomplished by high and low levels of contraction of the antagonist muscles or by a third signal acquired from a motion switch (Jacobson, 1982). The ultimate goal of the prostheses research is to develop artificial limbs which are controlled naturally by the amputee's motor intents and are functionally responsive (like the natural human limbs). Progress in prostheses requires a more intimate cybernetic interface between the amputee and the artificial limb, and a clearer understanding of the neuro-muscular-skeletal system which controls the natural limbs.

The purpose of this paper is to report on the problems of improving the control aspects in human–prostheses systems. Figure 1 suggests that the amputee and the limb prosthesis should be regarded as a man–machine control system where the controlled element is the powered prosthesis. Therefore, when designing the human–prosthesis system, it is necessary to pay attention to the dynamic properties of the man–machine system as a whole. Force, position, velocity and impedance can be used in such a system as the controlled variables. Here, only the position control, which is one of the most basic problems in arm manipulation, is discussed. It is proposed that a bilinear mechanism of the musculoskeletal system be built into the human–prosthesis interface so that the amputee can change the dynamic characteristics of the powered prosthesis continuously by his intents. Consequently, the myoelectric powered arm can be simulated on the computer, and the position control through the surface EMG signal can be evaluated from the viewpoint of the dynamic properties. Finally the tracking tasks can be analysed experimentally in order to predict to what degree the amputee's controllability can be improved in the position control.

Bilinear interface

The skeletal muscle of the human body is activated by impulses of the motor-neurones. The relationship between the nerve impulses and the contractile force of a muscle constitutes an ingenious and a very complex mechanism. It is well known, however, that the macroscopic mechanical properties of the muscle can be represented using two fundamental functions of length–tension curves and force–velocity curves (Dowben, 1980).

Assuming that the muscle force is in proportion to the level of activation ($0 \leqslant \alpha \leqslant 1$; normalized by the maximum), the muscle force F can be given by:

$$F = \alpha \cdot g(L,V) \tag{1}$$

where $g(L,V)$ is a nonlinear function under the maximum level of activation. Approximating $g(L,V)$ by the Taylor expansion around rest length $L = l_0$ and the velocity of contraction $V = 0$, and neglecting the second and higher terms, the following linear relation can be obtained:

$$g(L,V) = f_0 - k_1 x - b_1 \dot{x} \tag{2}$$

where f_0 is the maximum tension at isometric contraction ($V = 0$), x is the relative length of muscle ($x = 0$ at rest and $x > 0$ is shortening), $\dot{x}$ is velocity of shortening, and k_1 and b_1 are positive constants. Substituting (2) into (1) yields:

$$F = u - k' ux - b' u\dot{x} \tag{3}$$

where $u = \alpha \cdot f_0$, $k' = k_1/f_0$, and $b' = b_1/f_0$. This is the visco-elastic model of a muscle, where the viscous and elastic coefficients are not constant but are in proportion to the contractile force u.

By assuming that the forearm and hand can be regarded as a rigid link rotating about a fixed axis, the flexor and extensor have the same properties, and the moment arm d is angle-independent, the muscle torques T_f and T_e about the joint are given by:

$$T_f = d \left(u_f - k u_f \theta - b u_f \dot{\theta} \right) \tag{4}$$

$$T_e = d\,(u_e + ku_e\theta + bu_e\dot{\theta}) \tag{5}$$

where $k = k'\,d$, $b = b'\,d$ and θ is defined as 0 in the right angle for the upper arm and is positive toward flexion. u_f and u_e are the contractile forces of the flexor and extensor.

Based on the above, the dynamic model equation for the horizontal movements of the forearm can be obtained as follows:

$$I/d \cdot \ddot{\theta} = u_f - u_e - (u_f + u_e)k\theta - (u_f + u_e)b\dot{\theta} \tag{6}$$

where, I is the moment of inertia. The visco-elastic linear model proposed hitherto is as follows:

$$I/d \cdot \ddot{\theta} = u_f - u_e - k\theta - b\dot{\theta} \tag{7}$$

Equation (7) is based on the assumption that the stiffness and viscosity of the muscle are invariant and independent of the contractile forces. On the other hand, Equation (6) represents the fact that the driving torque about the joint and the stiffness and viscosity about the joint can be controlled independently through the sum and difference of the controlled forces, respectively. Since the difference $u_f - u_e$ controls the input, while sum $u_f + u_e$ controls the system parameters, Equation (6) is nothing but a bilinear system. Thus, one of the important phenomena of the natural limb is that the parameter adaptation can be performed by co-activation of antagonistic groups of muscles without proprioceptive feedback. The control advantages of the bilinear structure are given in Hogan (1980) and Ito and Tsuji (1985).

If one adds the bilinear structure to the human–prosthesis interface, the amputee might be able to adjust not only the driving torque, but also the mechanical impedance about the joint of the prosthetic arm. For example, when starting to move the forearm toward the desired position, it will be more desirable to make the mechanical impedance as small as possible, because it becomes easy to move the arm. But conversely, when bringing the arm to rest, it is desirable to make the impedance large to brake the arm. Thus, the prosthetic arm with the bilinear interface performing like the natural limb would give an amputee an essential component of the natural adaptive capability despite the severe sensory loss due to amputation. The effects of bilinear interface on the amputee's control manners are discussed in the next section.

Position control

Experimental procedures

The experimental arrangement for position control is shown in Figure 2. The subject's forearm was fixed on the horizontal table by keeping the elbow at a right angle. The EMG signals were taken from biceps and triceps under isometric contraction. The pair of differential electrodes on each muscle were 15 mm diameter discs placed 20 mm apart. After full wave rectification the EMG signals were processed by a couple of low-pass analogue filters ($f_c = 1\,\mathrm{Hz}$, the first-order). The filtered signal, called integrated EMG, is proportional to the muscle force (Hogan and Mann, 1980). The filtered outputs are the command signals u_f and u_e.

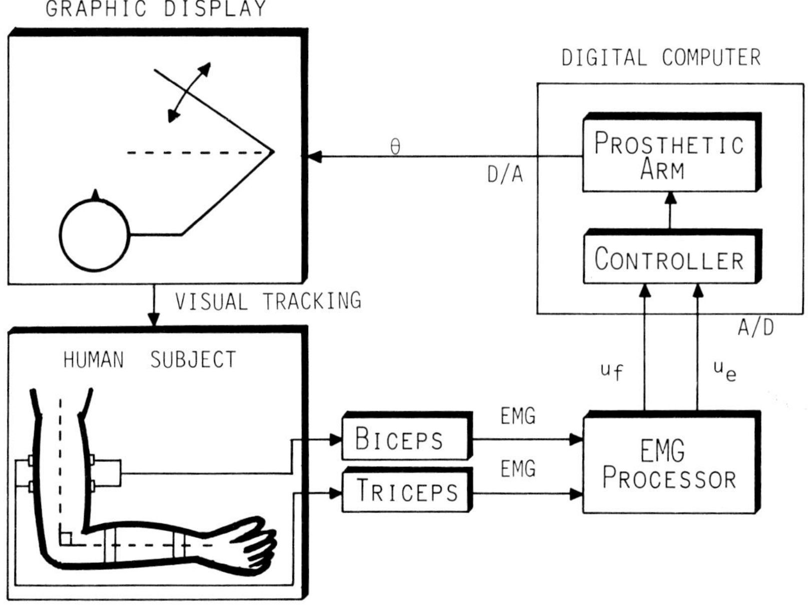

Figure 2.   Experimental arrangement for position control.

The forearm was drawn on the graphic screen. The solid line represented the current position and the broken line, the desired position. The computation time for the controller and mathematical model of the prosthesis was insignificant. The forearm shown on the computer display was rotated corresponding to the output (joint angle) θ from the computer. Therefore, the human subject was able to rotate the forearm with the aid of the EMG signals generated by biceps and triceps.

The dynamic equation, which simulated the horizontal movement about the elbow joint of a prosthetic arm, is given as follows:

$$I\ddot{\theta} + B_J\dot{\theta} = T \tag{8}$$

where

I : the moment of inertia.
B_J : the coefficient of viscosity about the joint.
T : the input to the actuator.
θ : elbow joint angle, ($\dot{\theta}$ = angular velocity; $\ddot{\theta}$ = angular acceleration).

Two interface controllers were tested. The first one was bilinear as shown in Figure 3; the second one was linear, shown within the broken line. In the bilinear controller, the driving force and system parameters can be controlled independently through the difference and sum of the flexor and extensor.

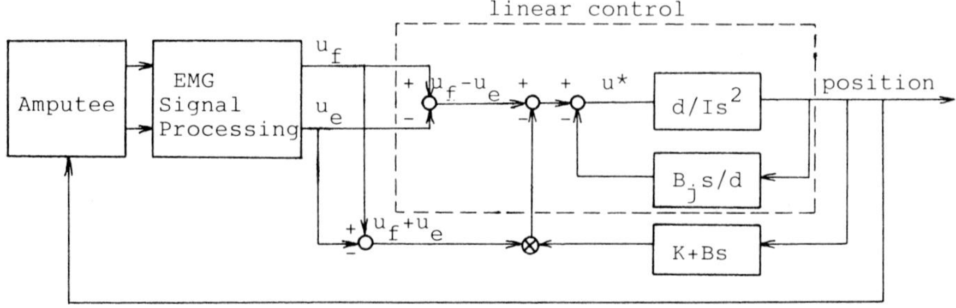

Figure 3.   Interface controllers for position control.

Arm positioning control

Some computer-controlled tracking tests were performed. The human subject sat in front of the graphic display and was instructed to move the forearm on the display from the current position to the desired position as fast as possible. The step responses were superimposed (Figure 4). The solid lines denote the results for the bilinear system; broken lines denote results for the linear system. The parameters were set at $d/J = 2.0$, $B_J/d = 0.4$, $K = 0.1$, $B = 0.2$. Figure 4(a) shows the time responses of the joint angle (which was the controlled variable). It can be seen that the rising time of the bilinear system is shorter than that of the linear system. u_f (flexor) and u_e (extensor) are shown in Figure 4(c). The control of the linear system is a bang-bang form. First, the flexor accelerates the forearm and then the extensor is activated to decelerate it, synchronizing with the relaxation of the flexor. On the other hand, in the bilinear system, co-activation of the flexor and extensor is observed in the latter half. Since co-activation gives an increase of $u_f + u_e$ and consequently makes the viscous coefficient larger, the prosthesis control system can be switched to a system with larger damping. This shows that the human subject skilfully utilizes the variable structure of the bilinear system.

The input u^* to the inertia element d/Is^2 makes the difference of both position controls clearer. As shown in Figure 4(b), both inputs are bang-bang forms. However, it is seen that the linear system yields loose switching and small amplitude, while the bilinear system yields sharp switching and large amplitude, i.e., about three times as large as the linear one. This leads to the differences in speed of response between both systems.

Linear systems never allow the control to change the system parameters. Therefore, when the large input is added to get a short rising time, it becomes more difficult to predict the switching point, which will lead to the overshoot. In contrast, the multiplicative control of bilinear systems can be used to change the system to a large damping system at any time, which may result in large input in the first half.

In the next phase of the experiment, the desired positions were shifted by angles (within 80 degrees) and time spans (4–8 seconds) supplied by the random number generator of the computer. Duration time was 50 seconds and three normal subjects

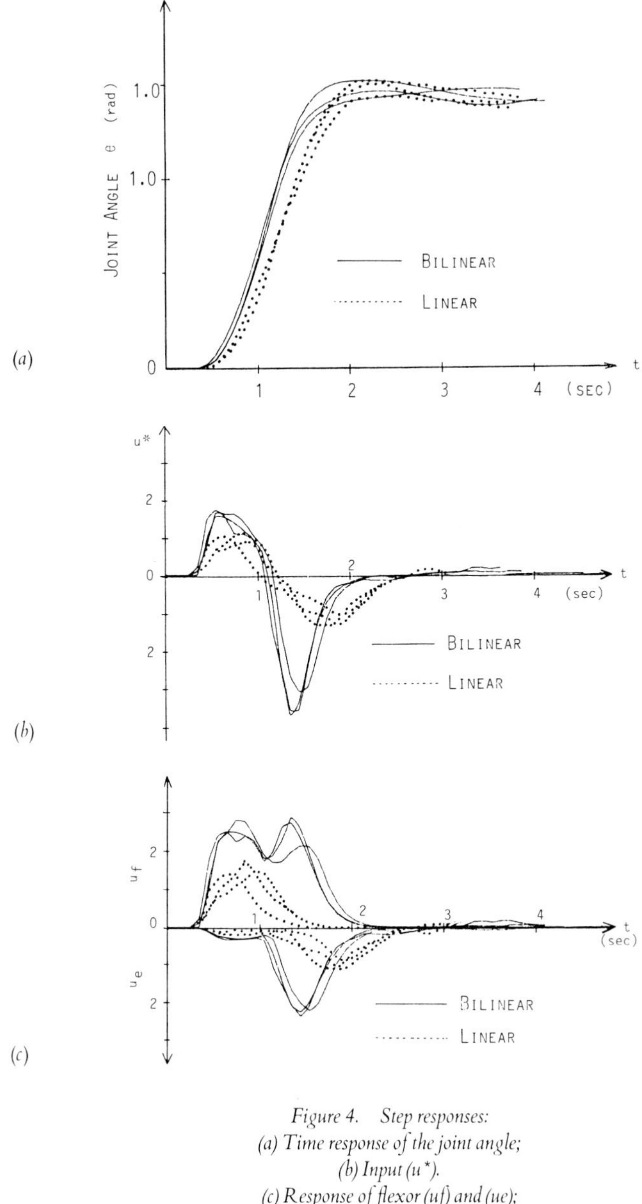

Figure 4.   Step responses:
(a) Time response of the joint angle;
(b) Input (u).*
(c) Response of flexor (uf) and (ue);

were trained to a stable level of performance scores. In general, at least twenty trial runs were carried out before the runs were recorded.

Figure 5 shows time scores of: (a) linear and (b) bilinear system, respectively. From the top, r denotes reference input, θ the joint angle, EMG_f the myoelectric surface signal of the flexor, u_f the filtered EMG signal, EMG_e the myoelectric surface signal of

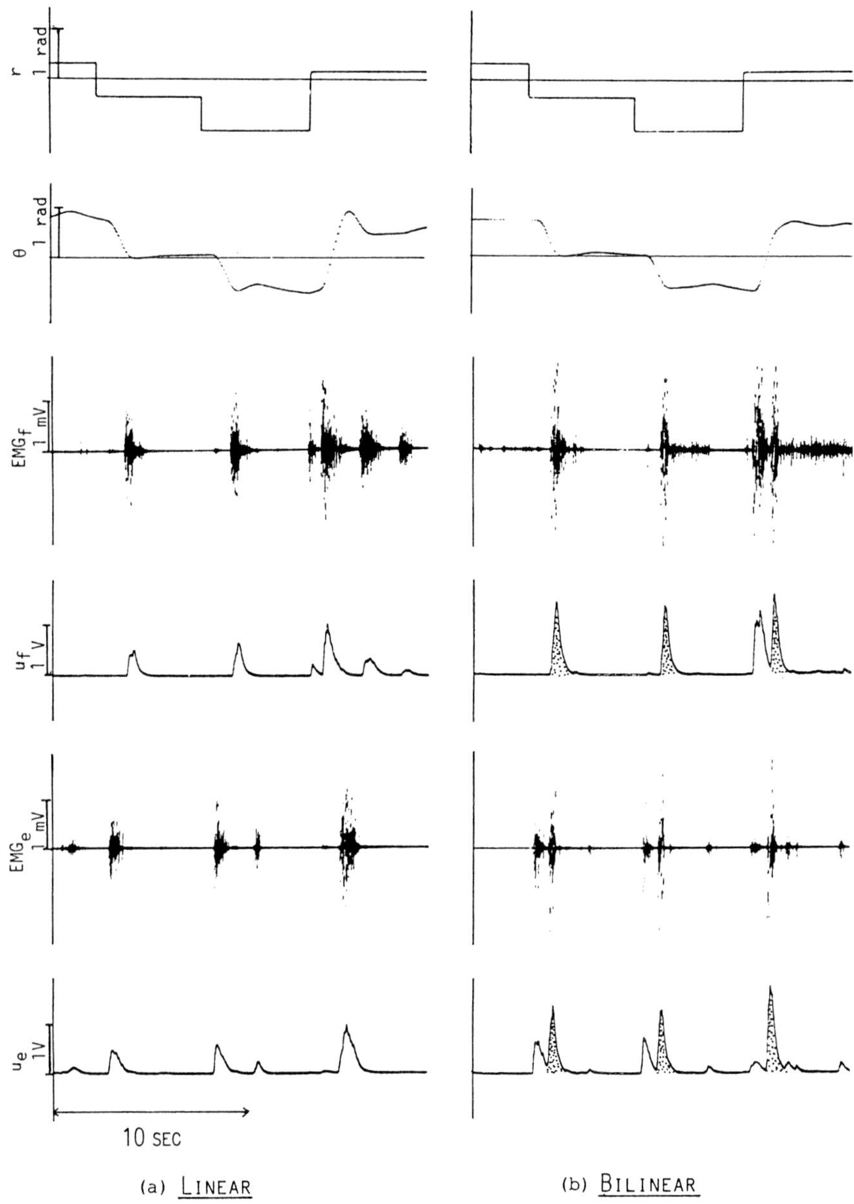

Figure 5.   Tracking time scores of (a) linear and (b) bilinear systems for random step inputs.

the extensor, and u_e the filtered EMG signal. It can be seen that the control patterns between both systems exhibit striking contrasts. In the linear system, the flexor and extensor are activated reciprocally. In contrast, the bilinear system allows the subject to co-activate both muscles.

The following performance scores were also computed;

1) Integrated squared error

$$\text{ISE} = \sum_{i=1}^{N} [\int_0^{T_i} e_i^2(t)dt / \int_0^{T_i} r_i^2(t)dt] \tag{9}$$

2) Integrated absolute error

$$\text{IAE} = \sum_{i=1}^{N} [\int_0^{T_i} |e_i 1(t)|dt / \int_0^{T_i} |r_i(t)|dt] \tag{10}$$

3) Integrated time absolute error

$$\text{ITAE} = \sum_{i=1}^{N} [\int_0^{T_i} t|e_i(t)|dt / \int_0^{T_i} t|r_i(t)|dt] \tag{11}$$

4) Integrated control cost

$$\text{ICC} = \sum_{i=1}^{N} [\int_0^{T_i} (u_{fi}(t) + u_{ei}(t))^2 dt / \int_0^{T_i} |r_i(t)|dt] \tag{12}$$

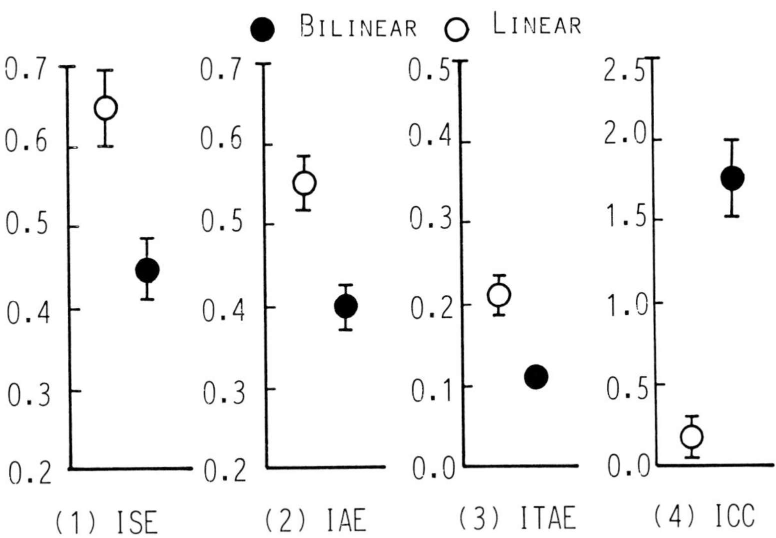

Figure 6. Performance scores.

The averaged scores and standard deviations of ten trials after training are shown in Figure 6. Three scores (ISE, IAE, ITAE) of the bilinear controller, except ICC, were lower by 30 per cent–50 per cent than the linear controller. This suggests that the bilinear controller can improve the amputee's position control. Although the control cost (ICC) increased substantially, the subject's EMG levels did not exceed 40 per cent of the maximum voluntary contraction.

Continuous arm movements

It may be concluded that the bilinear controller is not a good one for continuous and smooth arm movements. To examine this assumption, another tracking test was performed using the continuous and smooth inputs. The experimental conditions were the same except for the input and display.

The reference input was a random-appearing signal composed of ten sine waves of arbitrary phases and different frequencies in the range of $0 \cdot 309$ to $9 \cdot 865$ rad/sec ($w_c = 2 \cdot 265$ rad/sec). The visual display was an oscilloscope. The controlled element, which simulated the horizontal movement about the elbow joint of a prosthetic arm, is given in Equation (8). The parameters were given by $d/J = 2 \cdot 0$, $B_J/d = 0 \cdot 4$, $K = 0 \cdot 1$, $B = 0 \cdot 2$. The subjects, three healthy persons (non-amputees), were instructed to keep the error as small as possible. Each subject was trained to reach a stable level again.

The tracking time scores for the bilinear system are shown in Figure 7. Few co-activations of the flexor and extensor were observed. Such results are much different from that depicted in Figure 5. It was not necessary to stop the prosthetic arm rapidly since the input was continuous and smooth.

The frequency characteristics for the above experiment were computed using the following relations:

$$G(jw) = P_{rc}(jw)/P_{re}(jw) \tag{13}$$

where $G(jw)$ is the open-loop frequency responses of the overall system, and $P_{xy}(jw)$ is the cross-spectral density function between x and y and estimated by FFT method. The open-loop frequency responses from the error e to the controlled variable θ for both systems are shown in Figure 8. These are averages and standard deviations of ten trials. There is a little difference between both systems in terms of the amplitudes and phase characteristics.

The transfer characteristics in normal manual control with visual feedback was summarized as the crossover model and represented as follows:

$$G_h(s)G_c(s) = w_c e^{-\tau_e s} \tag{14}$$

where $G_h(s)$ denotes the human operator, $G_c(s)$ denotes the controlled element, and the crossover frequency w_c is equivalent to the loop gain and lies between 3–5 rad/sec. The effective time delay τ_e is an accumulation of additional lags due to the transport delays and the high frequency neuromuscular dynamics, and lies between $0 \cdot 2$–$0 \cdot 4$ seconds. The crossover model suggests that the subject adapts himself to the variation of the controlled elements by adjusting his own transfer characteristics.

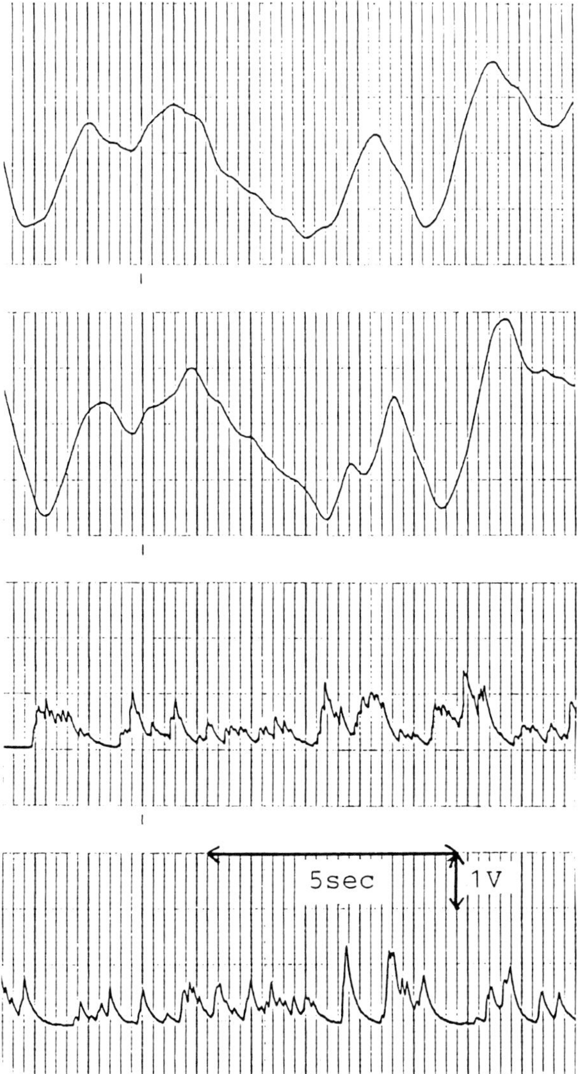

Figure 7.   Tracking time scores of bilinear systems for continuous random-appearing inputs.

The solid line in Figure 8 indicates the crossover model. The frequency responses in myoelectric control can be fitted by the crossover model within the range near the crossover frequency w_c (5 rad/sec). From this, it can be seen that tracking characteristics of the myoelectric command signal with a bilinear controller can be maintained at almost the same level as in the manual control.

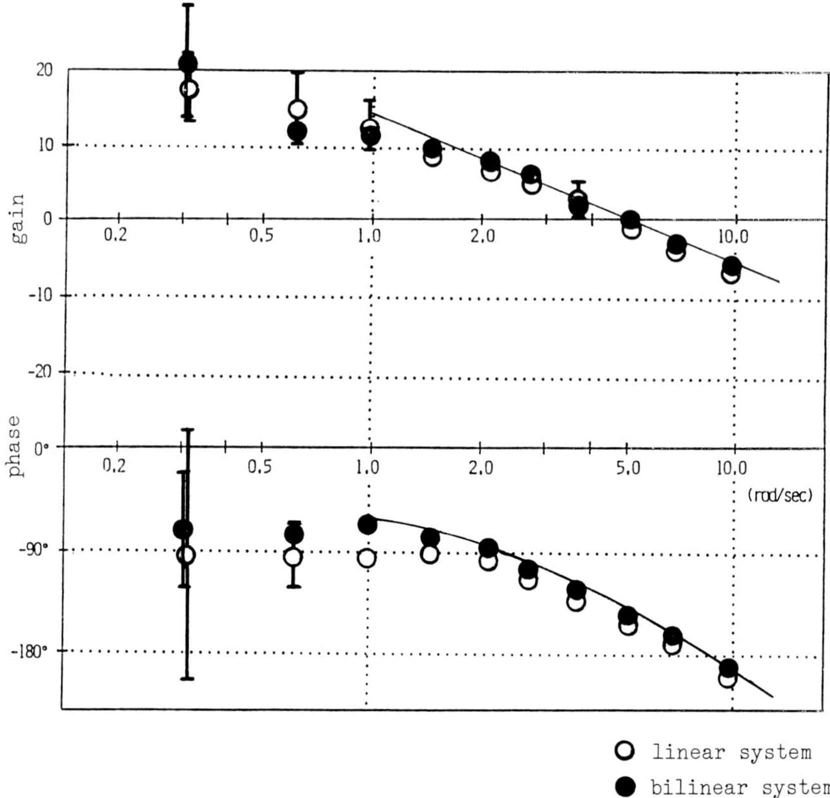

Figure 8.   Open-loop frequency responses for linear and bi-linear systems.

Conclusion

It is an important property of the bilinear system that the control input is able to vary the system structure, where no feedback compensatory loop to the amputee is added. Modification of viscoelastic property about the joint by co-activation of the antagonist muscles increases the flexibility of the system and plays a significant role in facilitating the open loop control of movement (Houk, 1979; Hogan, 1984). In this study, it was shown that the position control could be improved substantially by adding the bi-linear structure to the present interfaces in human–prosthesis system.

Acknowledgements

The work presented in this paper was supported in part by the Scientific Research Foundation of the Ministry of Education under Grant #60550300 and the Mazda Foundation's Research Grant.

References

Dowben, R.M., 1980, Contractility. *Medical Physiology*, 14th edition (edited by V.B. Mountcastle), (St. Louis: C.V. Mosby).

Hogan, N., 1980, Mechanical impedance control in assistive devices and manipulators. *Proceedings of the 1980 JACC*.

Hogan, N., 1984, Adaptive control of mechanical impedance by coactivation of antagonist muscles. *IEEE Transactions on Automatic Control*, AC-29, pp. 681–690.

Hogan, N. and Mann, R.W., 1980, Myoelectric Signal Processing: Optimal Estimation Applied to Electromyography Parts I, II. *IEEE Transactions on Biomedical Engineering*, BME-27, pp. 382–410.

Houk, J.C., 1979, Regulation of stiffness by skeletomotor reflexes. *Annual Review in Physiology, 1979*, pp. 99–104.

Jacobson, S.C., Knutti, D.F., Johnson, R.T. and Sears, H.H., 1982, Development of the Utah artificial arm. *IEEE Transactions on Biomedical engineering*, BME-29, pp. 249–269.

Ito, K. and Tsuji, T., 1985, Control Properties of Human–Prosthesis System with Bi-linear Variable Structure, *Proceedings of 2nd IFAC Conference on Analysis, Design, and Evaluation of Man–Machine Systems*.

19.

Biomechanical and ergonomic considerations of hand dynamic splints

Kai-Nan An
Laura A. Mildenberger[1]
and Stephen P. Chesher[2]
Department of Orthopedics
Biomechanical Research Laboratory
Mayo Clinic/Mayo Foundation
Rochester, MN 55905, USA

[1]430 Nevada
Frankfort, IL 60423, USA

[2]Louisville Hand Surgery
1001 Doctors' Office Building
250 East Liberty Street
Louisville, KY 40202–1578, USA

Abstract

Dynamic splinting is an important technique for the hand therapist in rehabilitation of patients following trauma or joint replacement. Despite this fact, data supporting a systematic selection and design based on biomechanical principles have been limited. Biomechanical and ergonomic analyses were therefore undertaken to identify factors that may be of importance for the proper fabrication of dynamic splints. Functional anatomy of the hand, including force transmission and joint motion and tendon excursion, was reviewed. The mechanical properties of springs and rubber bands used for splints were determined experimentally. Mathematical models were established for various types of dynamic splint design. Parametric analyses were performed to verify the influence of varying mechanical characteristics of components and their geometric arrangement.

Introduction

A splint or orthosis is a device that facilitates healing of a pathologic condition. Orthosis is a Greek word meaning ''to make straight''. In books several hundred years old, splinting devices were described for hand treatment. Not until the late nineteenth century were various designs for special applications systematically applied. In the early twentieth century, the application of splints for hand problems was primarily static, as in the resting splint for rheumatoid arthritis. In the early 1940s, reconstructive surgical work was initiated for acute hand lesions which resulted in the advent of some dynamic splints.

The greatest advances in hand splinting have occurred in the past fifteen years with the development of hand rehabilitation centres. Adequate orthotic management of the hand requires significantly more knowledge and planning than manual skill. The variations of patients' conditions are innumerable, as are the number of approaches and materials available. Therefore, much thought and planning prior to splint construction is necessary. Understanding the basic biomechanical and ergonomic principles is important and beneficial in dynamic splint application.

Splinting is primarily adopted to treat hand pathology due to trauma, rheumatoid arthritis, and central nervous system disease. Splinting is effective to immobilizing joints or bones in an effort to reduce the inflammatory process and to improve or reduce the tendency to contracture. More recently, dynamic splinting has been used for post-surgical rehabilitation of patients with joint replacements and tendon repairs.

The purpose of this paper is to review the basic biomechanical concepts involved in hand dynamic splint design and illustrate ergonomic concepts by practical examples. First, biomechanics of the hand is reviewed to provide the basic requirements and goals of dynamic splinting. Second, the mechanical characteristics of various rubber bands and springs used for traction are analysed and discussed. Third, a mathematical model of the dynamic splint is formulated to illustrate the effect of the geometric parameters of splint design on performance. Fourth, simulation of various types of dynamic splints and the expected results for various applications are presented. Finally, examples of clinical appliations are used to illustrate ergonomic concepts.

Biomechanics of the hand

The complexities of the function and anatomy of the human hand have long been recognized. From a biomechanical point of view, the human hand can be considered a linkage system of intercalated bony segments. The joints between each of the phalanges and metacarpals are spanned by ligaments, tendons, and muscles. When the muscles contract, these joints are moved in a characteristic manner constrained by the interposing soft tissues and the bony articulations. The relationship between joint motion and tendon excursion, joint constraint forces and tendon forces, and the mechanisms of joint constraint which are relevant to the implementation of dynamic splinting are therefore reviewed.

Joint motion and tendon excursion

In moving a finger or thumb, the tendons slide a certain distance to execute the movement. Excursions of the extensor and flexor tendons of the hand during joint movement have been measured by numerous investigators and found to be quite important in hand surgery, especially in dealing with tendon transfer procedures (Bunnell, 1956; Brand *et al.*, 1975; Armstrong and Chaffin, 1978; An *et al.*, 1983). It has also been hypothesized that the potential tendon excursion of a muscle is closely related to its fibre length.

Representative tendon excursions of the seven index finger muscles versus the flexion angle of the metacarpophalangeal (MCP) joint have been delineated (Figure 1).

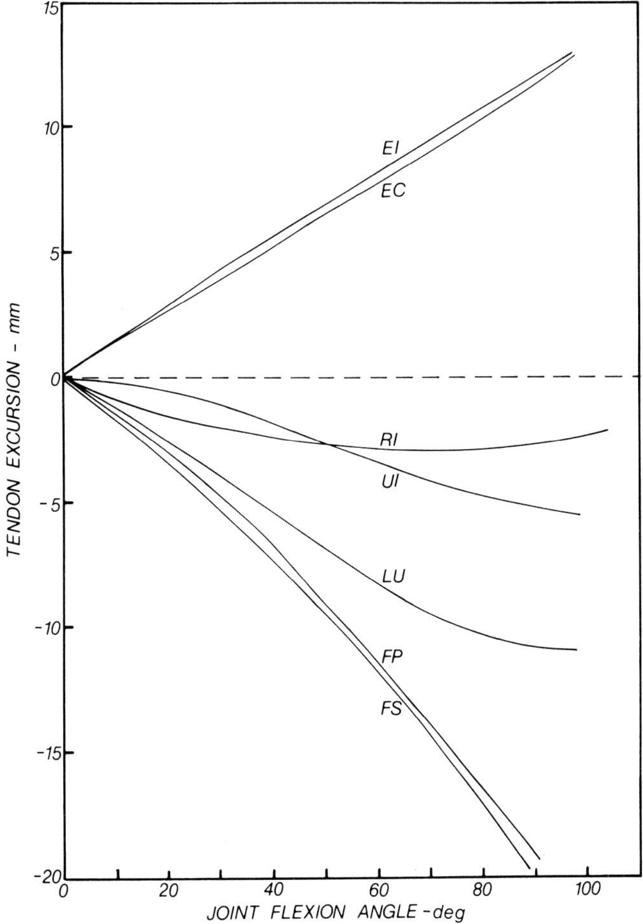

Figure 1.   Tendon excursion of the seven index finger muscles versus the flexion angle of the MCP joint. A positive value represents elongation or distal displacement of the tendon; a negative value corresponds to shortening or proximal retraction of the tendon.

Key to muscles:
FP = Flexor Profundus;
FS = Flexor Sublimis;
EI = Extensor Indicis;
EC = Extensor Digitorum Communis;
RI = Radial Interosseous;
UI = Ulnar Interosseous;
LU = Lumbrical.

Excursions of the four extrinsic muscles are relatively large and almost linearly increase with the flexion angle. The flexors have larger excursions than the extensors. For intrinsic muscles, both interosseous muscles have relatively smaller excursions throughout the range of MCP joint flexion, and most of the excursion occurs in the middle range of motion. However, the lumbrical muscle has a fairly large excursion which is almost equivalent to that of extrinsic muscles.

It has also been verified that the slope of the curve of the tendon excursion versus the joint rotation is equal to the moment arm of that particular tendon or muscle in the plane of motion of the joint (An *et al.*, 1983). The representative moment arms of the seven index finger muscles at the MCP joint as a function of the flexion angle are shown in Figure 2. Extension moment arms of the extensor indices (EI) and the extensor communis (EC) are fairly constant throughout the range of motion. The flexion moment arms of the flexor profundus (FP) and flexor superficialis (FS) are slightly increased with increasing flexion angles. For the three intrinsic muscles, the flexion–extension moment arms at the MCP joint are fairly constant in the middle range of motion but tend to decrease at both extremes of flexion and extension.

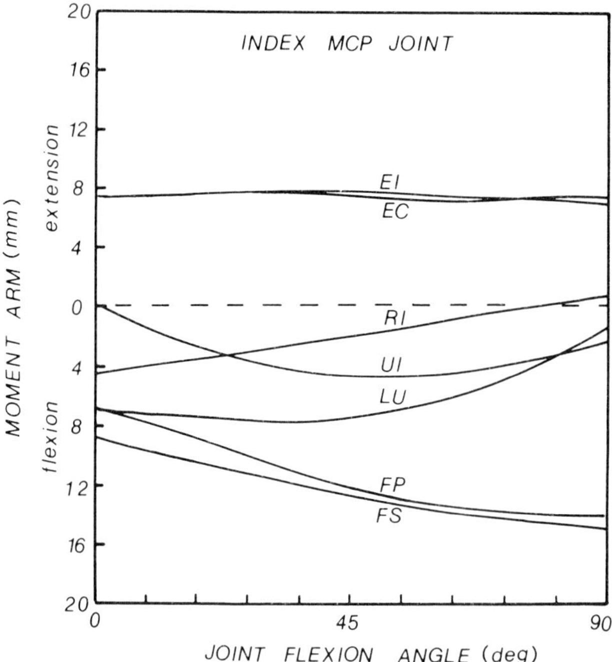

Figure 2.   Moment arms of the index finger muscles at the MCP joint as function of the joint flexion angle. Abbreviations as Figure 1.

The excursions of the two flexors (FP and FS) per 100 degrees of flexion of the three index finger joints are summarized in Table 1. This information is quite important for measuring the gliding motion between the FP and FS when faced with the development of adhesions after tendon injury and repair. The amount of relative gliding can be calculated by multiplying the amount of proximal interphalangeal (PIP) joint rotation by the difference of the moment arms of these two muscles at the PIP joint. For example, with 90° of PIP joint flexion, the total amount of relative gliding will average 3·2 mm. On the other hand, if the distal interphalangeal (DIP) joint is simultaneously flexed, the relative glide between these two flexors will be increased even more by an amount up to 7 mm.

Table 1.   Excursions of index finger flexor tendons.

Joint	Excursion (mm)/100° of joint rotation	
	FP	FS
DIP	7·3 (1·17)	
PIP	14·1 (2·03)	10·8 (1·92)
MCP	19·0 (2·58)	21·1 (2·50)

Mean (S.D.)

Joint force and tendon force

Forces across the joint and tendon forces of the fingers and thumb under various isometric functions have been studied extensively (Chao *et al.*, 1976; Cooney and Chao, 1977; An *et al.*, 1985).

In most isometric functions, the FP and FS have consistently high force values compared to other muscles. In finger tip and pulp pinch, the FP and FS resist an average of 2 to 3 times the applied force. In these pinch functions, the forces of the intrinsic muscles range from 0·5 to 1·5 times the applied force.

The joint forces in pinch functions are also of interest. Axial compressive forces in the IP and MCP joints range from 2 to 5 times the applied force during tip pinch functions, respectively. The volar-dorsal shear forces on the distal segment at all hand joints are directed dorsally and are of substantial magnitude. This implies that under most activities, for example — at the MCP joint, the proximal phalanx has a tendency to sublux volarly on the metacarpal head. The radial-ulnar shear forces at the MCP joint are directed radially. That is, the proximal phalanx has a tendency to shift in the ulnar direction, loading the radial support structures. These findings closely parallel clinical impressions of the patterns of joint deformities. When designing splints for the correction or retardation of such deformities, these forces should be kept in mind.

Most recently, the major flexor tendon forces involved in hand movement and functions were measured *in vivo* by using a specially designed tendon transducer on patients during carpal tunnel surgery (Cooney *et al.*, 1986). The results demonstrate that during active extension and flexion with resistance, the flexor tendons have about 200 to 400 g of force, respectively. Active flexion with mild resistance produces up to 1000 g force in the flexor tendons. In firm pinch, the forces in the flexor tendons range up to 12 000 g. This information along with data on suture strength is quite useful in considering the design of rehabilitation programmes for patients after tendon repair.

Capsuloligamentous structures

Joint stability to restrain isometric pinch and grasp forces as well as during movement is provided by the joint articular surfaces, capsuloligamentous structures, and musculotendinous units. In general, the muscles and tendons assume the primary role of these constraints. The capsuloligamentous structures, on the other hand, provide initial stability to the instantaneous force and moment and provide the second line of defense in maintaining stability.

Quantitative anatomic studies of various units of the collateral ligaments of the finger joints have been performed (Hakstian *et al.*, 1967; Schultz and Storace, 1978; Smith and Kaplan, 1967; Hagert, 1981; Minami *et al.*, 1984). The location and length changes as a function of joint position were measured. It was noted that the fibres furthest from the centre of rotation undergo the greatest change in length, and this eccentricity is crucial to stability at progressive angulations of the joint.

The apparent distance between the origin and insertion of the dorsal and middle portions of the collateral ligaments proportionally decreases with the increase of joint hyperextension. During joint flexion, however, the length of the dorsal portion progressively increases from 0° to 50° and remains constant between 50° and 70°. The apparent length of the volar portion of the ligament decreases when the MCP joint is flexed from 20° to 80°; however, when the MCP joint is extended to a neutral position, further increase occurs and continues with hyperextension.

Table 2. Laxity of the MCP joint of the index nger.

MCP Joint Angle (degree)	Distraction (mm)	Pronation/Supination (degree)	Dorsal-Volar Subluxation (mm)	Radial-Ulnar Rotation (degree)
		Movement		
0	2·3 (0·4)	37·8 (6·8)	10·1 (0·6)	17·0 (2·7)
45	1·0 (0·3)	25·7 (6·1)	8·1 (0·8)	10·2 (3·9)
75	0·8 (0·3)	18·8 (5·0)	4·4 (1·0)	2·8 (1·5)

Mean (S.D.)

Changes in ligament length during motion suggest the role each portion of the ligament plays on restraint of the MCP joint. Ususally, for each joint, there are planes where the joint can be rotated with a minimum of resistance. These specific planes correspond to the degrees of freedom of that joint. For example, the MCP joint moves freely in flexion–extension and somewhat in abduction–adduction. Beyond the range of free motion, resistance from the capsuloligamentous and the musculotendinous structures is developed. The resistance and laxity of the MCP joint due to capsuloligamentous structures has been experimentally obtained (Minami *et al.*, 1985). The laxity of the index MCP joint and the stiffness of resistance beyond the range of free motions are summarized in Tables 2 and 3.

Table 3a.   Stiffness of translational displacement of index finger.

Joint Flexion Angle (degree)	Stiffness of Translational Displacement (N/mm)		
	Distraction	Dorsal	Volar
0	145·9 (39·4)	29·6 (4·0)	58·1 (11·1)
45	105·1 (29·3)	36·6 (3·1)	42·8 (4·6)
75	72·4 (8·2)	32·8 (4·6)	47·2 (4·0)

Mean (S.D.)

Table 3b.   Stiffness of rotational displacement of index finger.

Joint Flexion Angle (degree)	Stiffness of rotational displacement (N.cm/deg)			
	Pronation	Supination	Radial Rotation	Ulnar Rotation
0	5·0 (0·6)	4·3 (1·0)	11·9 (1·3)	12·4 (1·2)
45	5·4 (0·9)	4·8 (0·9)	18·1 (4·6)	21·2 (3·7)
75	6·2 (0·9)	6·0 (1·0)	20·6 (3·2)	22·1 (2·2)

Mean (S.D.)

Mechanical characteristics of splint components

Rubber bands and coil springs are the most commonly used components for providing traction in dynamic splinting of the hand. In order to have an adequate splint design, it is necessary to understand the basic physical and mechanical properties of these tractional elements.

Rubber band

Physically, it has been recognized that for the same amount of stretch, the stiffer rubber band will provide more traction force. This concept can be explained by the simple relationship:

$$F = K \times \Delta L / L_0$$

where F is the traction or tension of the rubber band, L_0 is the resting length of the rubber band (at no stretch), ΔL is the change of the length of the rubber band, and K is the stiffness or the spring constant of the rubber band. The spring constant, K, is further related to the cross-sectional area of the rubber band and the material property, namely, the modulus of elasticity. Therefore, when a rubber band is described as stiffer or stronger, it would have either a larger cross-sectional area or a higher modulus of elasticity. Additionally, if two rubber bands have the same spring constant, the shorter one will be stiffer than the longer one. It should be noted that when two or more rubber bands of similar type are used together in parallel, the cross-sectional area of the rubber band system and the stiffness should increase proportionally.

Values for the spring constant and the original length were obtained for nine sizes of rubber bands in clinical use. The values for the spring constant ranged from 134 to 531 g (0·3 to 1·2 pounds), and those for original resting length ranged from 1·78 to 8·13 cm (0·7 to 3·2 inches).

Once the rubber band has been selected and the amount of stretch has been measured, the amount of traction or tension of the rubber band can easily be calculated. For example, if rubber band #30 is elongated to 2·5 inches (6·35 cm), it will provide 87·1 g of force; at 2·75 inches (6·99 cm) it will yield 121 g of force.

The rubber bands tested were obtained from an office supply house rather than from a discount, grocery, or drug store. The uniformity of a given batch of bands was also examined and found to be quite satisfactory.

Coil spring

Coil springs made of steel have recently been used to provide traction for dynamic splints. The spring was designed so that when it is bent in a plane about the centre of the coil, a resistive moment will be generated. This relationship is described arithmetically as follows:

$$M = K_m \times \alpha = F_n \times r$$

where M is the moment, α is the angle of rotation, K_m is the coil spring constant in the unit of moment/degree, F_n represents the force component perpendicular to the arm of the coil spring, and r is the distance between the point of force application to the centre of the coil. There are two important concepts about the coil spring which should be considered in the application of splints.

If traction is to be generated from a coil spring, the further away the attachment is placed from the centre of the coil less tractional stiffness will be experienced. When the attachment is closer to the centre of the coil, more force will be experienced by the same amount of rotation. However, the moment generated about the centre of the coil will remain equal.

If the coil spring is used to provide traction through the cable, placement of the cable on the arm of the coil spring controls the excursion of the cable and the amount of the rotation of the spring, and thus develops the traction. The relationship between the traction and the excursion of the cable attached to the coil spring is a complex one, depending on the geometry of the arrangement. For a given amount of cable excursion, the amount of traction is approximately inversely proportional to the square of the distance from the centre of the coil spring and the attachment.

The spring constant, K_m, of the most commonly used coil spring has been obtained by testing four samples. The average spring constant is 6·71 ± 1·61 g cm/deg.

Modelling of the dynamic splint system

The design and structure of the dynamic splint appear at first glance to be quite simple. However, due to the infinite possible combinations of the basic components, the outcome will be different and sometimes difficult to predict. The advantage of mathematical simulation is that the important parameters in the geometric arrangement can be easily examined and illustrated. From a biomechanical point of view, it is beneficial to know the amount of moment generated about the joint centre and the compressive and shear forces created on the bony structures due to traction from the dynamic splint system. Inappropriate applications of these forces can be potentially harmful.

In a simplified dynamic splint for extension of the MCP joint (Figure 3), the tension generated by the rubber band is transmitted to the proximal phalanx through a pulley and cable arrangement. The parameters involved include those defining the location of the outrigger, the attachment on the proximal phalanx, and the mechanical properties of the rubber band or the spring used.

R = distance between the outrigger of the splint and the centre of the joint;

ϕ = angle between the line joining the centre of the pulley and the joint and line parallel to the dorsal surface of the hand;

S = distance between the splint attachment on the phalanx and the centre of the joint;

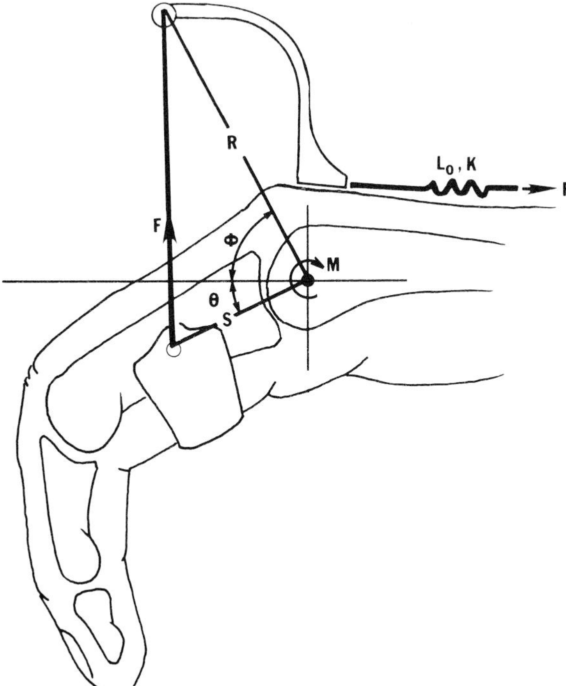

Figure 3.   Analytic model of the low profile dynamic splint of the hand about MCP joint.

θ = flexion angle of the MCP joint measured from the neutral extended position;

K = spring constant of the rubber band system;

L_0 = original unloaded resting length of the rubber band;

F = tension in the cable developed by the transaction of the rubber band;

M = moment about the joint centre due to the tension, F.

A formula for force and moment calculations as functions of the geometric parameters and joint angles has been derived (Mildenberger *et al.*, 1986). The length between the outrigger pulley and the attachment on the phalanx is changed during flexion of the joint and, as a result, stretches the rubber band and generates tension. Based on the trigonometric analysis, the length at a given joint angle, θ, can be calculated as,

$$L = [(R\cos\phi - S\cos\theta)^2 + (R\sin\phi + S\sin\theta)^2]^{\frac{1}{2}}$$

The magnitude of the traction due to the stretch of the rubber band is

$$F = F\,|\,_{\theta=0} + K\cdot[L - L\,|\,_{\theta=0}]/L_0,$$

where $|\,_{\theta=0}$ indicates the value at $\theta = O$.

The moment about the centre of the MCP joint due to the traction F is

$$M = F\cdot R\cdot S\cdot \sin(\theta + \phi)/L$$

The compressive force, F_c, along the proximal phalanx to the joint is

$$F_c = F\cdot\cos\psi$$

The shear force, F_s, perpendicular to the shaft of the proximal phalanx, is

$$F_s = F\cdot\sin\psi$$

where ψ is the angle between the cable attachment and the phalangeal shaft,

$$\psi = a\tan\{\sin(\theta + \phi)/[S/R - \cos(\theta + \phi)]\}$$

By using the above model, the effect on the moment and force created throughout the range of MCP joint flexion was examined by modifying the following parameters: (a) outrigger height; (b) splint pulley angle; (c) location of the splint attachment on the phalanx; (d) the resting length of the rubber band; and (e) the spring constant of the rubber band. In each simulation, the values used for the parameters were varied from:

R = $2\cdot0$ inches;

L_0 = $2\cdot0$ inches;

S = $1\cdot0$ inches;

ϕ = $60°$;

K = $0\cdot5$ pound/unit change of the rubber band length;

$F\,|\,_{\theta=0}$ = o pound

Effect of outrigger height, R

With L_o, S, ϕ, and K kept constant, the traction and moment about the joint centre at three different outrigger heights were obtained as the function of the MCP joint flexion angle (Figure 4). In general, the traction increased with increasing joint flexion angle. However, the moment did not increase monotonically with the joint angle, simply because the direction of traction in reference to the phalangeal shaft changed. The greater the height of the outrigger, i.e., the larger R, the larger the traction and moment about the joint centre.

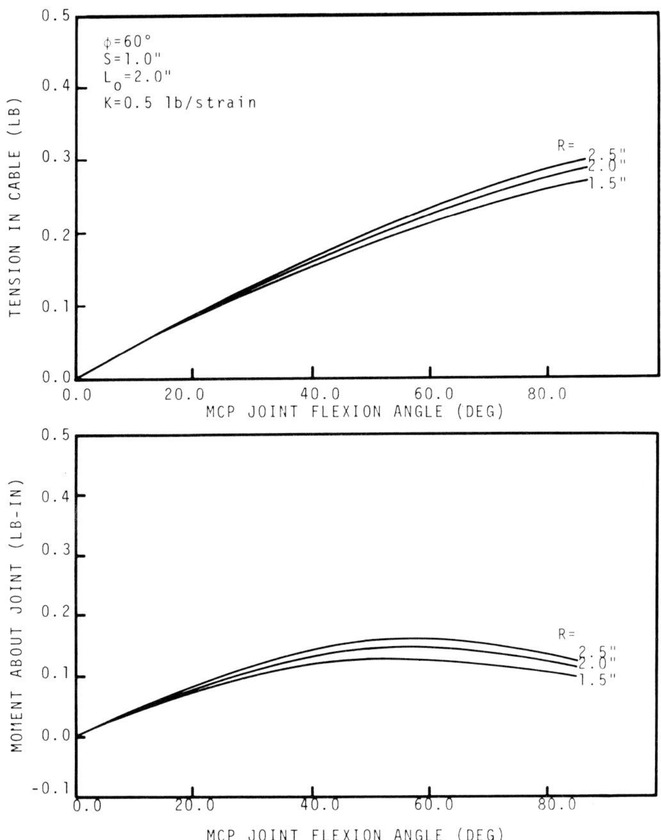

Figure 4. Effect of the outrigger height, R, on the tension generated in the cable and moment about the joint throughout the range of joint flexion.

Effect of splint pulley angle, ϕ

The relative location of the pulley and the attachment on the phalanx with respect to the phalanx had a significant effect on the traction and moment (Figure 5). Both the traction and moment increased with the decrease of the angle, ϕ, i.e., as the splint

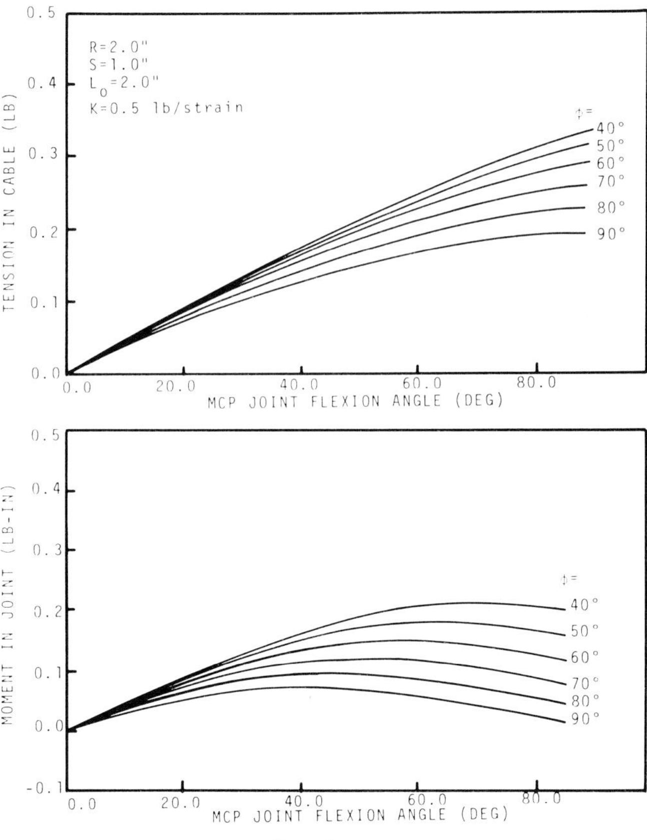

Figure 5.

Eects of the splint pulley angle, ϕ, on the traction and moment about the joint centre.

pulley is moved distally. It should be noted that when the pulley is located directly on top or even proximal to the joint centre, the moment generated is very small and can even cause a change in sign at the extreme flexion angle, i.e., flexion moment rather than extension moment.

Effect of attachment on the finger, S

The effects of the location of the sling attachment to the proximal phalanx on the traction and moment were quite similar to that of the pulley angle. When the attachment is further distal to the joint centre, the rate of variation of the traction and moment as a function of the joint angle were higher.

Effect of the rubber band resting length, L_o

The effect of the original, unloading length of the rubber band on the traction and the moment are quite obvious as that of the basic characteristics of the rubber band or spring. The shorter the resting length the stiffer or stronger the rubber band and the splint system. The traction and moment vary more throughout the range of joint flexion.

Effect of rubber band spring constant, K

The spring constant of the rubber band or spring has direct and proportional effects on the traction and moment generated. The stiffer rubber bands generate more traction and moment with flexion of the joint.

Considerations of special applications

Common applications of the dynamic splint in the rehabilitation of the hand are those of joint replacement and tendon repair (Bunnell, 1946, 1956; Bunnell and Howard, 1950; Brand, 1984; Colditz, 1983, 1984; Fess, 1984; Moberg, 1983, 1984; Mildenberger *et al.*, 1986; Peacock, 1952). In this section, considerations of these types of dynamic splint arrangements are discussed to illustrate the previously described biomechanical and ergonomic principles.

Low-profile versus high-profile MCP extension splint

Recently, the low-profile splint has gained popularity. Numerous authors have described the functional and cosmetic superiority of the low-profile splint design (Colditz, 1983; Moberg, 1983). Moberg stated, "in his experience, patients strongly object to outriggers on orthoses because they are very inconvenient, disturb sleep, require changes in dressing, and are too conspicuous". He thus recommended avoidance of the outrigger, or a reduction in their size to an absolute mimimum (Moberg, 1984). Colditz reports that a less cumbersome design may result in increased patient compliance. Bunnell, who first described the low-profile design, emphasized the convenience of being able to pass the splint through a coat sleeve (Bunnell, 1946, 1956; Bunnell and Howard, 1950).

A systematic approach for choosing the appropriate traction for a given clinical application is paramount in order to avoid harm to the patient and achieve the desired increase in motion (Brand, 1984). Generally, therapists tend to construct outriggers at arbitrary heights prior to selection of the appropriate rubber band for each specific clinical application. The rubber band chosen then is often selected to accommodate outrigger height rather than meeting the force requirement.

As discussed above, longer rubber band resting lengths are superior in providing more uniform traction throughout the range of the joint motion. Unfortunately, the superiority of the longer rubber band in a high-profile designed splint necessitates fabrication of a towering outrigger to accommodate the length, thereby compromising cosmetic and functional adaptability. Conversely, the low-profile splint can be constructed to readily accommodate a longer rubber band length with minimal impingement upon daily living activities. All outriggers could be constructed approximately one inch from the volar or dorsal surfaces of the hand. The actual traction is dissociated from the outrigger and provides for simplified adjustment.

However, the question remains whether the low-profile splint design is bio-mechanically equivalent or superior to the high-profile design. Based upon the above established model, these two types of splint designs were compared based on the traction, moment about the joint, axial compression, and dorsal-shear force when the phalanx is rotated from an extended to a flexed position.

The comparison of these two splint designs is shown in Figure 6. Both types of splints have identical force applied perpendicular to the phalanx and denoted by the vertical line when the joint is in 60° of flexion (Figure 7). In other words, at this specific position, these two types of splints exerted the same force and moment to the phalanx and joint. However, as the joint away from the specified position, the forces and moments were not the same.

The axial force applied along the phalangeal shaft toward the joint could either be compressive or distractive force. In this example, compression was found when the phalanx moved in the opposite direction of splint pull, and distraction was found when the phalanx moved in the same direction of splint pull. Distraction and compression ranged from 50 g to 75 g, respectively. The effect of these forces on the capsulo-ligamentous structures and the joint surfaces or on the fixation of implants should not be overlooked.

The results of this analysis indicate that if the dynamic splint is to be used statically, the low and high profile designs provide essentially the same forces. Therefore, because

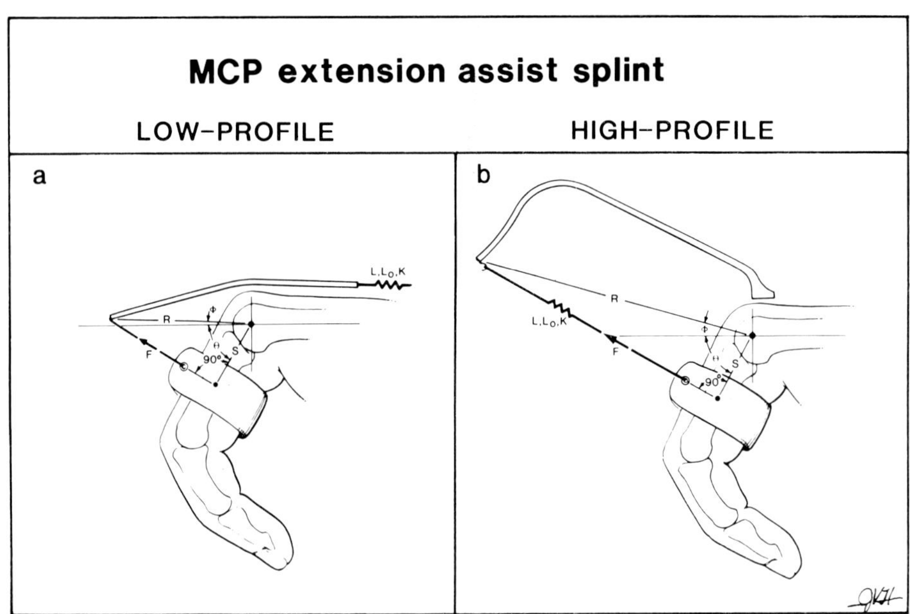

Figure 6.
Comparison of the low-profile and high-profile MCP joint extension assist splints. The values of geometric parameters and the rubber band property are: spring constant, $K = 142 \cdot 27$ g; resting length of rubber band, $L_0 = 5 \cdot 24$ cm; phalangeal attachment. $S = 2 \cdot 54$ cm; traction at 60° of flexion is $91 \cdot 18$ g; and outrigger distance, R, is $5 \cdot 12$ cm for low-profile (a) and $11 \cdot 09$ cm for high-profile (b).

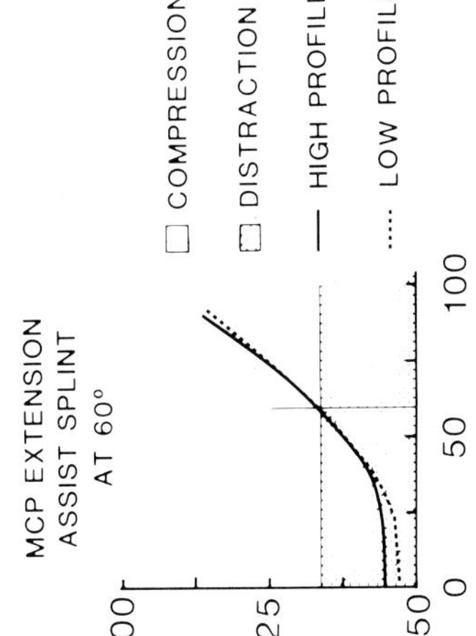

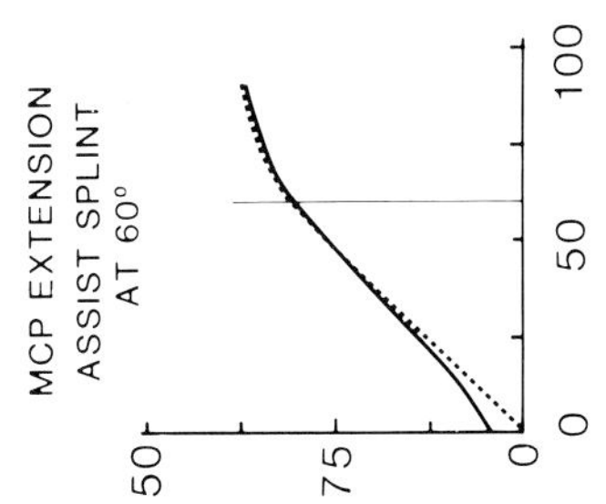

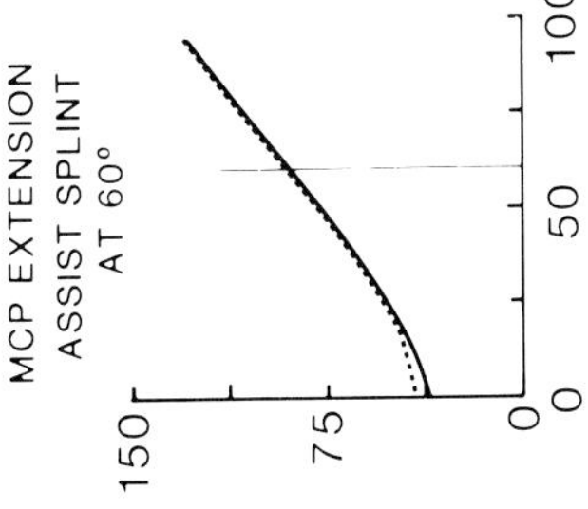

Figure 7.
Comparison of the low-profile versus high-profile splint.

of the cosmetic and functional superiority as well as increased ease in selecting elastic traction, the low-profile splint is superior for this application.

For applications that require the patient to move the digits throughout the range of motion, selection of the appropriate splint design becomes more complicated. Of particular interest is the fact that in this analysis, the low-profile design provided more distraction forces at the initial range of motion and less compressive force at the end range. This is of clinical importance, as distraction of the joint is favourable when attempting to increase joint range. The low-profile design also provided less shear force at the initial range of motion and slightly more shear force than the high-profile splint at the end. In most dynamic splint applications, shear force is an unfavourable component. These results confirm that the low-profile splint is not only superior cosmetically but also biomechanically.

A Protective splint for early care of proximal interphalangeal joint replacements

A variety of methods are used in the post-operative splinting of proximal interphalangeal joint (PIP) replacements. Evaluation of current splint designs disclosed a common shortcoming: the presence of unwanted lateral force against the proximal interphalangeal joint (Figure 8a). Modification of the conventional device can be done to minimize the lateral force applied and to reduce the amount of resistance throughout the joint range of motion. The modification is achieved by introducing an intermediate outrigger (Figure 8b).

The outrigger consists of a rigid wire bent in the shape of a rectangle with one of its four sides absent. Thus, there is a medial and lateral side which are connected by a transverse member. This transverse member is parallel to an axis which can be drawn through the MCP joints of the second through fifth digits. The absent side of the rectangle is the axis about which the outrigger pivots. The pivotal axis is dorsal to the axis through the MCP joints. The net result of this placement is to allow an arc of constantly decreasing radius between the tip of the outrigger to the pivotal axis of the digit as the hand is drawn into the fist. This decrease in radius between the tip of the outrigger and the joint centre matches well with the decrease in the distance between the end of the finger and the joint centre and results in more uniform traction throughout the range of motion.

A rubber band is attached on the proximal end to a thermoplast tower which is constructed as an integral extension of the splint protruding in a perpendicular direction from above the dorsal surface of the hand. The other end of this rubber band is secured to the centre of the transverse member of the intermediate outrigger.

A second rubber band is attached to the outrigger as well. Its other end is attached to the dorsal side of an orthoplast buddy splint individually wrapped around the DIP joints of the involved digit as well as its adjacent digits (Figure 9a). The adjacent digits assist in the passive flexion of the operated digit in addition to providing increased lateral stability. The exact position of the second rubber band is extremely important for the splint to function properly. The rubber band must be in the same geometric plane as the plane described by the digit moving throughout its range of motion. This

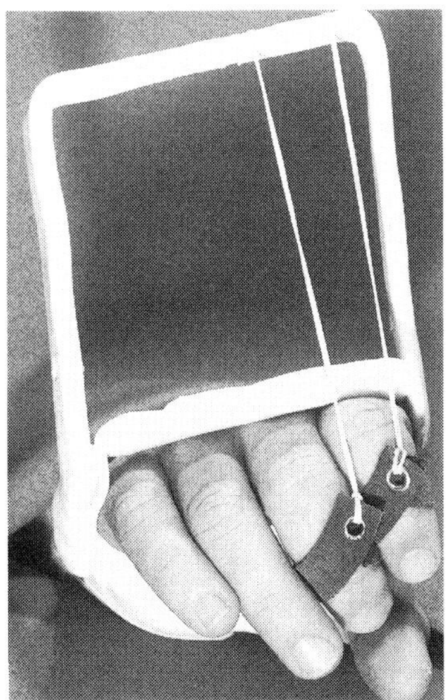

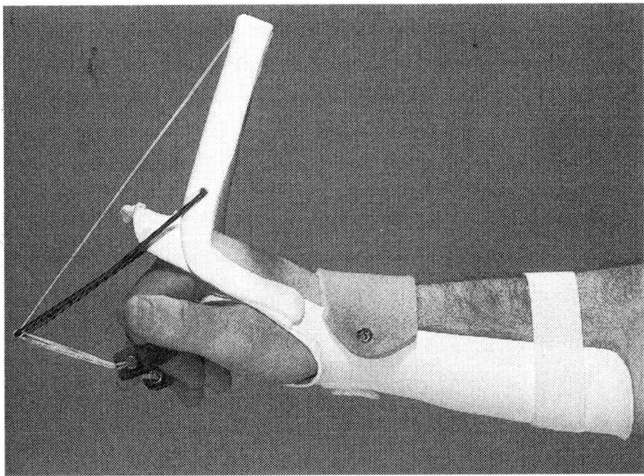

Figure 8.
Protective splint for early care of proximal interphalangeal joint replacement. In the traditional design (top), unwanted lateral force against the proximal interphalangeal joint is presented. This unwanted lateral force could be reduced by introducing an intermediate outrigger (bottom). Although the traditional high-profile splint is shown in the figure, the concept of the intermediate outrigger could also be well applied to the low-profile splint.

co-planar arrangement assures an absence of lateral forces on the repaired joint. In the traditional design, a progressively larger lateral force is exerted upon the joint as the hand is closed.

In instances in which a lateral load is necessary to re-align the digit back to its original or normal longitudinal axis, an additional outrigger device is used to apply this force against the digit to achieve independent control of lateral and resistive forces (Figure 9b).

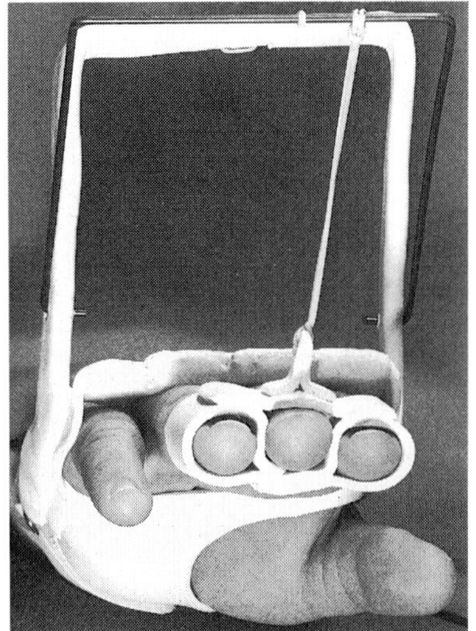
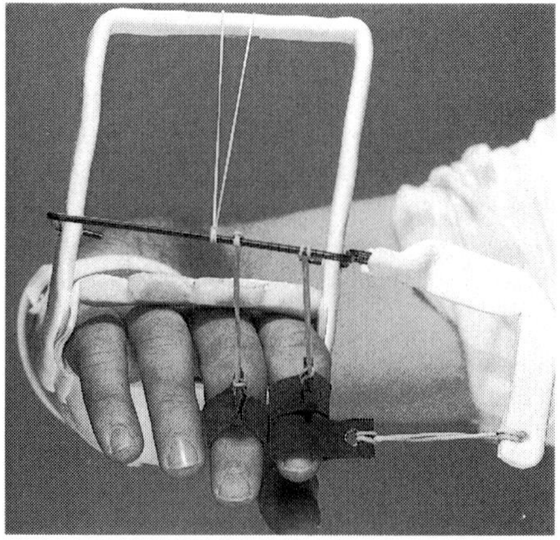

Figure 9.

Controlled planar motion of an individual digit can be achieved by using an orthoplast buddy splint wrapped around the joint and its adjacent digits (left). Although the high-profile splint is shown, this concept of attachment to the adjacent digits could be well adopted to the low-profile splint.

If lateral load is necessary to realign the digit back to its original plane of motion, an additional outrigger is used (right).

To date, 95 patients have been treated using the modified splint design during the last two years. The initial results obtained have been encouraging. This reduction in lateral stress provides a considerable advantage over the traditional technique because of the weakened state of the collateral ligamentous structures during the early post-operative period.

Earlier controlled motion may also enhance the alignment of the remodelled collagen fibres of the collateral ligaments and thus improve the joint stability.

Whole finger flexion assist split

Recently, based on laboratory results and clinical experiences, earlier controlled motion was advocated for post-operative rehabilitation of flexor tendon repair (Hunter *et al.*, 1987). Earlier controlled motion was found to stimulate optimal arrangement of the newly-formed and remodelled collagen fibres in the tendon at the repair site and thus the strength. Furthermore, earlier motion can prevent adhesion formation between the tendons themselves and with the surrounding soft tissue which may cause joint contracture.

In order to allow earlier motion after tendon repair, a strong suture is required as well as a sound splint device which will provide protective movement.

Traditionally, the splint was designed with the cable attached to the tip of the finger so that the traction would assist flexion of both the DIP and PIP joints. Recently, the traditional device has been modifid (Figure 10) by adding a roller bar at the palm region

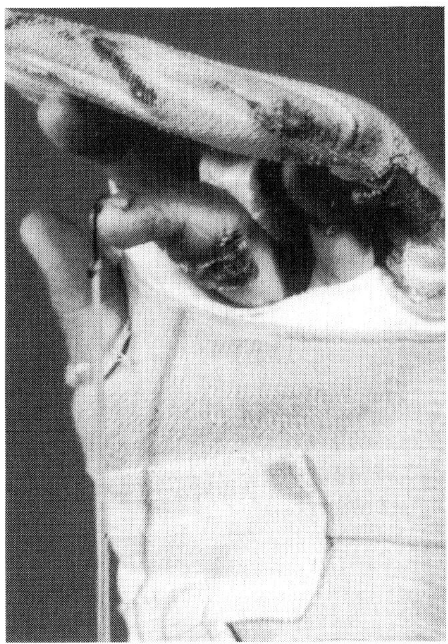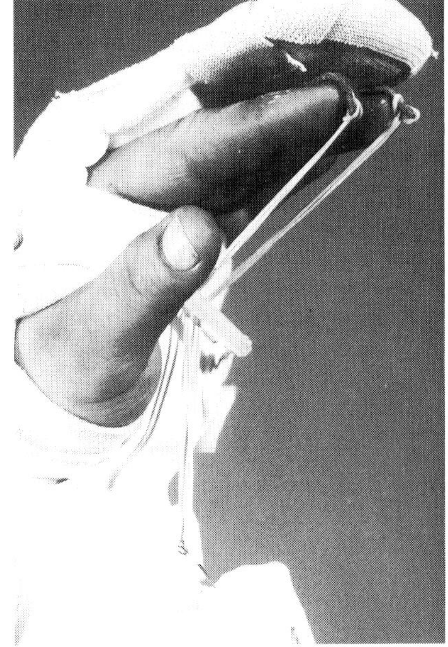

Figure 10.
Whole finger flexion assist splint for tendon repair patient. The traction is generated by a coil spring placed on the forearm and transmitted to the digits by either cables or rubber bands. Large bow-string effect of the cable taking place in the traditional design (left). In modified design, a roller bar is added at the palm to prevent the bow-string effect and increase the DIP and PIP joint range of motion and amount of sliding motion of the flexors.

to produce PIP and DIP flexion to the digit. In addition, a coil spring at the forearm provides more constant or uniform traction throughout the range of joint motion.

Addition of the roller bar at the palm can prevent the bow-stringing effect of the cable as seen in the traditional device. Therefore, larger excursion of the cable and thus increased range of DIP and PIP joint motion can be expected. It is agreed that it is the motion of both the DIP and PIP joints that produces the relative gliding motion between the flexor profundus and flexor superficialis tendons.

Summary

The exact design of a dynamic splint for hand rehabilitation seems at first to be trivial. However, to create an optimal design, a fundamental knowledge of bio-mechanical and ergonomic principles is required. This paper presents some of these principles and will hopefully be beneficial for those involved in the construction of dynamic splints.

Design and fabrication of the dynamic splint depends primarily upon the purpose for which it is employed. For example, when a splint is applied to correct a joint contracture, the appropriate design and arrangement would allow for adequate movement

Appendix:
Time-shared machine-readable databases: Their application to assistive device development for physically challenged persons

Hans Krobath and David Shafer

Research and Training Institute
At the Human Resources Center,
Albertson,
Long Island, USA

Currently available machine-readable databases accessible by subscription can greatly enhance the design, fabrication and test of assistive devices for physically challenged persons. These databases, easily accessed through telephone long lines, have greatly compressed the time and expense usually required to process a physical function-enhancing device so that it is ready for service to a person with a disability.

A strategy of useful database identification

Since ergonomics represents conceptual integration of human biology, physiology, psychology, mechanics, and anatomy, rehabilitation specialists must rely on experts in these disciplines to generate ideas in a biometric context. Access to these experts is now available through online machine-readable databases either by paying a fee or, in several cases, through a cost-free password (from this point on the term ''machine-readable databases'' will be shortened to ''database'').

Design

''Don't re-invent the wheel'' has never been more axiomatic than in these times of dwindling financial resources in rehabilitation. It is absolutely imperative to verify, through database search, that a needed assistive device does not exist before another is designed and fabricated, often at considerable cost.

One of the first ''stops'' in the database identification process is ''database of databases'', maintained on *DIALOG* by the Information Retrieval Laboratory in Urbana, Illinois. It's primary author, Martha Williams, is well known in the area of database selection.

An almost required next stop is at *ABLEDATA*, a database maintained by Catholic University's National Rehabilitation Information Center. This collection of over 13000 assistive devices has a learning curve of one hour in one of its formats on BRS Technologies of Latham, New York. In addition, the database is maintained by a large staff, who change over 1000 data elements per week.

While this is an article on machine-readable databases, it is important to note that one voice-based operation must be consulted. The University of West Virginia's Job Accommodation Network (*JAN*), reachable by telephone at (800) *JAN PCEH*, contains over 4000 local computer-based files listing job site modifications completed by companies in the United States for employees with handicapping conditions.

Properly identifying databases extant for pre-design work should involve persons in a network which represents many disciplines. By investing minutes in face-to-face meetings and telephone calls, one can learn of currently available databases in specific disciplines. Set out here are databases which have been found to aid the pre-design thinking process by listing devices and components that have an impact on a current challenge.

One major emerging database is maintained by the Veterans Administration (VA) on *COMPUSERVE*. Known as *VAR–4* it contains specific sections for persons engaged in ergonomics and rehabilitation. An online *Journal of Rehabilitation Engineering*, as well as *Technology Transfer Committee* and Current Progress Reports in Engineering bulletin boards are a few of the useful sections of this database. It is particularly useful because each VA station focuses on one part of the universe of ergonomics engineering.

Case study

A client, able to deliver a puff from the buccal cavity only measuring $2\,cm^2$, and without use of upper extremities, desired to operate a self advancing 35 mm camera. Positioning the camera on the client's wheelchair presented no especial challenges, but locating a leaf shutter release puff switch that would deliver the necessary force became a challenge.

Process for the case

Rehabilitation databases, viewed as the most likely candidates for the needed device were consulted, using thesauri indigenous to them. No product meeting specifications was discovered. Commercial product databases were then consulted using search strategies generic to the information utility that contained them. No products were discovered that met the required specifications. Medical databases were consulted using the logic that a $2\,cm^2$ output from the buccal cavity may have been dealt with in another context. No device was discovered. The final set of databases were in Patents. Many shutter release devices were either in an application phase, pending or actually granted, both domestically and internationally. Many came close to output requirements using a puff of air, but none, unfortunately, accepted a $2\,cm^2$ puff to operate the device. A device was designed and fabricated at considerable expense because of the first stage metal version, then injection mould process. The one value in the process beyond the immediate need of the client was the availability of the injection mould for replications. Unfortunately, contributing to future high costs was the need for automatic screw machine forming of the tapered thread mating shaft for camera applications.

One example of specificity can be found in *4–SIGHTS*, a database maintained by the Greater Detroit Society for the Blind. In the quest for a device cited above, it became useful to become very focused. In *4–SIGHTS*, a segment known as Technological Aids and Assistive Devices was helpful. It contained sections such as: audible aids and devices, tactile aids and devices, visual aids and devices, special modifications, and measuring devices. Access to this source currently requires an investment of $7·00 per contact hour — a bargain compared to many engineering databases.

Mining databases successfully

The optimal approach to so many useful databases is straightforward:

(a) Subscribe to major information utilities including *Dialog, Compuserv, BRS, Easynet, Newsnet, Mead Data Central* and others that contain database groups to which you need access.
(b) Acquire thesauri that contain exact or lexicologically acceptable term relationship to targeted databases. Master them.
(c) Take training by database producers either at their seminars, or minimally, using their on-line tutorials.
(d) Master the nuances of each database. Perfect practice makes perfect.
(e) Make a conscious effort to identify database organizational patterns that are similar. (Much akin to a pianist recognizing a piece that Bach wrote before seeing the composer's name.

Secrets of the trade

The old saw about a test containing most of the answers if one could but recognize them applies to database mastery in enhancing the design, fabrication and test of ergonomically-enhanced assistive devices. In the dataset on a device are items such as descriptors and other fugitive terms. They become the keys to information.

Descriptor gold

By carefully noting descriptors in each revealed data-set one is able to use them in expanding the search for an existing device. Terms not identified during study of the appropriate thesauri, now identified, serve as a springboard into potentially elusive files. In turn these newly-discovered files within a database may yield more terms that can serve as fresh ''ammunition'' to further explore the database.

Database reality

"You've got to be in it to win it," goes the commercial in the lottery. Devoting or rather investing time with the zeal of a new convert from physical databases to online databases yields new useful information sources. Previously passed-over databases now are added to one's repertoire of 'at the ready' information sources either for assistance in design, fabrication or test of an assistive device.

Run to stay in place

With three new machine-readable databases coming online every business day it behoves the user to "try-on" these new treasures. See if they "fit" and either place them "back on the rack" or add them to databases to be mastered.

Reading magazines published by information utilities such as *BRS (Bulletin), COMPUSERVE (Online Today),* and *DIALOG (Log on)* as well as *Database Monthly, Information Center, Datamation, Local Area Network Magazine,* and booklets published by the Veterans Administration — *Rehabilitation Research and Development,* and the Association for the Advancement of Rehabilitation Technology — *Rehabilitation Technology Review* are pristine sources for new databases.

Cost benefit

Within recent experience, prices quoted from commercial organizations to search for the existence of one ergonomically-enhanced assistive device ranged up to $1500·00. By careful thinking with people from several disciplines, including engineering, rehabilitation, learning and medicine, it is possible to formulate a search plan that can be executed in minutes on-line. A dedicated rehabilitation professional can further compress costs by waiting for least-cost clock hours to run a search.

Cost containment

What becomes increasingly attractive as one masters multiple databases is the immediate cost savings to a project. Discovering a component that would take an engineer two days to design, fabricate and test at a cost exceeding $600·00, and which retails for $40·00 at an aircraft supply house, means cost savings that are passed along to the authorizing agency. By sharing acquired information with other agencies and universities, additional savings are realized over time. The major national/international computer bulletin board for people working in the hard sciences at the university level is *SCIENCEnet* (avialable through *OMNET*). By its very nature ergonomics requires inter-disciplinary dialogue. *SCIENCEnet* allows this process to occur very efficiently. North Carolina and other universities are developing similar database concepts, but *SCIENCEnet* is in place and well known.

The future of multiple database access

One can assume that more and more expert and gateway systems will emerge that allow relative novices to access databases that reveal thousands of ergonomically-enhanced assistive devices.

Where to get further information on databases cited

Dialog
Dialog Customer Service 3460 Hillview Ave., Palo Alto, CA. 94304.
(800) 227–1960 outside California, (800) 982–5838 in California.

BRS

BRS Information Technologies 1200 Rte.7, Latham, NY 12110 (800) 345–4BRS or (518) 783–1161.

Compuserve
Compuserve, 5000 Arlington Centre Boulevard, Post Office Box 20212, Columbus, Ohio 43220. (800) 848–8990 or (614) 457–8650 in Ohio.

Directory of online databases

Directory of Online Databases Cuadro/Elsevier, 52 Vanderbilt Avenue, New York, NY 10017 (212) 370–5520

Database of databases

Database of Databases on *DIALOG*, file 230
Producer: Martha E. Williams, Information Retrieval Research Lab., Coordinated Science Lab. 1101 W. Springfield Ave., Urbana, Il 61801 (217) 333–8462.